Trends in Solid Mechanics 1979

PROF. DR. IR. W.T. KOITER

Trends in Solid Mechanics 1979

Proceedings of the Symposium dedicated to the 65th Birthday of W.T. Koiter

Delft University of Technology
June 13 - 15, 1979

J.F. Besseling/A.M.A. van der Heijden/*editors*

Delft University Press
Sijthoff & Noordhoff International Publishers
1979

Published by

Delft University Press
Mijnbouwplein 11
2628 RT Delft
The Netherlands

and

Sijthoff & Noordhoff
International Publishers
Stadhoudersplein 1
2404 BE Alphen aan den Rijn
The Netherlands

Cover by Ben Aalbers, Voorburg, The Netherlands

ISBN 90 286 0699 8

Contents

Preface

Already at the 14th IUTAM Congress at Delft in 1976 the idea to commemorate the sixty-fifth anniversary of Professor Warner Tjardus Koiter on June 16, 1979 was discussed in a group of friends and colleagues. That date should be also exactly thirty years after Koiter took up his professorship at Delft University of Technology. Afterwards it appeared that Koiter judged for himself that at his sixty-fifth birthday the right moment had come to retire from the University.

Meanwhile a group of colleagues former promovendi of Koiter, encouraged by the response received in 1976, planned a three days symposium 'Trends in Solid Mechanics 1979' intended for the presentation of an authoritative survey in the field to which Koiter himself contributed so much. The committee was fortunate to find in the Royal Institution of Engineers in the Netherlands (KIVI) the sponsor of the symposium.

Invitations to scientist prominent in their fields to act as speaker were all greeted with enthusiasm and they eagerly accepted to testify their friendship and respect for a leader in solid mechanics. The members of the organizing committee are very grateful that their efforts for the symposium in Koiter's honour met such a response, not only by the invited speakers but also by those invited as the other participants. The total number of listed participants was 191 and they came from Belgium, Denmark, France, Germany, Israël, Italy, Japan, the Netherlands, Poland, Switzerland, the United Kingdom, the USA and the USSR. But the perhaps most active participant of the symposium was not listed: Warner Tjardus Koiter himself.

This preface is not the place to evaluate the significance of the various contributions in this book: the reader is referred to the following chapters. Those who were present at the lectures will agree that the lively and meaningful discussions during the fifteen minutes discussion period were proof of the interest of the topics chosen and must have been a rewarding experience for the speakers.

What is the source of scientific endeavour? Surely it gives great satisfaction to put together, piece by piece, the endless puzzle that is presented to us by nature. To make this a common undertaking in the spirit of goodwill for all mankind, makes it more than an intellectual joy. However, like in all things of cultural value, one must be shown the way. It is the inspiration of great examples that is needed for the motivation and the perseverance of new workers in the field. Delft University has had in Professor Koiter for thirty years such an inspiring personality in the field of mechanics.

Organizing Committee

J.B. Alblas -- Eindhoven University of Technology
J.P. Benthem – Delft University of Technology
J.F. Besseling, Chairman – Delft University of Technology
J.W. Cohen – University of Utrecht
W.K.G. Floor – Delft University of Technology
A.M.A. van der Heijden, Secretary – Delft University of Technology
J.G. Lekkerkerker – Delft University of Technology
P. Meijers – Delft University of Technology
A. van der Neut – Delft University of Technology
A.D. de Pater – Delft University of Technology.

Acknowledgments

The organizing committee is indebted to many organizations for effective help in the preparation of the Symposium.
Delft University of Technology had provided its outstanding facilities for the Symposium. The Royal Institution of Engineers in the Netherlands (KIVI) had put its Congress Office at the disposal of the Symposium. The Ministry of Education and Science, the Department of Mechanical Engineering of Delft University of Technology, the Royal Institution of Engineers (KIVI) and industry had rendered financial support or guarantees to the Symposium. The industrial companies involved are listed in alphabetical order:

AKZO
Comprimo, Engineers and Contractors
DAF Trucks
Fokker V.F.W. B.V.
Hoogovens
KLM
NIVR
Philips
Shell-Nederland
TNO-IWECO
VMF-Stork

Furtheron the editors of this book wish to express their gratitude to the publishers Delft University Press and Sijthoff & Noordhoff International Publishers for their most helpful service and effective production of this book.

Biographical Note

Biographical Note

Warner Tjardus Koiter was born in Amsterdam on June 16, 1914. After his primary and secondary school education at Zutphen he matriculated into Delft University of Technology in 1931, where he obtained the degree of Mechanical Engineer with distinction on December 6, 1936. During his final year before graduation he was assistant to Professor C.B. Biezeno's chair in Applied Mechanics. In that capacity he designed a torsion-fatigue machine which is still in use in the Laboratory.

After graduation Koiter was appointed in the Structures Department of the 'Nationaal Luchtvaartlaboratorium' (NLL, National Aeronautical Research Institute) at Amsterdam, where he participated in airworthiness checking of aircraft structures and made an important contribution to the stress analysis of cantilever two-spar wings with shear resisting skin.

In 1938 he moved to the 'Octrooiraad' (Government Patent Office) and in 1939 joined the 'Rijksluchtvaartdienst' (Government Civil Aviation Office); until 1940 he was in charge of structural airworthiness checking and finally Director of the Engineering Division.

When during the war civil aviation became non-existent he was attached to NLL, working on problems of his own choice, which resulted in some important NLL-reports on postbuckling-behaviour of plates. These investigations were deepened and culminated in his thesis 'Over de stabiliteit van het elastisch evenwicht' (On the Stability of the Elastic Equilibrium), which led to the degree of Doctor of Engineering Science, with distinction, of Delft University of Technology on November 14, 1945, under the sponsorship of Professor Biezeno.

In 1949 Koiter was appointed Professor of Applied Mechanics in the Department of Mechanical Engineering in which capacity he was generally regarded as the successor of Professor Biezeno who would retire in 1958. He also became a member of the Department of Naval Architecture and Aeronautical Engineering, serving both Departments in several committees and as chairman of the first-mentioned Department from 1959 to 1961. In the early seventies the so-called democratization of the Dutch universities caused estrangement between Koiter and the university, as a result of which in 1973 he considered resignation. Fortunately the Ministry of Education and Sciences appreciated the significance of this loss and ensured his association with Delft by creating for him a personal chair, called the Chair of the 'Theory, Strength and Stability of Structures', which he held until his retirement, September 1st, 1979.

The influence of Koiter as an outstanding scientist and teacher is apparent from his correspondence with many scientists on their and his own publications, from the numerous visiting professors and research fellows the group of Applied Mechanics received, the invitations to lecture abroad and to collaborate in research, and from the foreign distinctions bestowed on him.

During the academic year 1961/62 Koiter was Visiting Professor of Applied Mathematics at Brown University, Providence, Rhode Island, and the academic year 1973/74 he spent as the Sherman Fairchild Distinguished Scholar at California Institute of Technology, Pasadena.

In addition to his main duties of teaching and research Koiter has served mechanics and auronautics in many formal capacities. He was the Secretary of the Government Committee on Netherlands Aircraft Industry in 1945 and 1946. From 1946 he was a member of the Netherlands Aircraft Development Board. From 1955 to 1968 he chaired the Scientific Advisory Committee for the National Aeronautical Research Institute and the Netherlands Aircraft Development Board. Since 1970 he served as a Member of the Advisory Council of the TNO Institute for Mechanical Constructions.

In 1956 Koiter became the first Chairman of the Section of Mechanics of the Royal Netherlands Institute of Engineers, a post he filled until 1959. From 1952 to 1964 he was the Secretary of the International Committee for the Congresses of Applied Mechanics. He served the International Union of Theoretical and Applied Mechanics in many capacities: Member of the Bureau (1956-1960), Treasurer (1960-1968), President (1968-1972) and Vice-President (1972-1976).

Koiter was elected a Member of the Royal Netherlands Academy of Sciences in 1959. He is also a member of two other Netherlands learned societies, viz. the 'Bataafsch Genootschap der Proefondervindelijke Wijsbegeerte' (Batavian Society of Experimental Natural Philosophy) at Rotterdam and the 'Hollandsche Maatschappij der Wetenschappen' (Dutch Society of Science) at Haarlem. From 1959 to 1968 he was a member of the International Council of the Aeronautical Sciences. He is a Foreign Member of the Danish Center for Applied Mathematics and Mechanics at Copenhagen; since 1977 Foreign Associate Member of the National Academy of Engineering, Washington and Member of the 'Deutsche Akademie der Naturforscher Leopoldina zu Halle' since 1976. Several Scientific journals may count Koiter among the members of their editorial board.

In an advisory capacity he served industrial concerns occasionally by solving their mechanics problems or making suggestions for improved structural design.

The flow of Koiter's publications, most of them of a fundamental nature, has been continuous up to this day. This is evident from the list appended to this biographical note.

Koiter's activities cover many fields of mechanics and of accompanying mathematical methods, in particular where they play a central role in mechanical and aeronautical engineering. His striking contributions to the development of stability theory, plasticity theory and shell theory are among the most frequently cited references in those fields. In particular his 1945 doctor's thesis 'On the Stability of the Elastic Equilibrium' opened a new avenue of attack for research on the non-linear post-buckling behaviour of structures. The international response to this thesis was retarded some 15 years because of a language barrier; it was written in Dutch. But when discovered, the initiative was taken twice to translate it into English. Up to present time, thus after more than 30 years, it continues to be a firm foundation for further contributions to stability theory, a situation not often encountered with a thesis.

Recipient of 1965 Von Kármán Medal of the American Society of Civil Engineers.
Recipient of 1968 Timoshenko Medal of American Society of Mechanical Engineers.
Honorary Degree of Doctor of Science from University of Leicester (1969).
Recipient of 1971 Modesto Panetti Prize, Academia delle scienze, Torino.
Foreign Honorary Member of American Academy of Arts and Sciences (elected 1974).
Honarary Member of the Polish Society of Theoretical and Applied Mechanics (1969).
Honorary Degree of Doctor of Laws from University of Glasgow (1978).
Dr. Ing. E.h. der Fakultät der Abteilung für Bauingenieurwesen der Ruhr-Universität, Bochum (1978).
Honorary Degree of Doctor of Applied Sciences from Ghent University (1979).

Publications

Publications

Publications of W.T. Koiter

1939

'Het torsiecentrum van balken bij belasting door dwarskrachten [The centre of shear of beams]', *Verslagen en verhandelingen van het Nationaal Luchtvaartlaboratorium,* **8** (1939) pp. 174-188 (Nat. Aeron. Res. Inst., Report No. S 156).

1940

a. 'Berekening van veerend gesteunde balken [Analysis of beams on elastic supports]', *De Ingenieur,* **55** (1940) pp.W.57-W.59.

b. *De invloed van het ribverband en de bekleeding op de sterkte van vliegtuigvleugels, Deel V* [*The influence of the rib structure and the pannelling on the strength of aircraft wings*] Nat. Aeron. Res. Inst. Report No. S 132, (Amsterdam, 1940), 29 pp. + tables and figures.

1943

a. 'De invloed van elastische ribvervormingen op de spanningsverdeling in vleugels met twee liggers en elastisch vervormbare bekleding [The influence of elastic deformations of the ribs on the stress distribution in two-spar stressed-skin wings]', *Verslagen en verhandelingen van het Nationaal Luchtvaartlaboratorium,* **12** (1943) pp. S1-S13 (Nat. Aeron. Res. Inst. Report No. S 278).

b. *De meedragende breedte bij grote overschrijding der knikspanning voor verschillende inklemming der plaatranden* [*The effective width at stresses far in excess of the critical stress for various boundary conditions*], Nat. Aeron. Res. Inst. Report No. 287(Amsterdam, 1943), 63 pp. + tables and figures.

1944

Het schuifplooiveld bij grote overschrijding van de knikspanning [*The behaviour of plates in shear at stresses far in excess of the critical stress*] Nat. Aeron. Res. Inst. Report No. S 295, (Amsterdam, 1944), 86 pp. + tables and figures.

1945

a. *Over de stabiliteit van het elastisch evenwicht,* Thesis Delft (Amsterdam, 1945), 7 + 223 pp. Two translations of this thesis appeared, viz.:

On the stability of elastic equilibrium (Washington, 1967), 6 + 202 pp. (NASA Technical Translation F-10, 833, Clearinghouse US Dept. of Commerce/Nat. Bur. of Standards N 67-25033);

The stability of elastic equilibrium (Palo Alto, Cal., 1970), 13 + 306 pp. (Stanford Univ., Dept. of Aeronaut. and Astronaut. Report AD 704124/Air Force Flight Dynamics Lab. Report TR-70-25).

b. *Elementaire elasticiteitstheorie* [*Elementary theory of elasticity*] Nat. Aeron. Res. Inst., Report No. S 302, (Amsterdam, 1945), 46 pp.

c. *Aanvulling op het rapport S266 'De meedragende breedte van vlakke platen, voorzien van langsverstijvers met open profiel'* [*The effective width of flat plates, provided with longitudinal open profile stiffeners*] Nat. Aeron. Res. Inst. Report No. S 294, (Amsterdam, 1945), 2 pp. Note that Report No. S 266 (of the Nat. Aeron. Res. Inst.) was published in 1943 by A. van der Neut and W.K.G. Floor.

1946

a. (jointly with U.F.M. Dellaert and A.S. Thomson) 'Het aantal vereiste baanrichtingen voor landingsterreinen [The required number of runways for airports]', *De Ingenieur*, **58** (1946) pp. L.1-L.5.

b. *Theoretical Investigation of the Diagonal Tension Field of Flat Plates,* Net. Aeron. Res. Inst. Report No. S 295, (Amsterdam, 1946), 7 pp. This is an English summary of the Dutch Report No. S 295, published in 1944.

1947

a. 'Gekromde staven [Curved beams]', published in *De Technische Vraagbaak deel A* (Deventer, 1947) on pp. 600-622.

b. 'De internationale luchtvaardigheidsvoorschriften [The international airworthiness requirements]', *De Ingenieur*, **59** (1947) pp. L.45-L.50.

1949

Vooruitzichten voor de ontwikkeling van de toegepaste mechanica, Inaugurele rede Delft [*Prospects for the developments of applied mechanics*], (Delft, 1949), 20 pp.

1950

'A Strain Energy Expression for Thin Elastic Shells (Discussion of a paper by H.L. Langhaar)', *Journ. Appl. Mechs.*, **17** (1950) pp. 107-108.

1951

a. 'On Grammel's linearisation of the equations for torsional vibrations of crankshafts', *Proc. Kon. Ned. Akad. Wet.*, **B54** (1951) pp. 464-467.

b. 'Niet-lineaire vraagstukken uit de toegepaste mechanica [Non-linear problems in applied mechanics]', *Ned. Tijdschr. Natuurk.*, **17** (1951) pp. 258-272.

c. 'Moeilijkheden en mogelijkheden van de Nederlandse vliegtuigindustrie [Difficulties and possibilities for the Netherlands aircraft industry]', *De Ingenieur*, **63** (1951) pp. L.33-L.38.

1953

a. 'Cornelis Benjamin Biezeno', *Anniversary Volume on Applied Mechanics dedicated to C.B. Biezeno*, Haarlem (1953) pp. 7-10.

b. 'Sress-strain relation, uniqueness and variational theorems for elastic-plastic materials with a singular yield surface', *Quart. Appl. Math.*, **11** (1953) pp. 350-354.

c. 'On partially plastic thick-walled tubes', *Anniversary Volume on Applied Mechanics dedicated to C.B. Biezeno*, Haarlem (1953) pp. 233-251.

d. 'Some remarks on plastic shakedown theorems', *Proc. eighth Int. Congr. on Theor. and Appl. Mechanics, August 20-28, 1952, Istanbul, vol. 1*, Istanbul (1953) pp. 220-221.

e. 'On the stress-strain relations and the general theorems of plasticity', *Fourth Symposium on Plasticity, Abstracts of Papers, September 1-3, 1953, Brown University, Providence, R.I., U.S.A.*, (1953) p. 13. Also published as an unnumbered Report of the Lab. of Engg. Mechs. (Delft, 1953), 24 pp.

f. 'Einige ergänzende Bemerkungen zum Aufsatz des Herrn Woinowsky-Krieger im Ingenieur-Archiv Bd XX, Seite 391', *Ingenieur-Archiv*, **21** (1953) p. 381.

1954

a. 'The flexural centre or centre of shear', *Journ. Roy. Aeron. Soc.*, **58** (1954) pp. 64-65.

b. (jointly with J.B. Alblas) 'On the bending of cantilever rectangular plates I-IV', *Proc. Kon. Ned. Ak. Wet.*, **B57** (1954) pp. 250-269, 549-557; **B60** (1957) pp. 173-181.

c. 'Approximate solution of Wiener-Hopf type integral equations with applications. I. General theory. II. Some plate bending problems and an example for hydrodynamics', *Proc. Kon. Ned. Ak. Wet.*, **B57** (1954) pp. 558-579.

1955

a. 'Approximate solution of Wiener-Hopf type integral equations with applications. Abstract'. *Proc. Kon. Ned. Ak. Wet.*, **A58** (1955) pp. 257-258. Also in *Indag Math.*, **17** (1955) 2.

b. 'On the diffusion of load from a stiffener into a sheet', *Quart. Journ. of Mechs. and Appl. Maths.*, **8** (1955) pp. 164-178.

c. *Inleidende beschouwingen over het slippen van motorvoertuigen bij geblokkeerde wielen* [*Introduction to the problem of skidding of motor vehicles with locked wheels*], Unnumbered Report of the Lab. of Engg. Mechs. (Delft, 1955), 22 pp.

1956

a. 'On the flexural rigidity of a beam, weakened by transverse saw cuts. I + II', *Proc. Kon. Ned. Ak. Wet.*, **B59** (1956) pp. 354-374.

b. *Spanningsconcentratie in een op trek belaste plaatstrook met een groot centraal cirkelvormig gat* [*Stress concentration in a strip with a large circular hole*], Unnumbered Report of the Lab. of Engg. Mechs. (Delft, 1956), 12 pp.

c. 'Buckling and Post-Buckling Behaviour of a Cylindrical Panel under Axial Compression', *Reports and Transactions National Aeronautical Research Institute*, **20** (1956) pp. 71-84 (Nat. Aeron. Res. Inst. Report No. S 476).

d. 'A new general theorem on shake-down of elastic-plastic structures', *Proc. Kon. Ned. Ak. Wet.*, **B59** (1956) pp. 24-34.

1957

a. 'An elementary solution of two stress concentration problems in the neighbourhood of a hole', *Quart. of Appl. Maths.*, **15** (1957) pp. 303-308.

b. (jointly with E. Sternberg), 'Thermal Stresses in Infinite Elastic Disks (Discussion of a paper by B. Sharna)' *Journ. Appl. Mechs.*, **24** (1957) pp. 487-488.

c. *Generalised plane stress in an infinite plate with a doubly-periodic set of equal holes. Part I. General theory for holes of arbitrary shape*, Unnumbered Report of the Lab. of Engg. Mechs. (Delft, 1957), 40 pp.

1958

a. 'Fail-Safe Structural Design', *Journ. Roy Aeron. Soc.*, **62** (1958) pp. 757-761.

b. (jointly with E. Sternberg), 'The Wedge Under a Concentrated Couple: A Paradox in the Two-Dimensional Theory of Elasticity', *Journ. Appl. Mechs.*, **25** (1958) pp. 575-581.

c. 'The Dutch Universities and South African Apartheid', *Science and freedom*, **10** (1958) pp. 29-31 (Apartheid and the World Universities; Report on a meeting held in London, November 1957).

1959

a. 'Some general theorems on doubly-periodic and quasi-periodic functions', *Proc. Kon. Ned. Ak. Wet.*, **A62** (1959) pp. 120-128. Also in *Indag. Math.*, **21** (1959) 2.

b. 'An infinite row of collinear cracks in an infinite elastic sheet', *Ingenieur-Archiv*, **28** (1959) pp. 168-172 (Festschrift Richard Grammel zum 70. Geburtstag).

1960

a. 'Stress Distribution in an Infinite Elastic Sheet with a Doubly-Periodic Set of Equal Holes', *Boundary problems in differential equations* (Madison, Wisconsin, U.S.A., 1960) pp. 191-213.

b. 'Complexe behandeling van het tweedimensionale elasticiteitsprobleem. I. Inleidende voordracht. III. Overzicht van bereikte praktische resultaten [Analysis of two-dimensional elasticity problems by means of complex function theory]', *De Ingenieur*, **72** (1960) pp. O.11-O.13, O.26-O.29.

c. 'Schalen [Schells]', *Kon. Ned. Ak., Wet., Verslag van de gewone verg. der Afd. Natuurk.*, **69** (1960) pp. 158-163.

d. 'A consistent first approximation in the general theory of thin elastic shells', *Int. Union of Theor. and Appl. Mechs. (I.U.T.A.M.). Proc. of the Symp. on the theory of thin elastic shells, August 24-28, 1959, Delft* (Amsterdam, 1960) pp. 12-33.

e. 'Ustojčivost' i zakritičeskoe povedenie uprugih sistem [Elastic stability and post-buckling behaviour], *Mehanika*, ? (1960) pp. 99-110. (Paper presented at the First All-Union Congress of Theoretical and Applied Mechanics, Moscow 1960).

f. 'General theorems for elastic-plastic solids', *Progress in solid mechanics* (Amsterdam, 1960) pp. 165-220.

1961

a. 'An infinite row of parallel cracks in an infinite elastic sheet', *Problems of continuum mechanics (Muskhelishvili Anniversary Volume)*, (Philadelphia, 1961) pp. 246-259.

b. 'A systematic simplification of the general equations in the linear theory of thin shells', *Proc. Kon. Ned. Ak. Wet.*, **B64** (1961) pp. 612-619.

c. (jointly with F.I.N. Niordson). On the linear theory of stability of thin elastic shells. Discussion, *Bygningsstatiske meddelelser*, **32** (1961) pp. 86-88.

d. 'Airworthiness requirements on fatigue strength', *Full-Scale Fatigue Testing of Aircraft Structures, Proc. Symposium June 9-11, 1959, Amsterdam, Neths.* (London, 1961) pp. 383-392.

1962

a. *'Stability of equilibrium of continuous bodies'*, Brown University, Div. of Appl. Maths. Techn. Report No. 79 (Providence, R.I., U.S.A., 1962), 2 + 110 pp.

b. *Theory of elastic stability and post-buckling behaviour*, Brown University, Div. of Appl. Maths. Techn. Report No. 80 (Providence, R.I., U.S.A., 1962), 2 + 3 + 82 pp.

c. *Correction of K-factor for cylindrical shell with grooves*, Lab. of Engg. Mechs. Report No. 227 (Delft, 1962), 2 pp.

d. *A spherical shell under point loads at its poles*, Brown University, Div. of Appl. Maths. Technical Report No. 75 (Providence, R.I., U.S.A., 1962), 2 + 20 pp. Also published in *Progress in Applied Mechanics. The Prager Anniversary Volume* (New York, 1963) pp. 155-169.

e. 'Het luchtvaartonderzoek op de tweesprong [The aircraft research at a cross-roads]' Avia/Maandblad voor Lucht- en Ruimtevaart, **11** (1962) pp. 622-623.

1963

a. *Solution of some elasticity problems by asymptotic methods*, Lab. of Engg. Mechs. Report No. 262 (Delft, 1963), 27 pp.

b. *Formules uit de algemene theorie van dunne elastische schalen (Samenvatting van het eerste gedeelte van het college 'Platen en schalen' in de cursus 1963/64)* [*Formulae from the general theory of thin elastic shells (Summary of the first part of the course of lectures on 'Plates and Shells', given during the academical year 1963/64)*], Lab. of Engg. Mechs. Report No. 267 (Delft, 1963), 13 pp.

c. 'Introduction to the post-buckling behaviour of flat plates', *Mémoires in -8° de la Société Royale de Sciences de Liège*, **5me serie, Tome VIII, Fascicule 5** (1963) pp. 17-40 (this issue 5 is also part 25 of the serial edition 'Les congrès et colloques de l'université de Liège', bearing the title 'Colloque sur le comportement postcritique des plaques utiliseés en construction métallique tenue à Liège les 12 et 13 novembre 1962', (Liège, 1963), 175 pp.

d. 'The effect of axisymmetric imperfections on the buckling of cylindrical shells under axial compression', *Proc. Kon. Ned. Ak. Wet.*, **B66** (1963) pp. 265-279. Also published as: Lockheed Missiles Space Company Techn. Report No. 6-90-63-86 (Sunnyvale, Cal., U.S.A., 1963), 3 + 21 pp.

e. 'Instability of Shells in the Light of the General Nonlinear Theory of Elastic Stability', *Abstracts of Papers of the Conference on shell theory and analysis. Lockheed Missiles & Space Comp. Res. Lab., August 21-23, 1963, Palo Alto, Cal., U.S.A.* (Palo Alto, Cal., U.S.A., 1963) pp. 12-17.

f. *A modified definition for the stability of equilibrium of continuous bodies*, Lab. of Engg. Mechs. Report No. 263 (Delft, 1963), 18 pp.

g. 'The concept of stability for continuous bodies', *Proc. Kon. Ned. Ak. Wet.*, **B66** (1963) pp. 173-177.

h. 'Elastic stability and post-buckling behaviour', *Proc. of a Symposium on Nonlinear Problems. April 30 – May 2, 1962, Univ. of Wisconsin, Madison* (Madison, Wisc., U.S.A., 1963) pp. 257-275.

1964

a. 'Couple stresses in the theory of elasticity. I + II', *Proc. Kon. Ned. Ak. Wet.*, **B67** (1964 pp. 17-44.

b. *Over de instabiliteit (in de zin van Liapounoff) van een evenwichtsconfiguratie van een conservatief mechanisch systeem in het geval dat de potentiële energie in de evenwichtsstand geen echt minimum heeft* [*On the instability (in the sense of Liapounoff) of an equilibrium configuration of a conservative mechanical system when the potential energy has not a true minimum in the equilibrium position*], Lab. of Engg. Mechs. Report No. 290 (Delft, 1964), 7 pp.

c. 'On the dynamic boundary conditions in the theory of thin shells', *Proc. Kon. Ned. Ak. Wet.*, **B67** (1964) pp. 117-126.

d. 'Knik van Schalen [Buckling of shells]', *Kon. Ned. Ak. Wet., Verslag gewone verg. Afd. Natk.*, **73** (1964) p. 58.

e. 'Knik van schalen [Buckling of shells]', *De Ingenieur*, **76** (1964) pp. O 33-O 41.

1965

a. *Note on the stress intensity factors for sheet string with cracks under tensile loads*, Lab. of Engg. Mechs. Report No. 314 (Delft, 1965), 12 pp.

b. 'On the instability of equilibrium in the abscence of a minimum of the potential energy', *Proc. Kon. Ned. Ak. Wet.*, **B68** (1965) pp. 107-113.

c. 'The energy criterion of stability for continuous elastic bodies. I + II', *Proc. Kon. Ned. Ak. Wet.*, **B68** (1965) pp. 178-202.

d. *Considérations elementaires sur la stabilité des systèmes non-lineaires avec quelques applications sur la stabilité de l'équilibre*, Lab. of Engg. Mechs. Report No. 302 (Delft, 1965), 16 pp. (Conférence à l'Université de Liège, mars 1965).

e. 'Solution of some elasticity problems by asymptotic methods', *Applications of the theory of functions in continuum mechanics. Proc. of the International Symposium, September 17-23, 1963, Tbilisi* (Moscow, 1965) pp. 15-31.

f. *Post-buckling analysis of a simple two-bar frame*, Lab. of Engg. Mechs., Report No. 312 (Delft, 1965), 25 pp.

1966

'On the nonlinear theory of thin elastic shells. I-III', *Proc. Kon. Ned. Ak. Wet.*, **B69** (1966) pp. 1-54.

1967

a. *On the thermodynamic background of elastic stability theory*, Lab. of Engg. Mechs. Report No. 360 (Delft, 1967), 16 pp.

b. 'Post-buckling analysis of a simple two-bar frame', *Recent Progress in Applied Mechanics. The Folke Odquist Volume* (Stockholm, 1967) pp. 337-354.

c. 'A sufficient condition for the stability of shallow shells', *Proc. Kon. Ned. Ak. Wet.*, **B70** (1967) pp. 367-375.

d. 'General equations of elastic stability for thin shells', *Proc. Symposium on the Theory of Shells, April 4-6, 1966, Houston, Texas, U.S.A.* (Houston, Texas, U.S.A., 1967) pp. 187-227.

1968

a. (jointly with H.B. Pacejka) 'Skidding of vehicles due to locked wheels', *Proc. Inst. of Mech. Engineers*, **183**, **3H** (1968-69) pp. 3-18 (Proc. Symposium 'Handling of vehicles under emergency conditions', Jan. 8, 1969, Loughborough).

b. *Error estimate of approximate solution to variational problem of elasticity theory*, Lab. of Engg. Mechs. Report No. 418 (Delft, 1968), 4 pp.

c. *Purpose and achievements of research in elastic stability*, Dept. of Mech. Engg. Report No. WTHD 8 (Delft, 1968), 3 + 20 pp. Also in *Recent Advances in Engineering Sciences. Proc. 4th Technical Meeting of the Society of Engineering Science, Inc. held at North Carolina State University, Raleigh, October 31 – November 2, 1966* (London, 1968) pp. 197-218.

d. *On the thermodynamic background of elastic stability theory*, Dept. of Mech. Engg. Report No. WTHD 7 (Delft, 1968), 3 + 12 pp. Also in *Problems of Hydrodynamics and Continuum Mechanics. Contributions in honor of the sixtieth birthday of L.I. Sedov. 14th November 1967* (Philadelphia, 1969) pp. 423-433.

e. *Summary of equations for modified, simplest possible accurate linear theory for thin circular cylindrical shells*, Lab. of Engg. Mechs. Report No. 442 (Delft, 1968), 7 pp.

f. *Foundations and basic equations of shell theory. A survey of recent progress*, Dept. of Mech. Engg. Report No. WTHD 5 (Delft, 1968), 3 + 14 pp. Also in *Theory of thin shells. Second IUTAM Symposium Copenhagen, September 5-9, 1967* (Berlin, Heidelberg, New York, 1969) pp. 93-105.

1969

a. 'A comparison between JOHN's refined interior shell equations and classical shell theory', *Zeitschr. angew. Math. Phys.*, **20** (1969) pp. 642-652.

b. 'The nonlinear buckling problem of a complete spherical shell under uniform external pressure', *Proc. Kon. Ned. Ak. Wet.*, **B72** (1969) pp. 40-123.

c. 'Enkele kanttekeningen bij de diskussie-nota Posthumus [Some marginal notes on the Posthumus discussion memorandum]', *Universiteit en Hogeschool*, **15** (1969) pp. 324-328.

1970

a. 'Error estimates for certain approximate solutions of problems in the linear theory of elasticity', *Zeitschrift angew. Math. Phys.*, **21** (1970) pp. 534-538.

b. 'Enkele opmerkingen over de invloed van de plaatbreedte op de zogenaamde spannings-intensiteitsfactor "Stress intensity factor" bij scheuren [Some remarks on the influence of the plate width on the so-called stress intensity factor in cracks]', *De Ingenieur*, **82** (1970) p. W 177.

c. 'On the foundations of the linear theory of thin elastic shells', *Proc. Kon. Ned. Ak. Wet.*, **B73** (1970) pp. 169-195.

d. (jointly with J.W. Hutchinson) 'Postbuckling theory', *Appl. Mechs. Reviews*, **23** (1970) pp. 1353-1366.

e. 'Enkele kanttekeningen bij het voorontwerp van de wet Universitaire bestuurshervorming 1970 [Some marginal notes on the draft Bill of the University Government Reform Act 1970]', *Universiteit en Hogeschool*, **16** (1970) pp. 451-454.

1971

a. 'On the mathematical foundation of shell theory', *Actes du congrès international de mathématiciens*. Vol. 3. 1-10 Septembre 1970, Nice (Paris, 1971) pp. 123-130.

b. 'Thermodynamics of Elastic Stability', *Proc. 3rd Canadian Congress of Applied Mechanics. May 17-21, 1971* (Calgary, 1971) pp. 29-37.

c. (jointly with G.D.C. Kuiken) 'The interaction between local buckling and overall buckling on the behaviour of built-up columns', *Dept. of Mech. Engg. Report No. WTHD 23* (Delft, 1971), 3 + 26 pp.

d. 'Rapport over een studiereis naar Canada en de Verenigde Staten van 11-28 mei 1971 [Report on a study trip to Canada and the United States from May 11 to 28, 1971]', *Rapporten inzake afvaardigingen naar bijeenkomsten ter beoefening of bevordering der wetenschappen, uitgebracht aan de Ned. regering of aan de Kon. Ned. Ak. v. Wet.*, **23** (1971) pp. 55-57.

1972

a. 'An appraisal of van der Neut's scientific work', *Contributions to the Theory of Aircraft Structures*, (Delft, 1972) pp. 1-9.

b. *Mechanica 4. Stijfheid en sterkte 1. Grondslagen* [*Mechanica 4. Rigidity and strength 1. Foundations*], (Haarlem, 1972), 70 + 240 pp.

c. (jointly with J.G. Simmonds). *Foundations of shell theory*, Dept. of Mech. Engg. Report No. WTHD 40 (Delft, 1972), 3 + 41 pp. Also in *Proc. 13th Int. Congress of Theoretical and Applied Mechanics. August 21-26, 1972, Moscow* (Berlin, Heidelberg, New York, 1973) pp. 150-176.

d. 'Vrije randen van platen en schalen [Free boundaries of plates and shells]', *Kon. Ned. Ak. Wet, Verslagen gewone verg. Afd. Natk.*, **81** (1972) p. 141.

1973

a. *On the principle of stationary complementary energy in the nonlinear theory of elasticity*, Dept. of Mech. Engg. Report No. WTHD 43 (Delft, 1973), 3 + 17 pp.

b. (jointly with J.P. Benthem) 'Asymptotic Approximations to Crack Problems', in *Methods of Analysis and Solution of Crack Problems* (Leiden, 1973) pp. 131-178.

1974

The influence of more or less localized short-wave imperfections on the buckling of circular cylindrical shells under axial compression (in a first approximation), Lab. of Engg. Mechs. Report No. 534 (Delft, 1974), 5 pp.

1975

a. 'Cornelis Benjamin Biezeno', *Kon. Ned. Ak. Wet., Verslagen gewone verg. Afd. Natk.*, **84** (1975) pp. 108-109.

b. *On the complementary energy theorems in nonlinear elasticity theory*, Dept. of Mech. Engg. Report No. WTHD 72 (Delft, 1975), 2 + 37 pp.

c. 'A basic open problem in the theory of elastic stability', *Joint IUTAM/ISU Symposium on Applications of Methods of Functional Analysis Problems of Mechanics. Résumés des conférences. September 1 to 6, 1975, Marseille. Part XXIX* (Marseille, 1975), 3 pp.

d. *A general theory for the interaction between local buckling and overall buckling and overall buckling of stiffened panels (Interim report)*, Lab. of Engg. Mechs. Report No. 556 (Delft, 1975), 14 pp.

1976

a. 'Obituary of C.B. Biezeno (March 2, 1888-September 5, 1975)', *Kon. Ned. Ak. Wet. Jaarboek 1975* (Amsterdam, 1976) pp.193-197.

b. (jointly with J.P. Benthem) 'On the Three-Dimensional Theory of Cracked Plates', *Journ. Appl. Mechs.*, **43 E** (1976) p. 374.

c. *Some comments on the so-called Timoshenko beam theory*, Lab. of Engg. Mechs. Report No. 597 (Delft, 1976), 7 pp.

d. 'Discussion of the paper "Maximum Stress in a Tensile Strip with a Large Hole" ', *Experimental Mechanics*, **16** (1976) p. 440.

e. 'On the complementary energy theorem in nonlinear elasticity', *Trends in Application of Pure Mathematics to Mechanics* (London, 1976) pp. 207-232.

f. 'Complementary energy, neutral equilibrium and buckling', *Proc. Kon. Ned. Ak. Wet.*, **B79** (1976) pp. 183-200.

g. 'Current trends in the theory of buckling', *Proc. IUTAM Symposium on Buckling of Structures. June 17-21, 1974, Cambridge, USA* (Berlin, Heidelberg, New York, 1976) pp. 1-16.

h. (jointly with M. Pignataro) 'An Alternative Approach to the Interaction between Local and Overall Buckling in Stiffened Panels, *Proc. IUTAM Symposium on Buckling of Structures. June 17-21, 1974, Cambridge, USA* (Berlin, Heidelberg, New York, 1976) pp. 133-148.

i. *Some properties of (completely) symmetric multilinear forms with an application to Ho's theorem for multi-mode buckling*, Lab. of Engg. Mechs. Report No. 587 (Delft, 1976), 7 pp.

j. *General theory of mode interaction in stiffened plate and shell structures*, Dept. of Mech. Engg. Report No. WTHD 91 (Delft, 1976), 3 + 44 pp.

k. *A general theory for the interaction of local buckling and overall buckling of stiffened panels*, Dept. of Mech. Engg. Report No. WTHD 83 (Delft, 1976), 3 + 49 pp.

1977

a. *On the so-called comparison theorem in the linear theory of elasticity*, Lab. of Engg. Mechs. Report No. 617 (Delft, 1977), 3 pp.

b. 'Timoshenko beam theory is not always more accurate than elementary beam theory. Discussion of paper by J.W. Nicholson and J.G. Simmonds, *Journ. Appl. Mechs.*, **44** (1977) pp. 357-358.

c. *A circular cylindrical shell under a uniform line load along a generator*, Lab. of Engg. Mechs. Report No. 620 (Delft, 1977), 12 pp.

d. *Basic principles of elastic stability*, Lab. of Engg. Mechs. Report No. 619 (Delft, 1977), 10 + 1 + 11 pp.

e. *Some comments on the non-linear theory of shells*, Lab. of Engg. Mechs. Report No. 608 (Delft, 1977), 6 pp.

f. (jointly with M. Pignataro) 'Sull' uso di metodi approssimati nell' analisi non lineare di pannelli rinforzati soggetti a compressione [On the use of approximate methods for the non-linear analysis of reinforced panels, subject to compression]' *Costruzioni Metalliche*, **29** (1977) pp. 148-152.

g. 'A general theory of interaction between local (panel) buckling and overall (Euler) buckling of stiffened plate and shell structures', *First Report, Second International Colloquium on Stability of Steel Structures April 13-15, 1977, Liège* (Liège, 1977?) pp. 253-254.

h. (jointly with M. Pignataro) 'A general theory for the interaction between local and overall buckling of stiffened panels, *Problem Attuali di Meccanica, Atti del Convegno Internazionale a ricordo di Modesto Panetti, November 10-12, 1975, Torino* (Torino, 1977) pp. 179-222.

1978

a. (jointly with Ph. Clément) 'On the so-called comparison theorem in the linear theory of elasticity', *Lab. of Engg. Mechs. Report No. 628* (Delft, 1978), 4 pp.

b. *Brief description of general non-linear theory of elastic stability, buckling, initial post-buckling behaviour and imperfection sensitivity*, Lab. of Engg. Mechs. Report No. 630 (Delft, 1978), 20 pp. (Text of paper read at GAMM Conference, March 28-31, 1978, Brussels).

c. *The intrinsic equations of shell theory with some applications*, Lab. of Engg. Mechs. Report No. 626 (Delft, 1978), 16 pp. + 9 tables.

d. *Some comments on an incomplete manuscript by R.S. Rivlin and K.N. Sawyers*, Lab. of Engg. Mechs. Report No. 634 (Delft, 1978), 16 pp.

e. *Engineering education at the cross-roads*, Lab. of Engg. Mechs. Report No. 637 (Delft, 1978), 16 pp.

Publications edited by W.T. Koiter

Proceedings of the IUTAM Symposium on the Theory of Thin Elastic Shells, August 24-28, 1959, Delft (Amsterdam, 1960), 11 + 496 pp.

(jointly with F. Rolla) *Proceedings of the Tenth International Congress of Applied Mechanics, August 31 – September 7, 1960, Stresa* (Amsterdam, 1962), 23 + 370 pp.

Anniversary Volume on Applied Mechanics dedicated to C.B. Biezeno by some of his friends and former students on the occasion of his sixty-fifth birthday March 2, 1953 (Haarlem, z.j. [1963]), 328 pp. + 1 fig.

(jointly with J.D. Achenbach e.a.) *Contributions to the Theory of Aircraft Structures presented to A. van der Neut by friends, colleagues and former students on the occasion of his sixty-fifth birthday, March 10, 1972* (Delft, 1972), 17 + 440 pp.

Theoretical and Applied Mechanics/Proceedings of the 4-th IUTAM Congress Delft, The Netherlands, August 30 – September 4, 1976 (Amsterdam, 1977), 10 + 491 pp. In addition, W.T. Koiter edited a volume of preprints (Amsterdam, 1976), 7 + 261 pp.

Lectures

Elastic Wave Propagation Problems in Non-destructive Evaluation

*J.D. Achenbach**

Summary

Methods to analyze diffraction of elastic waves by cracks are discussed. After a brief review of some elastodynamic theory, integral equations governing the crack-opening displacements are considered for both interior and surface breaking cracks. For specific geometries these integral equations have been solved numerically, and the results have been substituted in an appropriate representation theorem for the scattered displacement field. Particular attention has been devoted to amplitude spectra in the high-frequency range, which have been compared with approximate results obtained on the basis of elastodynamic ray theory. Some comparisons with experimental results are also included.

1. Introduction

Failure of structural components is generally caused by the presence of flaws which are inherent in the bulk material, or which appear either in the course of the manufacturing process or after a period of service. In a damage-tolerant design philosophy flaws are acceptable if they are sufficiently small that they will not lead to failure during the service life of the structure. As part of a damage-tolerant design philosophy it is important to have reliable methods of quantitative non-destructive evaluation (QNDE), not only to detect the presence and the approximate location of a flaw, but also to determine its size, shape and orientation.

Among the most useful QNDE methods are those based on the scattering of elastic (ultrasonic) waves by flaws in solids. Two general approaches to ultrasonic flaw detection have been taken. The imaging approach seeks to process the scattered field in such a manner that a visual outline of the object is produced on a display. The scattered-field approach attempts to infer geometrical characteristics of a flaw from either the angular dependence of its far-field scattering amplitude at fixed frequency, or from the frequency dependence of its far-field scattering amplitude at fixed angles. In this paper we will discuss analytical work for the scattered-field approach to crack-like flaws.

In experimental work on quantitative flaw definition by the ultrasonic pulse method either the pulse-echo method with one transducer or the pitch-catch method with two transducers is used. The transducer(s) may be either in direct contact with the specimen, or transducer(s) and specimen may be immersed in a water bath. Most experimental setups include instrumentation to gate out and spectrum analyze the signal diffracted by a flaw. The raw scattering data generally need to be corrected for transducer transfer functions and other characteristics of the system, which have been obtained on the basis of appropriate

* Department of Civil Engineering, Northwestern University, Evanston, Illinois 60201, USA

calibrations. After processing, amplitudes and phase functions, are available as functions of the frequency and the scattering angle. These experimental data can then be directly compared with theoretical results.

The solution to the direct diffraction problem, that is, the computation of the field generated when an ultrasonic wave is diffracted by a known flaw, is a necessary preliminary to the solution of the inverse problem, which is the problem of inferring the geometrical characteristics of an unknown flaw from either the angular dependence of the amplitude of the diffracted far-field at fixed frequency, or from the frequency dependence of the far-field amplitude at fixed angle. In recent years several analytical methods have been developed to investigate scattering of elastic waves by interior cracks as well as by surface-breaking cracks, in both the high- and the low-frequency domains. The appeal of the high-frequency approach is that the probing wavelength is of the same order of magnitude as the length-dimensions of the crack. This gives rise to interference phenomena which can easily be detected. The advantage of the low frequency approach is that useful approximations can be based on static results, and that the stress-intensity factor, which characterises near-tip stress fields, appears in the scattered far-field. Thus, conceptually the scattered field at low frequencies can directly provide information that is of interest from the point of view of fracture mechanics. It may be mentioned here that in a very well-known paper W.T. Koiter [1] was the first to analyze rigorously the stress-intensity factor for a surface breaking crack in a two-dimensional geometry. Methods to determine the width and particularly the depth of surface-breaking cracks are currently of great interest; and the results of Ref. [1] have proved useful in this context.

In the analytical work discussed in this paper we will primarily consider the direct problem. Diffraction of time-harmonic signals by cracks will be analyzed on the basis of linearized elasticity theory for a homogeneous, isotropic solid. The propagation of waves in such solids is a classical area of investigation. The equations governing elastodynamic theory, and several pertinent mathematical techniques to obtain solutions, have been discussed in detail in Ref. [2].

From the theoretical point of view a flat crack is a planar surface across which the displacement can be discontinuous. The exact mathematical formulation of the elastodynamic field generated by the presence of a crack is rather complicated if the displacement discontinuities are not known a-priori. That is the case in diffraction problems. Diffraction of an incident wave by a crack is a mixed boundary value problem, whose exact solution satisfies one or more generally singular integral equations for the displacement discontinuities. Only for a semi-infinite crack can an analytical solution conveniently be obtained, as shown by Maue [3]. The solution of the system of integral equations for a crack of finite dimensions requires a substantial amount of numerical analysis.

There is a considerable literature on diffraction of plane waves by slits and penny-shaped cracks. Much of this work was recently summarized in a review article by Kraut [4]. Another review article was recently published by Datta [5]. A great deal of useful information, and an extensive list of references is also contained in the book by Pao and Mow [6]. A comprehensive review of acoustic and electromagnetic scattering from discs and other simple shapes has been compiled by Bowman, Senior and Uslenghi [7]. Practical aspects, as well as analytical and experimental results on the application of elastic waves in the non-destructive testing of materials have recently been discussed in Ref. [8].

2. Elastodynamic theory

In this paper we consider time-harmonic motions, but the term $\exp(-i\omega t)$ is generally omitted. The equations governing elastodynamic theory relative to a system of Cartesian coordinates are stated in considerable detail in Ref. [2]. In a homogeneous, isotropic, linearly elastic solid the components of the stress tensor, $\tau_{ij}(\mathbf{x})$, are related to the gradients of the displacement components $u_i(\mathbf{x})$ by

$$\tau_{ij} = \lambda u_{k,k} \delta_{ij} + \mu(u_{i,j} + u_{j,i}) \tag{2.1}$$

where λ and μ are Lamé's elastic constants. Substitution of this relation into the balance equation of linear momentum yields the displacement equations of motion

$$\mathscr{L}[u_i] \equiv \mu u_{i,jj} + (\lambda + \mu) u_{j,ji} + \rho \omega^2 u_i = -f_i \tag{2.2}$$

where ρ is the mass density and f_i are the components of the body force distribution (per unit volume). These equations should be supplemented by

$$t_l = \tau_{kl} n_k . \tag{2.3}$$

At a surface S, Eq. (2.3) relates the components of the traction to the components of the stress tensor and the unit normal to S.

Basic singular solution

The basic singular solution for an unbounded solid is the solution of Eq. (2.2) when $f_i(\mathbf{x})$ is of the form

$$f_i = a_i \delta(\mathbf{x} - \mathbf{X}) \tag{2.4}$$

i.e., when **f** is a unit point load applied at $\mathbf{x} = \mathbf{X}$, and acting in the direction defined by the unit vector **a**. In this solution the components of **u** depend linearly on the components of **a**. To express this dependence we introduce

$$u_i^G = u_{i;m}^G a_m \quad \text{and} \quad \tau_{ij}^G = \tau_{ij;m}^G a_m \tag{2.5a,b}$$

which relate u_i^G and τ_{ij}^G to a_m through tensors of orders two and three, respectively. The solution for the Green's displacement tensor is well known. It is

$$u_{i;m}^G(\mathbf{x}-\mathbf{X}) = \frac{1}{\rho\omega^2} [-G_L(R) + G_T(R)]_{,im} + \frac{1}{\mu} G_T(R)\delta_{im} \tag{2.6}$$

where

$$G_\beta(R) = (4\pi R)^{-1} \exp(ik_\beta R) \qquad \beta = L, T \tag{2.7}$$

and

$$R = [(x_1 - X_1)^2 + (x_2 - X_2)^2 + (x_3 - X_3)^2]^{1/2} \tag{2.8}$$

$$k_L = \omega / c_L \qquad c_L^2 = (\lambda + 2\mu)/\rho \tag{2.9}$$

$$k_T = \omega / c_T \qquad c_T^2 = \mu/\rho. \tag{2.10}$$

The corresponding expression for $\tau^G_{ij;m}$ follows from (2.1) as

$$\tau^G_{ij;m}(\mathbf{x}-\mathbf{X}) = \lambda\, \delta_{ij}\, (u^G_{p;m})_{,p} + \mu\, [(u^G_{j;m})_{,i} + (u^G_{i;m})_{,j}]. \tag{2.11}$$

Let us define the length of the vector $\mathbf{x}$ by x, i.e.,

$$x^2 = x_1^2 + x_2^2 + x_3^2, \tag{2.12}$$

then for $X < x$, and neglecting terms of order $0(x^{-1})$, the far-field expressions for (2.6) and (2.11) reduce to, see e.g. [9],

$$u^G_{i;m}(\mathbf{x},\mathbf{X}) = A^{G;L}_{i;m}(\hat{\mathbf{x}},\mathbf{X})\, G_L(x) + A^{G;T}_{i;m}(\hat{\mathbf{x}},\mathbf{X})\, G_T(x) \tag{2.13}$$

$$\tau^G_{ij;m}(\mathbf{x},\mathbf{X}) = B^{G;L}_{ij;m}(\hat{\mathbf{x}},\mathbf{X})\, G_L(x) + B^{G;T}_{ij;m}(\hat{\mathbf{x}},\mathbf{X})\, G_T(x) \tag{2.14}$$

where

$$\hat{\mathbf{x}} = \mathbf{x}/x \tag{2.15}$$

and

$$A^{G;L}_{i;m} = (\lambda + 2\mu)^{-1}\, \hat{x}_i \hat{x}_m \;\exp\,(-ik_L\, \hat{\mathbf{x}}\cdot\mathbf{X}) \tag{2.16}$$

$$A^{G;T}_{i;m} = \mu^{-1}\, (\delta_{im} - \hat{x}_i \hat{x}_m)\; \exp\,(-ik_T\, \hat{\mathbf{x}}\cdot\mathbf{X}) \tag{2.17}$$

$$B^{G;L}_{ij;m} = i\,(\lambda + 2\mu)^{-1}\, k_L\, (2\mu\, \hat{x}_i \hat{x}_j + \lambda \delta_{ij})\, \hat{x}_m \;\exp\,(-ik_L\, \hat{\mathbf{x}}\cdot\mathbf{X}) \tag{2.18}$$

$$B^{G;T}_{ij;m} = i\, k_T\, (\delta_{im}\, \hat{x}_j + \delta_{jm}\, \hat{x}_i - 2\, \hat{x}_i \hat{x}_j \hat{x}_m)\; \exp\,(-ik_T\, \hat{\mathbf{x}}\cdot\mathbf{X}). \tag{2.19}$$

It is easily checked that the longitudinal and transverse parts of the far-field displacements have only components parallel and normal to $\mathbf{x}$, respectively.

Integral representation for the displacement field

Let V be a bounded domain and let S be its boundary surface. Let S_Γ be a sphere with radius Γ around P, and let V_Γ denote the domain interior to S_Γ, Γ is chosen so large that S_Γ completely surrounds S. By using the elastodynamic reciprocity relation and the appropriate radiation condition, the displacement can be expressed in terms of u_i and τ_{ij} on S by the following integral representation, see e.g. [10]

$$I_m(\mathbf{x}) = -\begin{cases} 0 & \text{when } P \in V \\ u_m(\mathbf{x}) & \text{when } P \in V_\infty \end{cases} \tag{2.20}$$

where $V_\infty = \lim_{\Gamma\to\infty} V_\Gamma - V$, and

$$I_m(\mathbf{x}) = \int_S [\tau^G_{ij;m}(\mathbf{x}-\mathbf{X})\, u_i(\mathbf{X}) - u^G_{i;m}(\mathbf{x}-\mathbf{X})\, \tau_{ij}(\mathbf{X})]\, n_j \, \mathrm{d}A(\mathbf{X}). \tag{2.21}$$

Now consider an isolated flat crack of area A in an unbounded solid. Suppose the crack faces A^+ and A^-(upper and lower) are subjected to tractions $t_j^+(\mathbf{x})$ and $t_j^-(\mathbf{x})$. If, as is typically the case no net tractions act on the crack, we have $t_j^+ + t_j^- = 0$, which implies $\Delta\tau_{ij} \equiv 0$. For a point in the body (2.20) and (2.21) then yield

$$u_m(\mathbf{x}) = -\int_{A^+} \tau^G_{ij;m}(\mathbf{x}-\mathbf{X})\, \Delta u_i(\mathbf{X})\, n_j^+ \, \mathrm{d}A(\mathbf{X}). \tag{2.22}$$

Here n_i^+ is still an outer normal, and is thus directed from the + to the – face of the crack.

In the far-field Eq. (2.22) further reduces to

$$u_m(\mathbf{x}) = -I^L_m(\hat{\mathbf{x}})\, G_L(x) - I^T_m(\mathbf{x})\, G_T(x) \tag{2.23}$$

where

$$I^L_m = ik_L(\lambda+2\mu)^{-1}(2\mu\hat{x}_i\hat{x}_j + \lambda\delta_{ij})\hat{x}_m \int_{A^+} e^{-ik_L\hat{\mathbf{x}}.\mathbf{X}}\, \Delta u_i(\mathbf{X}) n_j^+ \, \mathrm{d}A(\mathbf{X}) \tag{2.24}$$

$$I^T_m = ik_T(\delta_{im}\hat{x}_j + \delta_{jm}\hat{x}_i - 2\hat{x}_i\hat{x}_j\hat{x}_m) \int_{A^+} e^{-ik_T\mathbf{x}.\mathbf{X}}\, \Delta u_i(\mathbf{X})\, n_j^+ \, \mathrm{d}A(\mathbf{X}). \tag{2.25}$$

Here (2.18) and (2.19) have been used.

Integral equation

Suppose a displacement wave $u_i^{in}(\mathbf{x})$ is incident on a traction-free flat crack of area A. The components of the stress tensor corresponding to the incident wave are $\tau_{ij}^{in}(\mathbf{x})$. The incident wave generates a scattered field $u_i^s(\mathbf{x})$, such that the total displacement field is defined as

$$u_i^t(\mathbf{x}) = u_i^{in}(\mathbf{x}) + u_i^s(\mathbf{x}). \tag{2.26}$$

The scattered field does of course satisfy Eqs. (2.1)-(2.3). Since the crack is free of surface tractions, the surface tractions corresponding to the scattered field are

$$\mathbf{x}\in A: \quad t_i^s(\mathbf{x}) = -t_i^{in}(\mathbf{x}). \tag{2.27}$$

Returning to Eq. (2.22), we note that for $\mathbf{x}\in V_\infty$, the scattered field may be expressed as

$$u^s_m(\mathbf{x}) = \int_A \tau^G_{2i;m}(\mathbf{x}-\mathbf{X})\, \Delta u_i(\mathbf{X})\, \mathrm{d}A(X_1, X_3) \tag{2.28}$$

where the X_1 and X_3 have been chosen in the plane of the crack. The stresses according to Eq. (2.28) can be obtained by substitution of (2.28) into Hooke's law given by Eq. (2.1). Then by letting $\mathbf{x}$ approach A, the corresponding stresses should equal $-\tau_{ij}^{in}(\mathbf{x})$, according to Eq. (2.27). The resulting equation is an integral equation for Δu_i^s. This equation requires careful handling, because the integrand contains singularities. For a slit the integral equation has been stated, and solved numerically by Tan [11]. For a flat crack of general shape the integral equation has been derived by Budiansky and Rice [12].

For simple geometries, such as penny-shaped cracks and two-dimensional slits, it is convenient to use Fourier transform techniques to derive a governing system of dual integral equations, see e.g. Mal [13], [14], Sih and Loeber [15], [16] and Luong et al [17]. These kinds of equations can be reduced to a single singular integral equation, which can then be solved numerically. Examples of computations for the diffracted field based on this approach are included in Ref. [18].

3. The surface-breaking crack

The detection of a surface-breaking crack is an important problem in QNDE. It appears, however, that no exact solution to the direct scattering problem is available, not even for a crack normal to the free surface in a two-dimensional configuration. For experimental investigations we refer to the papers by Tittmann et al, see e.g. Ref. [19].

In this section we report on some recent work on the two-dimensional problem shown in Fig. 1. The incident wave can be either a time-harmonic surface wave, or a time-harmonic longitudinal or transverse body wave. The scattered field will be computed. An exact formulation will be followed whereby the two-dimensional elastodynamic problem is analyzed by Fourier transform techniques. The term $\exp(-i\omega t)$ is everywhere omitted.

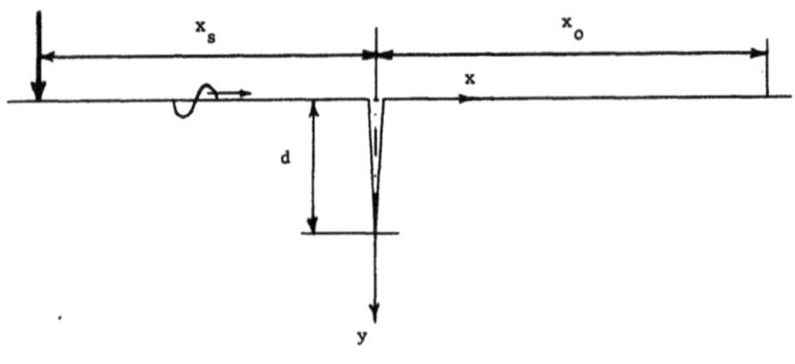

Fig. 1. Surface-breaking crack in 2-D geometry.

First the problem is decomposed into a symmetric and an antisymmetric problem relative to $x = 0$. The resulting mixed boundary-value problems are solved separately. Here we consider a few details of the antisymmetric problem. The complete analysis can be found in Ref. [20].

The boundary conditions for the antisymmetric problem are

$$y = 0, -\infty < x < \infty: \quad \tau_y = 0 \text{ and } \tau_{yx} = 0 \tag{3.1a,b}$$

$$x = 0, 0 \leq y < \mathrm{d}: \quad \tau_{xy} = -(\tau_{xy})_{\mathrm{inc}} \tag{3.2}$$

$$x = 0, \mathrm{d} \leq y < \infty: \quad u_y = 0 \tag{3.3}$$

$$x = 0, 0 \leq y < \infty: \quad \tau_y = 0. \tag{3.4}$$

Here $(\tau_{xy})_{\text{inc}}$ is the shear stress due to the incident wave. In addition it is required that the elastodynamic field is outgoing, and that the stresses and strains are integrable at the crack tips.

Equations (3.1)-(3.4) are considered as boundary conditions for the fields in the quarter-plane $x, y \geq 0$. In this quarter plane the following expressions for the displacements in terms of Fourier integrals satisfy the elastodynamic equations of motion:

$$u_x(x,y) = \frac{2}{\pi} \int_0^\infty [\xi k_1^{-2} A\, e^{-\alpha_1 y} - 2k_2^{-2} \alpha_2 C\, e^{-\alpha_2 y}] \cos(\xi x)\, d\xi$$

$$+ \frac{2}{\pi} \int_0^\infty [\alpha_1 k_1^{-2} B\, e^{-\alpha_1 x} + 2k_2^{-2} \xi D\, e^{-\alpha_2 x}] \sin(\xi y)\, d\xi \tag{3.5}$$

$$u_y(x,y) = \frac{2}{\pi} \int_0^\infty [-\alpha_1 k_1^{-2} A\, e^{-\alpha_1 y} + 2k_2^{-2} \xi C\, e^{-\alpha_2 y}] \sin(\xi x)\, d\xi$$

$$- \frac{2}{\pi} \int_0^\infty [\xi k_1^{-2} B\, e^{-\alpha_1 x} + 2k_2^{-2} \alpha_2 D\, e^{-\alpha_2 x}] \cos(\xi y)\, d\xi \tag{3.6}$$

where

$$k_1 = k_L = \omega/c_L \qquad k_2 = k_T = \omega/c_T \tag{3.7a,b}$$

$$\alpha_j = (\xi^2 - k_j^2)^{1/2} \qquad \xi > k_j \tag{3.8}$$

$$= -i(k_j^2 - \xi^2)^{1/2} \qquad \xi < k_j. \tag{3.9}$$

In Eqs. (3.5) and (3.6), A, B, C and D are functions of the transform variable ξ, which are to be determined. Equations (3.5) and (3.6) together with the boundary conditions (3.1)-(3.4) lead to a singular integral equation for $a(y)$, which is the tangential derivative (i.e., with respect to y) of the tangential crack-opening displacement for the antisymmetric problem. The manipulations leading to this equation are discussed in some detail in Ref. [20]. The result is

$$\int_0^d \frac{a(s)}{s-y}\, ds + \int_0^d a(s) \left[\frac{1}{s+y} + K_a(y,s) \right] ds = -\frac{\pi k_1^2}{4(k_2^2 - k_1^2)} (\tau_{xy})_{\text{inc}} \tag{3.10}$$

where

$$\frac{4K_a(y,s)}{k_2^2 - k_1^2} = \int_0^\infty \left[\frac{4\alpha_1 \alpha_2 \xi^2 - (\alpha_2^2 + \xi^2)^2}{4\alpha_2 \xi} - \frac{1}{2}(k_2^2 - k_1^2) \right] \sin(\xi s) \cos(\xi y)\, d\xi$$

$$+ \int_0^\infty \frac{f(\xi,s)}{4\alpha_2 \xi} [(\alpha_2^2 + \xi^2)^2 e^{-\alpha_2 y} - 4\alpha_1 \alpha_2 \xi^2 e^{-\alpha_1 y}]\, d\xi \tag{3.11}$$

$$f(\xi,s) = \frac{4\alpha_2 \xi [E_1 e^{-\alpha_1 s} + E_2 e^{-\alpha_2 s} + E_3]}{4\alpha_1 \alpha_2 \xi - (\alpha_2^2 + \xi^2)^2} \tag{3.12}$$

$$E_1 = -\xi^2 \tag{3.13}$$

$$E_2 = \alpha_2^{-2} \left(\xi^4 - k_2^2 \xi^2 + \frac{1}{4} k_2^4\right) \tag{3.14}$$

$$E_3 = -\frac{1}{4} \alpha_2^{-2} k_2^4. \tag{3.15}$$

The analogous equation for the symmetric problem is

$$\int_0^d \frac{b(s)}{s-y} \, ds + \int_0^d b(s) \left[\frac{1}{s+y} + K_s(y, s)\right] ds = \frac{\pi k_2^2}{2\mu(k_2^2 - k_1^2)} (\tau_y)_{inc} \tag{3.16}$$

Here $b(y)$ is the y- derivative of the normal crack-opening displacement, and $(\tau_y)_{inc}$ is the normal stress on $x = 0, 0 \le y < d$ due to the incident wave. Also

$$\frac{K_s(y, s)}{k_1^2 - k_2^2} = \int_0^\infty \left[\frac{(\alpha_2^2 + \xi^2)^2 - 4\alpha_1 \alpha_2 \xi^2}{\alpha_1 \xi} - 2(k_1^2 - k_2^2)\right] \sin(\xi s) \cos(\xi y) \, d\xi$$

$$+ \int_0^\infty \frac{g(\xi, s) k_2^2}{k_1^2 (\alpha_2^2 + \xi^2)} \left[(2\alpha_1^2 + k_2^2)(\alpha_2^2 + \xi^2) e^{-\alpha_1 y} - 4\alpha_1 \alpha_2 \xi^2 e^{-\alpha_2 y}\right] d\xi \tag{3.17}$$

$$g(\xi, s) = \frac{(k_1/k_2)^2 (\alpha_2^2 + \xi^2) [F_1 e^{-\alpha_1 s} + F_2 e^{-\alpha_2 s} + F_3)}{4\alpha_1 \alpha_2 \xi^3 - (\alpha_2^2 + \xi^2)^2} \tag{3.18}$$

$$F_1 = \alpha_1^{-2} [k_2^2 (k_2^2 - 2k_1^2) - 4\alpha_1^2 \xi^2] \tag{3.19}$$

$$F_2 = 4\xi^2 \tag{3.20}$$

$$F_3 = \alpha_1^{-2} k_2^2 (2k_1^2 - k_2^2). \tag{3.21}$$

Equations (3.10) and (3.16) have been solved numerically for a discrete set of values of $a(s)$ and $b(s)$, by using a quadrature based on expansions in terms of Chebyshev polynomials. For the required values of y and s the integrals in the kernels $K_a(y,s)$ and $K_s(y,s)$ are obtained by numerical integration in the complex ξ-plane. The computed values of $a(s)$ and $b(s)$ are subsequently integrated numerically to obtain the crack-opening-displacements in the normal and the tangential directions.

Once the crack-opening displacement has been computed, the scattered displacement field can be obtained by a representation integral of the general form given by Eq. (2.28). For the two-dimensional surface-breaking crack shown in Fig. 1 we can write the vertical displacement at a surface position x in the form

$$u_y^s(x, 0) = \int_0^d \tau_{xi;y}^G (x - Y) \, \Delta u_i(Y) \, dY \tag{3.22}$$

where we have converted to the x,y system shown in Fig. 1. Thus, the index i can be either x or y, and the summation convention is implied. In Eq. (3.22), Δu_x and Δu_y are the crack-opening-displacements computed earlier, and $\tau^G_{xi;y}(\mathbf{x}-\mathbf{Y})$ is the stress at a position defined by $x=0, y=Y$ due to a normal line load applied at $x=x, y=0$.

Line load excitation

In Fig. 1 a time-harmonic line load is applied at $y=0$, $x=-x_s$. The load generates wave motions which are incident on the surface-breaking crack located in the plane $y=0$. We will compute $u^s_y(x,0)$ at a point of observation $x=x_o$ for the case that both x_s and x_o are large. The boundary conditions on the plane $y=0$ are

$$\tau_{yy}=\delta(x+x_s)e^{-i\omega t} \qquad \tau_{yx}=0. \tag{3.23}$$

These boundary conditions define Lamb's problem, whose solution is available in the literature, see e.g. Ref. [2]. In particular for large $\omega\,|x+x_s|/c_L$ only the surface wave motions need be considered, and the stresses τ_{xx} and τ_{xy} assume the simple forms:

$$\tau_{xx}=-\frac{i\omega s_R}{\kappa_T}\{[2(1-\kappa_L^2)+\kappa_T^2]\,\Phi\exp[-\omega(s_R^2-s_L^2)^{1/2}y]$$
$$-(1-\kappa_T^2)^{1/2}\,\Psi\exp[-\omega(s_R^2-s_T^2)^{1/2}y]\}\exp[i\omega s_R(x+x_s)] \tag{3.24}$$

$$\tau_{xy}=-\frac{\omega s_R}{\kappa_T}\{2(1-\kappa_L^2)^{1/2}\,\Phi\exp[-\omega(s_R^2-s_L^2)^{1/2}y]$$
$$-\frac{1}{2}(2-\kappa_T^2)\,\Psi\exp[-\omega(s_R^2-s_T^2)^{1/2}y]\}\exp[i\omega s_R(x+x_s)] \tag{3.25}$$

where

$$s_L=1/c_L \quad s_T=1/c_T \quad s_R=1/c_R \tag{3.26a,b,c}$$

$$\kappa_L=s_L/s_R \qquad \kappa_T=s_T/s_R \tag{3.27a,b}$$

$$\Phi=\frac{\kappa_T(2-\kappa_T^2)(1-\kappa_L^2)^{1/2}(1-\kappa_T^2)^{1/2}}{2(4-\kappa_T^4)(1-\kappa_L^2)^{1/2}(1-\kappa_T^2)^{1/2}-(2-\kappa_L^2-\kappa_T^2)} \tag{3.28}$$

$$\Psi=\frac{\kappa_T(2-\kappa_T^2)^2(1-\kappa_L^2)^{1/2}}{2(4-\kappa_T^4)(1-\kappa_L^2)^{1/2}(1-\kappa_T^2)^{1/2}-(2-\kappa_L^2-\kappa_T^2)}\,. \tag{3.29}$$

In Eq. (3.26c) c_R is the velocity of Rayleigh surface waves.

Equation (3.25) provides the right-hand side of the boundary condition (3.2), while (3.24) provides the corresponding input to the boundary condition for the symmetric problem. With these conditions the integral equations (3.10) and (3.16) have been solved numerically as described earlier. Figures 2 and 3 show the crack-opening displacements versus y, for a few values of the dimensionless frequency. The normalizing constant A that has been used is

$$A = \frac{(2 - \kappa_T^2)}{2\mu\kappa_T}\,\Phi \qquad (3.30)$$

where Φ is defined by Eq. (3.28).

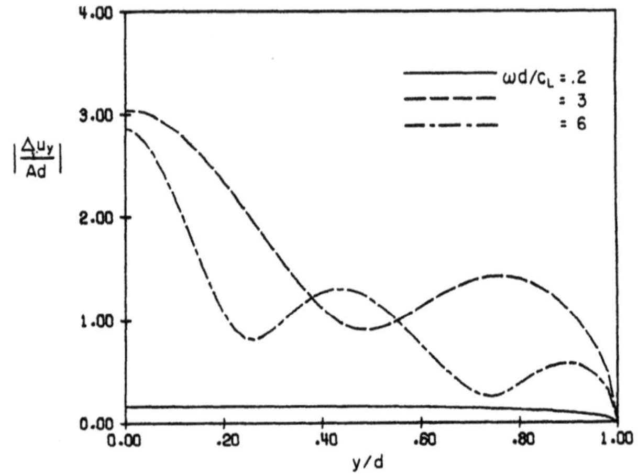

Fig. 2. Tangential crack-opening-displacement

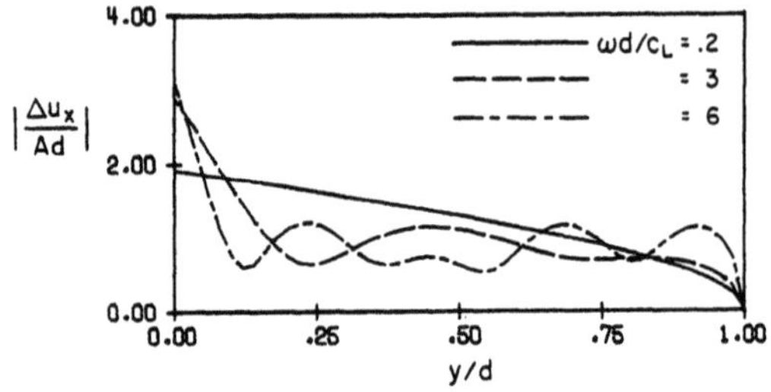

Fig. 3. Normal crack-opening-displacement.

When $\omega d/c_L$ increases the wavelength decreases, and the variation of the crack-opening-displacement with y becomes less even. The crack-opening displacements at the mouth of the crack (denoted by subscripts '*cm*') have been plotted in Figs. 4 and 5. The peaks at certain frequencies suggest resonance effects with surface motions on the faces of the crack.

Substitution of Δu_x and Δu_y in Eq. (3.22) yields the scattered displacement field at $x = x_0$. The required expressions for $\tau^G_{xi;y}(\mathbf{x} - \mathbf{Y})$ are given by Eqs. (3.24) and (3.25), respectively, provided that x_s is replaced by $-x_0$. The normal displacement has been plotted in Fig. 5. The spacing of maxima and minima is an important feature of this result.

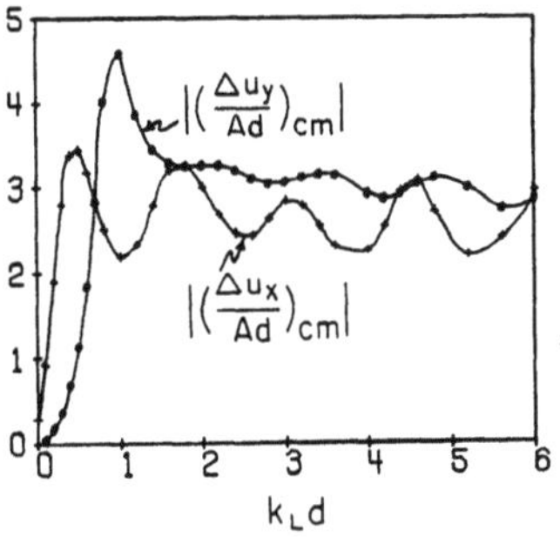

Fig. 4. *Crack opening displacements at the crack mouth versus* $k_L d = \omega d/c_L$.

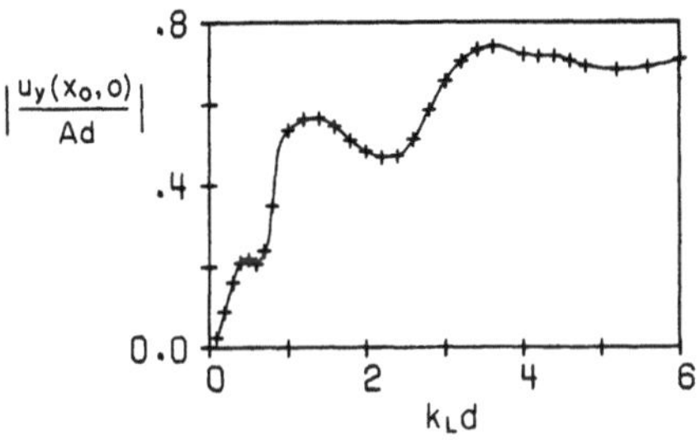

Fig. 5. *Normal component of scattered displacement at the free surface versus* $k_L d = \omega d/c_L$.

4. Elastodynamic ray theory

At high frequencies the diffraction of elastic waves by cracks can be analyzed conveniently on the basis of elastodynamic ray theory. The basic concepts of elastodynamic ray theory have been presented by Karal and Keller [21]. For time-harmonic wave motion, ray theory provides a method to trace the amplitude of a disturbance as it propagates along a ray. In a homogeneous, isotropic, linearly elastic solid the rays are straight lines, which are normal to the wavefronts. An unbounded solid can support rays of longitudinal and transverse wave motion. These rays are denoted as L-rays and T-rays, respectively. The free surface of a solid can, in addition, support rays of surface-wave motion, which are denoted as R-rays.

In analogy with geometrical optics, the simplest theory for diffraction of elastic waves by cracks may be called geometrical elastodynamics (GE). In GE a crack acts as a screen, which creates a shadow zone of no motion, and zones of reflected waves. The shadow zone is bounded by all rays passing through the source point and the edge of the crack. The geometrical reflections of these rays bound the zone(s) of reflected rays. The displacement field according to GE, which is denoted by $\mathbf{u}^g$ is of the same order of magnitude as the incident field. The GE-field is, however, physically unrealistic, because of discontinuities at the boundaries of the shadow zone and the zone(s) of reflected waves.

The geometrical theory of diffraction (GTD) provides a first correction to GE. This correction is valid for $\omega a/c_L >> 1$, and at points where $S/a > 1$. Here ω is the circular frequency, a is a length dimension of the crack, c_L is the velocity of longitudinal waves, and S is the distance from a crack edge. The correction provided by GTD is of order $(\omega a/c_L)^{-\frac{1}{2}}$. The basic ideas of GTD were introduced by Keller [22].

Basic to the elastodynamic version of GTD is the result that the incidence of a body-wave ray on the edge of a crack gives rise to two cones of diffracted body-wave rays, see Fig. 6, and two R-rays (one on each crack face). The surfaces of the inner and outer cones of body-wave rays consist of L-rays and T-rays, respectively. When an R-ray intersects the edge of a crack, a ray of reflected surface wave motion is generated, as well as cones of diffracted body-wave rays.

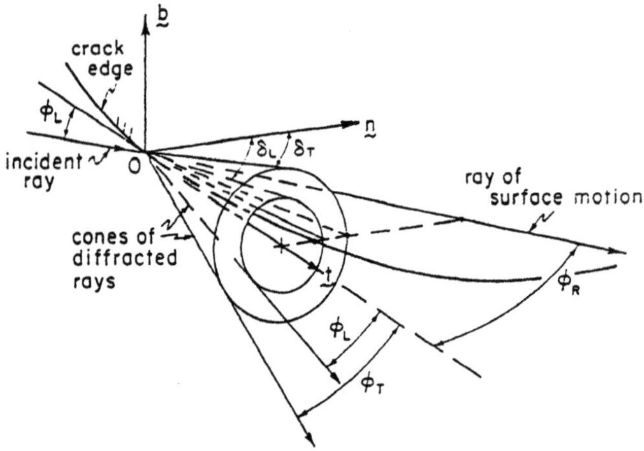

Fig. 6. Cones of diffracted L- and T-rays, and ray of surface motion generated by an incident L-ray.

For *plane* longitudinal and transverse waves, which are under arbitrary angles of incidence with a traction-free *semi-infinite* crack, the fields on the diffracted rays can be obtained by asymptotic considerations, as shown by Achenbach et al, [23] and [24]. The results can be expressed in terms of diffraction coefficients which relate the diffracted fields to the incident fields. Geometrical diffraction theory provides modifications to the semi-infinite crack results, to account for curvature of incident wave-fronts and curvature of crack edges, and finite dimensions of the crack. In the usual terminology the results for diffraction of plane waves by a semi-infinite crack are the canonical solutions.

With GE and GTD the total displacement field is of the form

$$\mathbf{u}^t = \mathbf{u}^g + \mathbf{u}^d. \tag{4.1}$$

This result is still not valid at the boundaries of the shadow zone and the zone(s) of reflected waves. In a further refinement which is called uniform asymptotic theory (UAT), the fields at these boundaries are corrected.

For incident waves with curved wavefronts and for curved diffracting edges, the cones of diffracted rays have envelopes, at which the rays coalesce and the fields become singular. The envelopes are called caustics, and GTD breaks down at caustics.

Within the context of the GTD theory of Refs. [23] and [24], the diffracted field at a point of observation Q is comprised of contributions corresponding to 'primary' diffracted body-wave rays, which are directly generated by incident body-wave rays, and contributions corresponding to 'secondary' diffracted body-wave rays. The latter are generated by rays travelling via the crack faces. Thus, the diffracted displacement field $\mathbf{u}^d$ at Q can be represented by

$$\mathbf{u}^d = \Sigma\, \mathbf{u}_\beta^\alpha + \Sigma \mathbf{u}_{\beta\gamma}^\alpha \tag{4.2}$$

where $\mathbf{u}_\beta^\alpha$ and $\mathbf{u}_{\beta\gamma}^\alpha$ represent the primary and secondary diffractions, respectively. In $\mathbf{u}_\beta^\alpha$, the symbol α defines the incident ray, i.e., $\alpha = L$ or $\alpha = T$, while β defines the diffracted ray, $\beta = L$ or $\beta = T$. In $\mathbf{u}_{\beta\gamma}^\alpha$ the symbol β defines the crack-face ray, i.e., $\beta = RS$ (surface-symmetric), $\beta = RA$ (surface-antisymmetric) or $\beta = TH$ (horizontally polarized transverse). The symbol γ defines the body-wave rays generated by diffraction of a crackface ray; thus $\gamma = L$ or $\gamma = T$. The summations in Eq. (4.2) are carried out over all rays of a particular type passing through Q.

Results obtained on the basis of Eq. (4.2) have been presented in Refs. [18], [25] and [26]. In Ref. [18] results obtained by elastodynamic ray theory have been compared with results obtained by numerical solution of a governing singular integral equation.

5. A comparison between theory and experiment

Experimental results in the high-frequency range that are suitable for comparison with theoretical results have been reported by Adler et al, see e.g. Refs. [27] and [28]. The experimental configuration is schematically indicated in Fig. 7. The sample is a circular disk (2.5 x 10 cm) of titanium alloy which contains a penny-shaped crack of radius 2500 μ parallel to the flat faces, and located at the center of the disk. The disk is immersed in water.

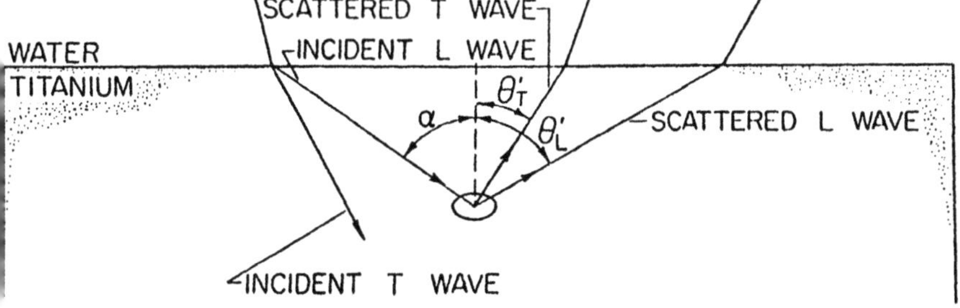

Fig. 7. Experimental set-up of Ref. [28].

A transmitter launches a longitudinal wave to the water-titanium interface under norma incidence. This wave is transmitted into the solid, diffracted by the crack, and the diffracte waves are transmitted back into the fluid, where they are received by a second transducer The experimental set-up and the processing of the data is discussed in some detail i Ref. [28].

In the experimental work the nature of the diffracted signals is determined by thei arrival times. Since the first arriving signals are related to longitudinal waves in the solid, it i possible to gate out and separate the purely longitudinal diffracted signals from subsequen signals. By appropriate processing of the experimental data, as described in Ref. [28], th amplitude-spectrum is obtained for the longitudinal diffracted waves only. Thus for th present comparison of analytical and experimental results we need to consider only th primary diffracted body-wave rays in our analytical work.

Diffracted L-rays in the solid

For normal incidence of a plane longitudinal wave on a penny-shaped crack, the displace ment vectors on the diffracted L-rays in the solid are

$$\mathbf{u}_L = \exp(i\omega S/c_L)(\omega S/c_L)^{-\frac{1}{2}}(1 + S/C)^{-\frac{1}{2}} D_L^L \mathbf{i}_L U_o. \qquad (5.1)$$

Here S is the distance along a diffracted ray from the point of diffraction to the point o observation, c_L is the velocity of longitudinal waves, and θ is the angle of the diffracted ray with the surface of diffraction. The geometrical quantities are indicated in Fig. 8.

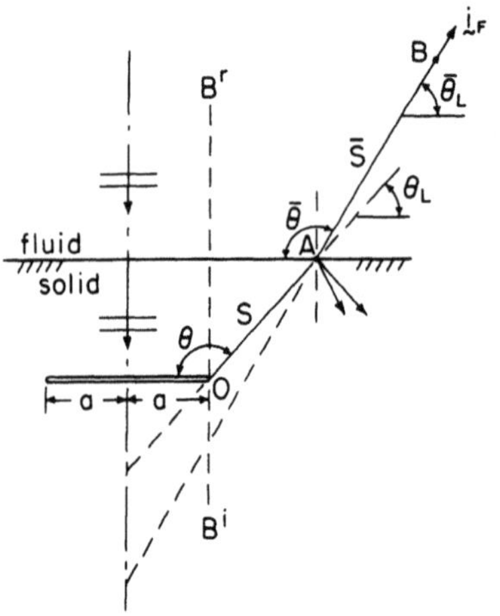

Fig. 8. Diffracted ray from a crack tip in an immersed specimen.

The amplitude U_0 defines the incident wave in the plane of the crack, and C is the distance to the caustic (for the problem considered here the caustic is the normal through the center of the crack), i.e.,

$$C = -a/\cos\theta \qquad (5.2)$$

where a is the radius of the penny-shaped crack. The vector $\mathbf{i}_L$, which defines the direction of the displacement is radial from the point of diffraction, as shown in Fig. 8. Finally, the dimensionless diffraction coefficient $D_L^L(\theta)$ is defined as

$$D_L^L(\theta) = [1-2(c_T/c_L)^2 \cos^2\theta][1+(c_R/c_L)\cos\theta] G_L(\theta) \tag{5.3}$$

where

$$G_L(\theta) = \frac{1}{\sqrt{\pi}} \left[\frac{1-\cos\theta}{1-(c_T/c_L)^2}\right]^{1/2} \frac{c_R(2c_T\cos\theta)^{-1}}{[1-(c_R/c_L)^2\cos^2\theta]} \frac{\exp(i\pi/4)}{K^+(-c_L^{-1}\cos\theta)}. \tag{5.4}$$

It is noted that $D_L^L(\theta)$ is symmetric in θ. In Eqs. (5.3) and (5.4), c_T and c_R are the velocities of transverse waves and Rayleigh waves, respectively, and $K^+(\xi)$ is the plus part of the modified Rayleigh function. An expression for $K^+(\xi)$ follows from the function $K^+(\xi,\eta)$, which is defined in the Appendix of Ref. [23], by setting $\eta = 0$. The result is

$$\pi \log K^+(\xi) = -\int_{s_L}^{s_T} \tan^{-1}\left[\frac{4s^2(s^2-s_L^2)^{1/2}(s_T^2-s^2)^{1/2}}{(s_T^2-2s^2)^2}\right] \frac{ds}{s+\xi} \tag{5.5}$$

where

$$s_L = 1/c_L, \qquad s_T = 1/c_T. \tag{5.6a,b}$$

For Poisson's ration $\nu = .352$, which corresponds to titanium, $|D_L^L(\theta)|$ and arg $[D_L^L(\theta)]$ have been plotted in Fig. 2 of Ref. [18]. It is noted that $|D_L^L(\theta)|$ is unbounded at $\theta = \pi/2$, i.e., at the shadow boundary where GTD cannot be applied. In the application of this paper θ is in the range $\pi/2 < \theta < \pi$, sufficiently far away from the shadow boundary.

Transmission across the solid-fluid interface

The incidence of an L-ray from the solid side on a solid fluid interface generates an L-ray of transmitted motion in the fluid and reflected L- and T-rays in the solid. The geometry for a single ray is shown in Fig. 8. Along the L-ray in the fluid we have

$$\mathbf{u}_L = \exp(i\omega\bar{S}/c_F)(1+\bar{S}/\bar{C})^{-1/2}(1+\bar{S}/\bar{E})^{-1/2}\, T(\theta_L)\, U_A\, \mathbf{i}_F. \tag{5.7}$$

Here c_F is the velocity of longitudinal waves in the fluid, $\bar{S}$ is the distance from the point of transmission A to the point of observation B, U_A defines the incident field at A, and the unit vector $\mathbf{i}_F$ defines both the direction of the transmitted ray and the displacement direction on that ray:

$$\mathbf{i}_F = (\cos\bar{\theta}_L,\ \sin\bar{\theta}_L). \tag{5.8}$$

The terms $(1+\bar{S}/\bar{C})^{-1/2}$ and $(1+\bar{S}/\bar{E})^{-1/2}$ follow from considerations of energy conservation in a bundle of rays. In Ref. [28] C, E and $\bar{\theta}_L$, as well as the transmission coefficient $T(\theta_L)$ have been determined as functions of the angle of incidence θ_L. Note that in the present geometry $\theta_L = \pi - \theta$ and $\bar{\theta}_L = \pi - \bar{\theta}$.

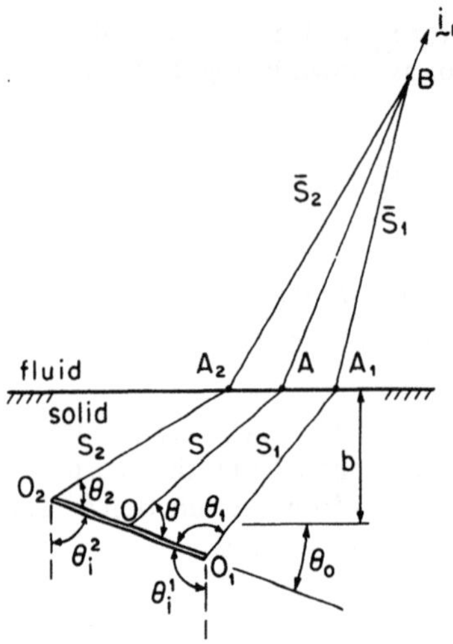

Fig. 9. Geometry in the plane of symmetry of a penny-shaped crack.

Far-field expressions

The interference patterns for the first arriving longitudinal waves in the fluid are generated by phase differences and amplitude differences on the direct rays from the two crack tips, see Fig. 9. The differences in the incident fields are represented by

$$U_0^1 = U_0 \exp[i(\omega a/c_L)\sin\theta_0]; \; U_0^2 = U_0 \exp[-i(\omega a/c_L)\sin\theta_0]. \tag{5.9a,b}$$

Now adding the primary diffracted longitudinal fields from the points O_1 and O_2 we obtain

$$\begin{aligned} \mathbf{u}_L = {} & H_1 \exp[i\omega(S_1/c_L + \bar{S}_1/c_F + a\sin\theta_0/c_L) + i\pi/4]\,U_0\,\mathbf{i}_F^1 \\ & + H_2 \exp[i\omega(S_2/c_L + \bar{S}_2/c_F - a\sin\theta_0/c_L) + i\pi/4]\,U_0\,\mathbf{i}_F^2 \end{aligned} \tag{5.10}$$

where

$$H_j = \frac{sgn(\cos\theta_j)\,T(\theta_L^j)\,|\,D_L^L(\theta_j;\theta_i^j)\,|}{(\omega S_j/c_L)^{1/2}(1+S_j/C)^{1/2}(1+\bar{S}_j/\bar{E})^{1/2}(1+\bar{S}_j/\bar{C})^{1/2}}\,. \tag{5.11}$$

One of the terms H_j is imaginary, since the ray has crossed a caustic.

We now introduce rays $\overline{OA}$ and $\overline{AB}$, as S and $\bar{S}$, respectively, which connect the center of the crack with the point of observation in the fluid. Sufficiently far from the interface Eq. (5.10) can then be reduced to, see Ref. [28].

$$\mathbf{u}_L \sim F(\theta,\theta_0)\exp[i\omega(S/c_L + \bar{S}/c_F) + i\pi/4]\,U_0\,\mathbf{i}_F \tag{5.12}$$

where

$$F(\theta,\theta_0) = H_1 \exp[-i(\omega a/c_L)(\cos\theta - \sin\theta_0)]$$

$$+ H_2 \exp\left[i(\omega a/c_L)(\cos\theta - \sin\theta_0)\right]. \tag{5.13}$$

Of particular interest is the absolute magnitude of F,

$$|F| = \{ |H_1|^2 + |H_2|^2 + 2|H_1||H_2| \sin[2(\omega a/c_L)(\cos\theta - \sin\theta_0)]\}^{1/2}. \tag{5.14}$$

Here we have taken into account that either H_1 or H_2 is imaginary.

Comparisons with experimental data

Theoretical results obtained from Eq. (5.14) have been plotted together with experimental data in Fig. 10. The frequency varies from 2 MHz to about 14 MHz. The angles in the solid are $\theta_L' (= \pi/2 - \theta_L) = 35°, 45°, 55°$ and $60°$, respectively. The amplitudes of the first few cycles agree well. At higher frequencies (above 6 MHz) the experimental results are lower than predicted by theory. One possible explanation is the effect of attenuation which is not accounted for in the theory. In all cases the positions of maxima and minima of the spectra agree well. The locations of the maxima are significant for the inversion process.

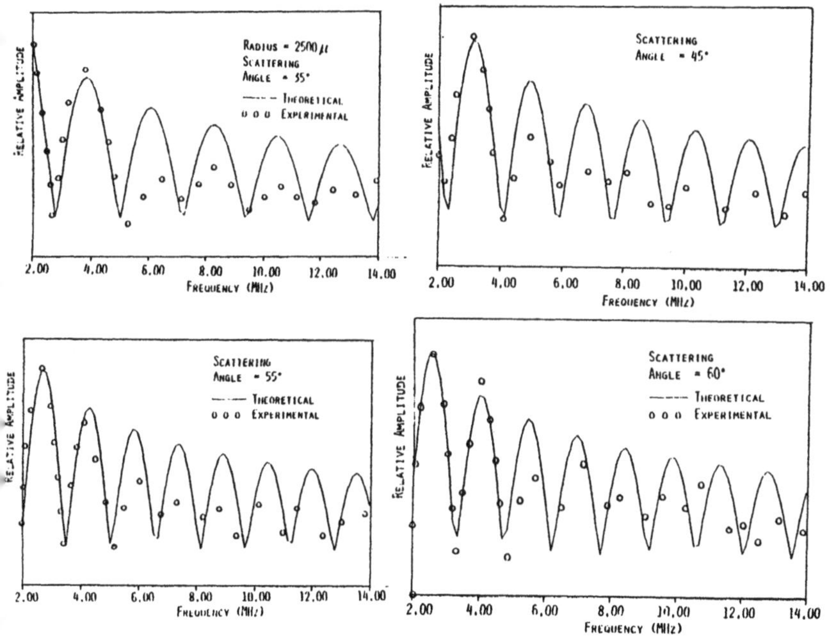

Fig. 10. Comparison of Eq. (5.14) with experimental results in Ref. [28].

The inverse problem

The discussion of the previous sections has been concerned with the direct problem, that is the computation of the scattered field when the size, shape and orientation of the crack are known. We will conclude with a few comments on the inverse problem for plane waves incident on penny-shaped cracks, for the special case that the incident wave is known to be in a plane of symmetry of the crack. The geometry in the plane of symmetry is then as shown in Fig. 9, where the incident wave is under an oblique angle with the plane of the crack. For a given point of observation, say the point B in Fig. 9, the unknowns then are θ_0 a and b.

The theoretical expression for the amplitude spectrum given by Eq. (5.14) implies that the amplitude of the primary diffracted field is modulated with respect to ω/c_L with period

$$P = \pi/a \mid \cos\theta - \sin\theta_0 \mid . \tag{5.15}$$

It is of interest to apply Eq. (5.15) to the experimental measurements. Since we know that $\theta_0 = 0$ each amplitude spectrum will give a number for a from the periodicity of the modulation. We have

$$a = \frac{c_L}{2\sin(\theta'_L)\,\Delta f_{ave}} \tag{5.16}$$

where $\theta'_L = \frac{1}{2}\pi - \theta_L$ and Δf_{ave} is the average frequency spacing between two consecutive maxima.

The results of the size determination are given in Table 1. The agreement between actual crack radius ($a = 2500\ \mu$) and the predicted values is excellent.

Table 1: Crack radius a computed from Eq. (5.16) for a penny-shaped crack in titanium ($c_L = 6330$ m/s)

$\theta'_L = \pi/2 - \theta_L$	Δf_{ave}	computed a in μ
35°	2.18	2530
40	1.87	2630
45	1.83	2450
50	1.68	2460
55	1.60	2410
60	1.47	2500
65	1.39	2510

6. Acknowledgements

I should like to acknowledge the cooperation of Dan Mendelsohn (surface-breaking crack) and Harry McMaken (ray-theory approach). This paper was written in the course of research sponsored by the Center for Advanced NDE operated by the Science Center, Rockwell International, for the Advanced Research Projects Agency and the Air Force Materials Laboratory under Contract F33615-74-C-5180.

7. References

1. W.T. Koiter, 'On the Flexural Rigidity of a Beam Weakened by Transverse Saw Cuts. II,' *Proc. Royal Neth. Acad. of Sci.*, **B59**, 365-374 (1956).
2. J.D. Achenbach, *Wave Propagation in Elastic Solids*, North-Holland, Amsterdam-New York, 1973.
3. A.W. Maue, 'Die Beugung elastischer Wellen an der Halbebene,' *Z. für Angew. Math. und Mech.*, **33**, pp. 1-10 (1953).
4. E.A. Kraut, 'Review of Theories of Scattering of Elastic Waves by Cracks,' *IEE Trans.*, **Su-23** p. 162 (1976).
5. S.K. Datta, 'Scattering of Elastic Waves,' *Mechanics Today, Vol. 4*, Pergamon Press, Oxford, 1978.
6. Y.H. Pao and C.C. Mow, *Diffraction of Elastic Waves and Dynamic Stress Concentrations*, Crane Russak, New York, 1973.
7. J.J. Bowman, T.B.A. Senior and P.L.E. Uslenghi, *Electro-Magnetic and Acoustic Scattering by Simple Shapes*, North-Holland, Amsterdam-New York, 1969.
8. Y.H. Pao (editor), *Elastic Waves and Non-Destructive Testing of Materials*, AMD-Vol. 29, American Society of Mechanical Engineers, 1978.
9. V.D. Kupradze, 'Dynamic Problems in Elasticity,' *Progress in Solid Mechanics*, North-Holland, Amsterdam-New York, 1963.
10. T.H. Tan, 'Scattering of Elastic Waves by Elastically Transparent Obstacles (Integral-Equation Method),' *Appl. Sci. Res.* **31**, pp. 29-51 (1975).
11. T.H. Tan, Diffraction Theory for Time-Harmonic Elastic Waves, Doctoral Dissertation, Delft, 1975.
12. B.A. Budiansky and J.R. Rice, 'An Integral Equation for Dynamic Elastic Response of an Isolated 3-D Crack,' WAVE MOTION **1**, in press.
13. A.K. Mal, 'Interaction of Elastic Waves with a Griffith Crack,' *Int. J. Eng. Sci.*, **8**, 1970, p. 769.
14. A.K. Mal, 'Diffraction of Elastic Waves by a Penny-Shaped Crack,' *Q. Appl. Math.*, **26**, pp. 231-238 (1968).
15. J.F. Loeber and G.C. Sih, 'Diffraction of Antiplane Shear Waves by a Finite Crack,' *J. Acoust. Soc. Am.*, **40**, pp. 90-98 (1968).
16. G.C. Sih and J.F. Loeber, 'Normal Compression and Radial Shear Waves Scattering at a Penny-Shaped Crack in an Elastic Solid,' *J. Acoust. Soc. Am.*, **46**, pp. 711-721 (1969).
17. W.C. Luong, L.M. Keer and J.D. Achenbach, 'Elastodynamic Stress Intensity Factors of a Crack near an Interface,' *Int. J. Solids Structures*, **11**, pp. 919-925 (1975).
18. J.D. Achenbach, A.K. Gautesen, and H. McMaken, 'Application of Elastodynamic Ray Theory to Diffraction by Cracks,' *Modern Problems in Elastic Wave Propagation*, Wiley-Interscience, New York, 1978.
19. B.R. Tittmann, F. Cohen-Ténoudji, M. deBilly, A. Jungman and G. Quentin, 'A Simple Approach to Estimate the Size of Small Surface Cracks with the use of Acoustic Waves,' *Appl. Phys. Lett.*, **33**, pp. 6-8 (1978).
20. J.D. Achenbach, D. Mendelsohn and L.M. Keer, '2-D Scattering by a Surface-Breaking Crack,' submitted for publication.
21. F.C. Karal and J.B. Keller, 'Elastic Wave Propagation in Homogeneous and Inhomogeneous Media,' *J. Acoust. Soc. Am.*, **31** p. 694 (1959).
22. J.B. Keller, 'A Geometrical Theory of Diffraction,' *Calculus of Variations and Its Applications*, McGraw-Hill, 1958.

23. J.D. Achenbach and A.K. Gautesen, 'Geometrical Theory of Diffraction for Three-D Elastodynamics,' *J. Acoust. Soc. Am.*, **61** pp. 413-421 (1976).
24. A.K. Gautesen, J.D. Achenbach and H. McMaken, 'Surface Wave Rays in Elastodynamic Diffraction by Cracks,' *J. Acoust. Soc. Am.*, **63**, p. 1824 (1978).
25. J.D. Achenbach, A.K. Gautesen and H. McMaken, 'Diffraction of Point-Source Signals by a Circular Crack,' *Bull. Seism. Soc. Am.*, **68**, pp. 889-905 (1978).
26. J.D. Achenbach, A.K. Gautesen and H. McMaken, 'Diffraction of Elastic Waves by Cracks-Analytical Results,' in *Elastic Waves and Non-Destructive Testing of Materials* (Y.H. Pao, editor), AMD-Vol. 29, American Society of Mechanical Engineers, 1978.
27. L. Adler and H.L. Whaley, 'Interference Effect in a Multifrequency Ultrasonic Pulse Echo and its Application to Flaw Characterization,' *J. Acoust. Soc. Am.*, **51**, pp. 881-887 (1972).
28. J.D. Achenbach, L. Adler, D. Kent Lewis and H. McMaken, 'Diffraction of Ultrasonic Waves by Penny-Shaped Cracks in Metals: Theory and Experiment,' submitted for publication.

Continuum-thermo-mechanics

*E. Becker**

Summary

In contrast to the to day much presented axiomatic treatment of the foundations of continuum thermodynamics a survey is given which is attached to the classical theory of Caratheodory and Born. Starting from the hypothesis of local equilibrium, irreversible processes in continuum mechanics may be dealt with. Timescales of different order of magnitude are distinguished. Then visco-elasticity, creep and such-like are discribed with the aid of internal statevariables like changes of apparent equilibrium on a time-scale comparable with the time-scale of observation. Also viscous stresses may be explained by the theory of internal statevariables if the to this purpose introduced relaxation time is much smaller than the characteristic time of the flow.

1. Introduction

When I was asked to present a survey on such a wide field as 'continuum mechanics' within the short time of one hour I decided that the best I could hope to do is to discuss one important aspect of continuum mechanics, and as such I chose the thermodynamic aspect. Thermodynamics has always played a decisive role in the mechanics of gases, a subject which has occupied me for many years. In solid mechanics the importance of thermodynamics has grown only more recently so that this growth might be considered a present 'trend in solid mechanics'. Certainly, thermodynamics is fundamental to all branches of continuum mechanics and is, therefore, a strongly unifying feature connecting the diverse branches. To elucidate such unifying aspects of continuum mechanics in a simple manner is the main didactic aim of a book which I wrote, together with my colleague W. Bürger, some years ago [1]. For good reasons the chapter on thermodynamics is a central one in this book. I would like to mention that some important notions of that chapter grew out of a discussion I had with Professor Koiter on problems of stability [2]. Ever since I had written this chapter I felt it was somewhat unsatisfactory as an introduction, and I have tried to simplify, improve and polish the treatment. The following necessarily brief and sketchy outline is the present, incomplete result of these modest attempts.

There is a steady stream of expository papers and textbooks on thermodynamics. I readily admit that many of the modern axiomatic treatises do not appeal very much to me. My own endeavours led me back to expositions of thermodynamics which are close to the classical treatment of the subject by Caratheodory and Born (for a succinct review see [3]). Therefore the following treatise may be considered old-fashioned. Nevertheless, two recent books by H. Buchdahl have greatly inspired me [4, 5]; I should also mention the book by

* Institut für Mechanik, Technische Hochschule Darmstadt, D-1600 Darmstadt

L.C. Woods [6]. Moreover, B.T. Chu has influenced my ideas about the internal variable theory [7], a subject close to my heart because I have used it in my own work on gasdynamics [8]. A brief outlook on internal variables in fluid dynamics is therefore a natural conclusion of this paper.

2. From mechanics to thermomechanics

Let us start with a constitutive equation for a homogeneously deformed elastic body in equilibrium:

$$\sigma = \mathbf{f}(\mathbf{F}). \tag{2.1}$$

According to (2.1) the – suitably defined – stress tensor σ is a unique function of the deformation gradient $\mathbf{F}$, defined by choosing a suitable reference configuration. Careful experimentation reveals that (2.1) is not true under all circumstances: Repeating experiments with the same material body shows that stress does not uniquely depend on deformation, or, in other words, that deformation does not uniquely define the 'state' of the body. In particular, the results of deformation experiments with the same body will differ if they are performed in different environments. These differences are more pronounced for gaseous bodies than, say, for elastic solids. However, the differences exist in both cases and they are the bridge which leads from mechanics to thermomechanics.

A further experience has to be mentioned: The relation between deformation and stress may depend not only on the environment but also on the time scale of observation in the following sense: When performing a deformation experiment in a fixed environment the body will settle down after a certain time in an 'apparent' equilibrium, with a definite relation of form (2.1) between σ and $\mathbf{F}$. On a larger time scale this equilibrium may slowly change without changes of the environment. This change may lead to another apparent equilibrium on the larger time scale, which again might slowly change on a still larger time scale. Effects like viscoelasticity, creep etc. are such changes of apparent equilibria on a time scale comparable with the time scale of observation.

How are we to incorporate the two experiences described above in a constitutive equation? The thermodynamic method of coping with the influence of the environment consists in introducing an additional scalar variable in (2.1) which at present will be denoted by θ, such that

$$\sigma = \hat{\mathbf{f}}(\mathbf{F}, \theta) \tag{2.2}$$

gives stress σ as a unique function of the deformation $\mathbf{F}$ and the variable θ. The physical meaning of the 'thermal' variable θ is left open at present; definite choices of θ will emerge later on. In my opinion, a quite natural way to incorporate the possibility of slowly changing equilibria is the straight-forward generalization of the thermodynamic method by introducing a finite set of additional 'internal' variables $q_i, \ldots, q_n$ in (2.2):

$$\sigma = \hat{\hat{\mathbf{f}}}(\mathbf{F}, \theta, q_i, \ldots, q_n) \tag{2.3}$$

such that σ is again uniquely determined by the independent variables on the right hand side of (2.3). In many cases the q_i have well-defined physical meanings; for example, they might be concentration variables in a body consisting of different chemical species which react with each other. In other cases they are more of the nature of an artifice.

On a sufficiently short time scale the internal variables remain constant and (2.3) reduces to (2.2) with the q_i playing the role of constant parameters. On the largest time scale on which (2.3) is applicable an equilibrium, for which a relation of the form (2.2) is valid, again prevails, because on this largest time scale the internal variables assume 'equilibrium values' $\bar{q}_i$ ($\mathbf{F}$, θ), so that (2.3) again reduces to an equation in the form of (2.2). The equilibrium on this largest time scale will be called an unconstrained equilibrium. In what follows the classical thermodynamic theory of such unconstrained equilibria of *homogeneously deformed bodies* will be studied first (chapter 3-7). Deformation $\mathbf{F}$ and thermal variable θ are sufficient as independent variables in that case. However, the results obtained for unconstrained equilibria apply also for apparent equilibria with fixed values of q_i, ..., q_n.

The variables $\mathbf{F}$, θ span the state space of the body under consideration. Each equilibrium state of the body corresponds to a point in an open domain of state space and vice versa. A 'change of state' is an ordered set of two states, $\{A, B\}$, with A the 'initial' and B the 'final' state; 'initial' and 'final' denote a sequence in time. Changes of state are effected by 'processes'. During such processes the system need not be in equilibrium states. Hence, in general, processes do not correspond to a sequence of points, a 'path', in state space.

In the following chapters the existence of three important functions of state, i.e. functions of the independent variables $\mathbf{F}$ and θ, will be postulated. These postulates are the abstractions from age-old experiences. In part 3 the empirical entropy η, in part 4 the internal energy U, and in part 6 the empirical temperature Θ will be defined. These three quantities have the common property that they are monotonic functions of the thermal variable θ. Therefore each of the quantities η, U, Θ, is itself a possible choice for the variable θ.

3. Entropy

Natural processes are irreversible! In order to precisely formulate the essence of that experience in a simple manner we assume that a material body can be isolated from its surroundings in such a way that the state of the body can be changed only by mechanical action on the body but not by a change in the environment of the body. (It is tacitly assumed that electromagnetic interactions, for example, are excluded, thereby restricting the theory to *thermomechanics*.) Of course, such an 'adiabatic body', or an 'adiabatic system' of bodies in general, can be realized, by enclosing the body or system with an 'adiabatic wall', only in an asymptotic sense. In this and the two following chapters, adiabatic bodies are studied.

Let us assume that an equilibrium state A of an adiabatic body is changed by a certain process ('adiabatic process') and that finally the body settles down to an equilibrium state B. Depending on the details of the process various states B can be reached from a given state A; (for brevity's sake 'state' stands for 'equilibrium state'). However, as a consequence of the irreversibility of natural processes, not every state B is attainable from A. This is made precise in the following statement, usually called Caratheodory's Principle:

1. In every neighborhood (definable by introducing a metric in state space) of a state A there are states B which cannot be reached from A by any adiabatic process.
2. If state B is unattainable from state A then A is attainable from B.

Caratheodory's Principle holds not only for homogeneously deformed bodies, with which we are concerned presently, but for arbitrary systems, for example for the 'combined systems' which will be introduced in part 6. Furthermore, the principle is also true under the restriction to isometric changes of state. These are changes of state with the same value of the deformation **F** in the initial and final state: $\mathbf{F}_A = \mathbf{F}_B$. Therefore, initial state A and final state B differ only by the value of θ : $\theta_A \neq \theta_B$. It is tacitly implied that isometric changes of state are possible for every initial state A. Caratheodory's Principle is certainly satisfied if a continuous and unique function $\eta(\mathbf{F}, \theta)$, called the *empirical entropy*, exists with the following properties:

$$\eta_B \geqslant \eta_A \quad \text{if state } B \text{ is adiabatically attainable from state } A, \qquad \eta_B < \eta_A \text{ otherwise.} \tag{3.1}$$

The hypersurface $\eta = \eta_A$ through point A in the state space is the boundary between those states which are attainable from A and those which are unattainable from A. A point B which lies on the hypersurface $\eta = \eta_A$ is attainable from A and, vice versa, A is attainable from B. Special processes which lead from one point of a hypersurface η = const. to another point on the same hypersurface are therefore reversible. The existence of an empirical entropy function with properties (3.1) is a sufficient condition for the validity of Caratheodory's Principle. It is possible to show, with weak additional assumptions concerning the topology of paths in state space, that the existence of an entropy function is a necessary consequence from Caratheodory's Principle. Without discussing this point let us *postulate* at this stage the existence of the empirical entropy function η. Sufficient smoothness and differentiability of $\eta(\mathbf{F},\theta)$ and of all other relevant functions is tacitly assumed from now on.

The validity of Caratheodory's Principle under the restriction to isometric changes of state implies that $\partial\eta(\mathbf{F}, \theta)/\partial\theta$ does not change sign. Otherwise there would be special states A such that either no isometrically unattainable states were in the neighborhood of A or no isometrical changes starting from A were possible. By suitably choosing the sign of the variable θ it is, therefore, always possible to have $\partial\eta/\partial\theta \geqslant 0$. We further restrict this by omitting the = – sign and hence postulate

$$\frac{\partial\eta}{\partial\theta} > 0. \tag{3.2}$$

4. Internal energy

The existence of the second fundamental function of state mentioned in part 2, the *internal energy* $U(\mathbf{F}, \theta)$, is derived as usual from the experience that the work W_{AB}, which has to be done on a body to change its state adiabatically from state A to state B, depends only on A and B and is independent of the process by which this change of state is effected. Hence the work can be written as the difference of the internal energies $U(\mathbf{F}_B, \theta_B$ and $U(\mathbf{F}_a, \theta_a)$:

$$W_{AB} = U_B - U_A. \tag{4.1}$$

The experience leading to the internal energy function is supplemented by the further experience that work cannot be extracted isometrically from a body. In view of (3.1) and (3.2) the variable θ grows for possible isometric changes and, hence, we may postulate:

$$\frac{\partial U(\mathbf{F}, \theta)}{\partial \theta} > 0. \tag{4.2}$$

Inequality (4.2) guarantees a unique solution of $U = U(\mathbf{F}, \theta)$ for θ, viz. $\theta = \theta(\mathbf{F}, U)$. Substitution into the entropy function yields:

$$\eta = \eta(\mathbf{F}, U), \tag{4.3}$$

$\eta(\mathbf{F}, U)$ is a unique function of its variables; (4.3) is a possible form of the 'canonical equation of state' of the body. In view of (3.2) and (4.2):

$$\frac{\partial \eta}{\partial U} > 0. \tag{4.4}$$

5. Maximum entropy principle; Gibbs' Relation

We now assume an adiabatic body to be in an equilibrium state A, with homogeneous deformation, under the action of surface tractions $\sigma^*_{ik}\, dA_{0k}$. Here dA_{0k} denotes the surface element in the reference configuration and the σ_{ik} are the Kirchhoff-Piola stresses. If the surface tractions are suddenly changed from $\sigma^*_{ik}\, dA_{0k}$ to $\sigma_{ik}\, dA_{0k}$ a process will commence which will finally lead to a new equilibrium state B. (Of course, the tractions $\sigma_{ik}\, dA_{0k}$ have to be chosen as an equilibrium system in such a way that the body in its final state is again homogeneously deformed). The assumption that the body finally always settles down to an equilibrium state B means that all equilibrium states are assumed *asymptotically stable*.

Because the process $A \to B$ is a natural one, the following holds true:

$$\eta_B - \eta_A > 0. \tag{5.1}$$

Assuming state B fixed and varying state A (by varying the initial tractions $\sigma^*_{ik}\, dA_{0k}$) we infer from (5.1) that the entropy of state B is higher than the entropy of all neighboring states A. This immediate consequence of the properties of entropy is called the 'maximum entropy principle'. With the denotation $\delta\mathbf{F} = \mathbf{F}_B - \mathbf{F}_A, \delta U = U_B - U_A$ relation (5.1) can be written as follows:

$$\frac{\partial \eta}{\partial F_{ik}}\, \delta F_{ik} + \frac{\partial \eta}{\partial U}\, \delta U + o(\delta \mathbf{F}^2, \delta U^2) > 0. \tag{5.2}$$

The derivatives are to be evaluated for the fixed state B.

During the process leading from state A to state B the constant surface tractions $\sigma_{ik}\, dA_{0k}$ perform the following amount of work:

$$\delta W_{AB} = V_0\, \sigma_{ik}\, \delta F_{ik}. \tag{5.3}$$

Here, V_0 is the volume of the body in the reference configuration. Because the body is adiabatically isolated, the equation $\delta W_{AB} = \delta U$ holds, according to the definition of internal energy. Inserting (5.3) for δU into (5.2) one obtains:

$$\left(\frac{\partial \eta}{\partial F_{ik}} + \frac{\partial \eta}{\partial U} \sigma_{ik} V_0\right) \quad \delta F_{ik} + o(\delta \mathbf{F}^2, \delta U^2) > 0. \tag{5.4}$$

The δF_{ik} can be varied independently and therefore (5.4) can be satisfied only if:

$$\sigma_{ik} = - \frac{1}{V_0} \frac{\partial \eta(\mathbf{F}, U)/\partial F_{ik}}{\partial \eta(\mathbf{F}, U)/\partial U} = \frac{1}{V_0} \frac{\partial U(\eta, \mathbf{F})}{\partial F_{ik}} . \tag{5.5}$$

From this important relation, together with the identity $d\eta = (\partial \eta / \partial U)\, dU + (\partial \eta / \partial F_{ik})\, dF_{ik}$ the following result is immediately derived:

$$\frac{\partial U(\mathbf{F}, \eta)}{\partial \eta} \, \mathrm{d}\eta = \mathrm{d}U - V_0 \, \sigma_{ik} \, \mathrm{d}F_{ik}. \tag{5.6}$$

This is the so-called 'Gibbs' Relation' for the body under consideration. In part 7 the 'metrical entropy' $S(\eta)$ will be introduced as a monotonically increasing function of the empirical entropy. It will also be shown that $\partial U(\mathbf{F}, S)/\partial S = T > 0$ is the absolute temperature. Gibbs' Relation then assumes the standard form

$$T\mathrm{d}S = \mathrm{d}U - V_0 \, \sigma_{ik} \, \mathrm{d}F_{ik}. \tag{5.7}$$

We now define an *ideal process* as a process which leads through a continuous sequence of equilibrium states. Such processes can be approximated by real ones as closely as desired. During an ideal process the surface tractions, which are the only forces assumed to act on the body, always coincide with their equilibrium values belonging to the instantaneous deformation of the body. The work done to change the deformation infinitesimally is given by $dW = V_0 \, \sigma_{ik} \, dF_{ik} = dU$. Hence, (5.7) yields $d\eta = 0$. This means that an ideal process in an adiabatic body effects an isentropic change of state and therefore, according to the definition of entropy, is reversible (by another ideal process).

6. Thermal contact; temperature

If two bodies (*a*) and (*b*) are in 'thermal contact' but are adiabatically isolated from the remaining environment, the state of body (*b*) usually changes when the state of body (*a*) is changed. Or, in other words: With a given state of (*a*) not every state of (*b*) is compatible if the 'combined system', consisting of bodies (*a*) and (*b*), is to be in equilibrium, that is if bodies (*a*) and (*b*) are to be in 'thermal equilibrium' with each other. Omitting arguments which may be found in many textbooks [3, 4, 6], it can be said in short that study of thermal equilibrium shows the existence of a function $\Theta(\mathbf{F}, \theta)$ for each body such that bodies (*a*) and (*b*) are in thermal equilibrium if and only if:

$$\Theta_a(\mathbf{F}_a, \theta_a) = \Theta_b(\mathbf{F}_b, \theta_b). \tag{6.1}$$

If a body (a) is in thermal equilibrium with (b), and if (b) in turn is in thermal equilibrium with (c) then (a) and (c) are also in thermal equilibrium (transitivity of thermal equilibrium). Therefore, one can use one body as standard, or 'thermometer', for defining the *empirical temperatures* Θ of all other bodies, and any bodies are in thermal equilibrium with each other if their temperatures Θ thus defined are equal.

If in an adiabatically isolated two-body system the state of body (a) is changed isometrically, then observation shows that thermal equilibrium between the two bodies no longer persists. Hence, the temperature of body (a) must have been changed by the isometric change of state. At the same time θ must have changed (cf. part 3). Therefore $\partial\Theta/\partial\theta$ cannot change sign. By suitably chosing the sign of Θ we can therefore assume $\partial\Theta/\partial\theta \geqslant 0$. Restricting this again by omitting the = – sign we finally postulate

$$\frac{\partial\Theta(\mathbf{F}, \theta)}{\partial\theta} > 0. \tag{6.2}$$

7. Metrical entropy; absolute temperature

For a system consisting of two homogeneously deformed bodies (a) and (b) in thermal contact the number of independent variables is, in view of (6.1), the sum of the number of independent variables of each system minus one. The validity of Caratheodory's Principle is postulated for combined systems, which are adiabatically isolated from their environment, too. Therefore, the existence of an empirical entropy η can be established for such systems in the same way as was done for a single body in part 3. Likewise, the internal energy U of the combined system is defined through work done on the adiabatic system. The internal energy can be assumed as function of the deformation $\mathbf{F}_a$, $\mathbf{F}_b$ of the two constituent bodies and of η : $U = U(\mathbf{F}_a, \mathbf{F}_b, \eta)$. Because of the additivity of work, internal energy is additive; hence:

$$U(\eta, \mathbf{F}_a, \mathbf{F}_b) = U_a(\eta_a, \mathbf{F}_a) + U_b(\eta_b, \mathbf{F}_b). \tag{7.1}$$

When performing an ideal process with the combined system, the relation $(\partial U/\partial\eta)\, d\eta = dU - dW$ (cf. (5-6)) is valid for bodies (a) and (b) and, as can be shown, for the combined system too. This leads to the validity of

$$\frac{\partial U}{\partial\eta}\, \mathrm{d}\eta = \frac{\partial U_a}{\partial\eta_a}\, \mathrm{d}\eta_a + \frac{\partial U_b}{\partial\eta_b}\, \mathrm{d}\eta_b \tag{7.2}$$

for all values η_a, η_b, $\mathbf{F}_a$, $\mathbf{F}_b$, for which $\Theta_a = \Theta_b$.

By a standard argument [4, 5] one shows that as a consequence of (7.2) the following is true for each of the two bodies (a) and (b):

$$\frac{\partial U_z}{\partial\eta_z} = T(\Theta_z) \cdot S_z'(\eta_z). \tag{7.3}$$

Here z denotes either (a), (b) (or any other body which could be used as part of a combined system). $T(\Theta)$ is a *universal* function of empirical temperature; that is, T is the same function for all bodies. Integrating the function $S'(\eta)$ yields $S(\eta)$, the 'metrical entropy'. Inserting the expression (7.3) for $\partial U/\partial\eta$ into (5.6) produces the Gibbs' Relation (5.7).

By studying ideal gas systems one can show that T is proportional to the ideal gas temperature. Because the factorization of $\partial U/\partial\eta$ is unique only up to a constant factor, the temperature T can be identified with the ideal gas temperature; in this case T is called the absolute temperatue. Since, by this choice, $T > 0$ and $\partial U/\partial\eta > 0$ (cf. (3.2), (4.2)) the function $S'(\eta)$ is positive. Therefore, the metrical entropy $S = \int S' \, d\eta$ is a monotonically increasing function of the empirical entropy η and, hence, is as suitable as the empirical entropy as a measure of irreversibility (part 3).

A combined system, consisting of two bodies (a) and (b) in thermal contact, can itself be brought into thermal contact with a third body (c). The reasoning leading to (7.3) can be repeated for this three-body-combination with the result, that (7.3) is valid also for the combined subsystem (a) + (b). In that case, T denotes the common absolute temperature of bodies (a) and (b) and $\mathbf{F}$ denotes the set of independent variables $\{\mathbf{F}_a, \mathbf{F}_b\}$. Therefore, (7.2) can be written in the form:

$$T \, dS = T(dS_a + dS_b). \tag{7.4}$$

Integration yields (omitting an irrelevant constant of integration);

$$S = S_a + S_b. \tag{7.5}$$

Eq. (8.2) shows that the metrical entropy of the combined system is additive in the same way as its internal energy is. This shows the way to generalizing the theory thus far expounded to non-homogeneously deformed, isothermal bodies by introducing the *fields* of *specific internal energy* u and *specific entropy* s together with the deformation field. ('Specific' quantities are defined as quantities per unit mass).

The next step towards a general continuum thermodynamics consists in justifying (7.5) also for non-isothermal bodies, in which the temperature varies with position $\mathbf{x}$. This can be done by studying combined systems for which the connection between the constituent bodies can be 'switched' at will from thermal contact to adiabatic (i.e. no thermal contact between the bodies). For such a system $S = S_a + S_b$ is a proper measure of entropy, as is shown, for example, in [4]. Of course, energy is also additive in that case, and the introduction of the fields u and s together with the deformation field and the temperature field is possible.

By now a point has been reached which is close to the starting point of chapter 4 of [1]. Eq. (4.3) of [1], which was made the basis of subsequent derivations in [1] is a simple consequence of the maximum entropy principle for a system consisting of a body that is thermally connected with a 'heat reservoir'. The well known principles of continuum mechanics, for example the 'Principle of Minimum Potential Energy' and 'Castigliano's Principle' are shown to be consequences of eq. (4.3) in [1]; hence, they are consequences of the maximum entropy principle.

8. Internal variable theory

The ideas put forth in the previous chapters can be generalized to homogeneously deformed bodies which are in apparent equilibrium states. Since it will simplify the following discussion without seriously diminishing the generality, let us first assume that *one* internal variable q suffices to describe these states. The state space of the body is then spanned by the independent variables $\mathbf{F}$, θ, q. By assuming these variables as 'independent', we imply that, by suitable processes performed with the body, apparent equilibria with any desired values of these variables (within an open domain of state space) can be realized.

By processes with an adiabatic body an empirical entropy function $\eta(\mathbf{F}, \theta, q)$ and an internal energy function $U(\mathbf{F}, \theta, q)$ can be established in principle in the same manner as explained earlier (chapters 3 and 4). Again $\partial\eta/\partial\theta > 0$ and $\partial U/\partial\theta > 0$ is assumed, the justification for these assumptions being given along the same lines as previously. Isometric changes are now changes of state with $\mathbf{F}$ *and* q fixed. The transfer of the argumentation employed in previous chapters for unconstrained equilibria to the apparent equilibria studied now hinges upon the assumption that the characteristic time t_1 for establishment of apparent equilibria is *much smaller* than the chracteristic time t_2 for establishment of unconstrained equilibria. In that case one can contemplate processes on a time scale t_0, with $t_1 \ll t_0 \ll t_2$. By such processes the system can be conveyed from one apparent equilibrium to another one without changing the value of q: During such processes the system behaves as a 'frozen' system. For frozen systems the thermodynamics theory developed in parts 3-7 is immediately applicable.

The empirical temperature $\Theta(\mathbf{F}, \theta, q)$ is defined in complete analogy to the situation for unconstrained equilibria (part 6), and finally, the arguments of part 7, applied to the present case, lead to the metrical entropy $S(\mathbf{F}, \theta, q)$ and the absolute temperature $T(\mathbf{F}, \theta, q)$. By suitable elimination of the variable θ one obtains $U = U(\mathbf{F}, S, q)$. Relation (5.7) is valid for a frozen system ($dq = 0$); therefore, Gibbs' Relation now assumes the form (in specific variables):

$$T ds = du - \frac{1}{\rho_0}\, \sigma_{ik}\, dF_{ik} - \frac{\partial u}{\partial q}\, dq. \tag{8.1}$$

From (8.1) one derives:

$$\sigma_{ik} = \rho_0 \frac{\partial u(s, \mathbf{F}, q)}{\partial F_{ik}}, \quad \text{or:} \quad \tau_{ik} = -\rho \frac{\partial u}{\partial F_{il}} F_{kl}, \tag{8.2}$$

Here σ_{ik} denotes the Kirchhoff-Piola-stresses, τ_{ik} the Cauchy-stresses, ρ_0 is the density in the reference configuration and ρ the density in the deformed configuration.

Keeping U and $\mathbf{F}$ fixed, the body considered will slowly change its apparent equilibrium by a naturally proceeding process till finally an unconstrained equilibrium is reached. During the process, leading through a succession of apparent equilibrium states to the final equilibrium, the entropy of the body cannot decrease. After the final equilibrium has been attained no further changes are possible. This means that the entropy has reached a maximum value as a function of q for $\mathbf{F}$ and u fixed. Therefore, unconstrained equilibrium is characterized by $\partial s(\mathbf{F}, u, q)/\partial q = 0$ and $\partial^2 s/\partial q^2 < 0$. In view of (8.1) this is equivalent to

$$\frac{\partial u(\mathbf{F}, s, q)}{\partial q} = 0, \quad \frac{\partial^2 u(\mathbf{F}, s, q)}{\partial q^2} > 0. \tag{8.3}$$

Solving the first relation (8.3) for the internal variable q yields the equilibrium value $\bar{q}(\mathbf{F}, S)$.
Substituting $q = \bar{q}$ in the energy function $u(\mathbf{F}, s, q)$ leads to:

$$\overline{u}(\mathbf{F}, s) := u(\mathbf{F}, s, \overline{q}(\mathbf{F}, S)). \tag{8.4}$$

This is the energy function for unconstrained equilibria. For states close to unconstrained equilibria, the energy function can be expanded as follows:

$$u(\mathbf{F}, s, q) = \overline{u}(\mathbf{F}, s) + \frac{1}{2} \frac{\partial^2 u}{\partial q^2} (q - \overline{q})^2 . \tag{8.5}$$

A term which is linear in $q - \overline{q}$ does not appear because (8.3) is valid.

The generalization of the theory to nonhomogeneously deformed, non-isothermal bodies by introducing the fields u, s, F, Θ, q is straightforward and follows the lines indicated at the end of part 7.

9. Fluids and viscosity

The class of bodies – or materials – considered here is wide enough to encompass solids as well as fluids. For a fluid the depency of u on $\mathbf{F}$ reduces to the dependency on $\Delta = \det \mathbf{F} = \rho_0/\rho$. In that case (8.2) yields [1]:

$$\tau_{ik} = -\rho^2 \frac{\partial u(\rho, s, q)}{\partial \rho} \delta_{ik} =: -p\,\delta_{ik}. \tag{9.1}$$

Eq. (10.3) shows that the Cauchy stress tensor in a fluid is spherically symmetric; p (ρ, s, q) denotes the pressure. The remaining part of the discussion is exclusively concerned with fluids; however, most of the conclusions can be easily transferred to the case of solids.

The generalization of the thermodynamic theory to nonhomogeneous bodies, as explained above, suggests a further step of generalization, namely the generalization to moving bodies, i.e. to flow fields. The thermodynamic quantities of each material particle in such a flow field will change with time. The generalization of the previous theory to that case leads to the thermodynamic theory of processes, or 'irreversible thermodynamics'. The basic assumption needed for this last generalization is a 'local equilibrium hypothesis', which can be formulated as follows [9]: Take a small volume of fluid, isolate it instantaneously from its surroundings by a (fictitious) rigid and adiabatic wall and move this adiabatic body with a constant velocity equal to its mean mass velocity at the time of isolation. The smaller the volume, the faster it will settle down to equilibrium. In the limit of an infinitesimally small volume, equilibrium will be attained instantaneously. The *local* thermodynamic quantities are defined as those belonging to the equilibrium state attained by the fictitious act of isolating an infinitesimally small volume. These local quantities depend on position in the flow field and, in unsteady flow fields, also on time. In its simplest form the local equilibrium hypothesis now states: 1) The local equilibrium state thus defined is an unconstrained equilibrium; 2) the stress at any point in the flow field at any time is given by the stress belonging to that equilibrium state. The theory based on this hypothesis is usually called 'ideal fluid flow theory'.

Ideal fluid flow theory has a number of consequences which contradict experience: Rigid bodies moving through a fluid will be found without drag. Intimately related to this absence of drag is the fact that ideal fluid flow can be easily shown to be isentropic: the specific entropy of individual gas particles does not change. This too is contrary to experience. The usual method of fluid dynamics to avoid such contradictions consists in introducing 'viscous' stresses in addition to the equilibrium pressure. (Apart from viscous stresses other dissipative mechanisms like heat conduction usually play a role; such mechanisms will be neglected here). However, at least some of the deficiencies of ideal fluid flow theory can also be removed by assuming that the equilibria attained in the infinitesimally small particles immediately after isolation are no unconstrained but apparent equilibria, which necessitate the introduction of internal variables. As a matter of fact, viscous stresses can, in principle, be explained by an internal variable theory.

It has been known since many years that the viscous stresses caused by 'volume viscosity' can be explained as an internal variable effect in the following way: If the characteristic time t_0 of the flow, i.e. the time scale for changes of state of the moving fluid particles, is considerably larger than the characteristic time t_2 introduced in part 8 the particles will never be far from unconstrained equilibrium. Therefore, one may use the following equation of state (cf. (8.5))

$$u(\rho, s, q) = \bar{u}(\rho, s) + c(\rho, s)\,(q - \bar{q})^2 . \tag{9.2}$$

Here

$$c := \frac{1}{2}\,\partial^2 u/\partial q \,\Big|_{q=\bar{q}} > 0, \quad \bar{q} = \bar{q}(\rho, s). \tag{9.3}$$

According to (9.3) the stress in the fluid is given by

$$p = \bar{p} - 2\,c\,\rho^2\,(q - \bar{q})\,\frac{\partial \bar{q}}{\partial \rho} + o((q - \bar{q})^2). \tag{9.4}$$

$\bar{p} = \rho^2\,\partial\bar{u}/\partial\rho$ is called the 'unconstrained equilibrium pressure'. Introducing the so-called 'Damköhler parameter' $D := t_2/t_0 \ll 1$ one can show that $q - \bar{q} = O(D)$. In the following derivations only terms which are linear in D will be retained. The pressure $\bar{p}$ is – apart from terms $O(D^2)$ – the pressure which a fluid particle attains after its fictitious isolation if one waits till it has reached unconstrained equilibrium. This is so because the density of the particle does not change after isolation and the entropy changes during the process leading from the apparent to the unconstrained equilibrium are of order $O(D^2)$.

Using a simple relaxation equation for the time rate of change of q, namely

$$\tau\,\dot{q} = \bar{q} - q \tag{9.5}$$

where $\tau = O(t_2)$ is a 'relaxation time', and also using the mass balance, $\dot{\rho} = -\rho\,\operatorname{div}\vec{v}$, eq. (9.4) is reduced to the following result (terms $O(D^2)$ are omitted):

$$p = \bar{p} - \lambda\,\operatorname{div}\vec{v}, \quad \text{with } \lambda := 2c\tau\rho^3\left(\frac{\partial \bar{q}}{\partial \rho}\right)^2 > 0. \tag{9.6}$$

λ is the volume viscosity in the usual sense.

To describe the effects of shear viscosity in a fluid by introducing internal variables is a much harder problem, which was recently tackled and solved by my coworker L. Pfeifroth. From previously published papers on internal variable theory [7, 10] one can infer that, in principle, viscous shear stresses are predicted by such theories if the material is assumed to be always close to unconstrained equilibrium. However, although these stresses emerge as linear functions of **A**, the symmetric part of the gradient tensor of the velocity field (or 1st Rivlin-Ericksen-tensor, cf. [1]), they also depend on **F** and hence on the reference configuration. If the theory is to describe a 'newtonian fluid', the dependency of the viscous shear stresses on **F** must reduce to a depency on $\Delta = \det \mathbf{F}$, and hence on the density of the fluid only.

In order to achieve this goal Pfeifroth postulates nine internal variables q_{ij} $(i, j = 1, 2, 3)$ and the following internal energy function (for states close to unconstrained equilibrium, cf. (8.5)):

$$u(\mathbf{F}, s, q_{ij}) = \overline{u}(\mathbf{F}, s) + M_{ijkl}(q_{ij} - \overline{q}_{ij})(q_{kl} - \overline{q}_{kl}) \tag{9.7}$$

with $M_{ijkl}(\mathbf{F}, s) = M_{klij}$. For reasons of frame indifference the depency of u on **F** reduces to a depency on the right-Cauchy-Green tensor $\mathbf{C} = \mathbf{F}^T \mathbf{F}$ (cf. [1]). The relaxation equation are postulated as follows:

$$\tau \dot{q}_{ij} = \overline{q}_{ij} - q_{ij}, \quad i, j = 1, 2, 3 \tag{9.8}$$

Onsager's relations [8] are automatically satisfied by (9.7) and (9.8). With two arbitrary functions $f(\Delta^2)$ and $g(\Delta^2)$, the equilibrium values $\overline{q}_{ij}$ and the coefficients M_{ijkl} are identified with

$$\overline{q}_{ij} = \frac{\partial f(\Delta^2)}{\partial C_{ij}}, \quad M_{ijkl} = g(\Delta^2) \left[\frac{\partial f(\Delta^2)}{\partial \mathbf{C}\, \partial \mathbf{C}} \right]^{-1}_{ijkl} \tag{9.9}$$

With these choices the following Cauchy stresses are obtained:

$$\tau_{ij} = \overline{\tau}_{ij} + 2\mu A_{ij} + \lambda A_{kk}\, \delta_{ij}. \tag{9.10}$$

$\overline{\tau}_{ij}$ are the equilibrium stresses; for a fluid: $\overline{\tau}_{ij} = -\overline{p}\, \delta_{ij}$. λ is a volume viscosity; the shear viscosity μ is given by

$$\mu = -4\tau \rho g \cdot \Delta^2 \cdot \frac{\mathrm{d}}{\mathrm{d}\Delta^2} \left(\Delta^2 \frac{\mathrm{d}f}{\mathrm{d}\Delta^2} \right) = \mu(\rho, s). \tag{9.11}$$

Pfeifroth has shown that $\mu > 0$ and $\lambda + \frac{2}{3}\mu > 0$ if the 'stability condition' $M_{ijkl}\, x_{ij}\, x_{kl} > 0$, for $x_{ij} \neq 0$, is satisfied (cf. (8.3, 2)). It is of interest to remark that the internal variables q_{ij} can be interpreted as internal Piola stresses; the corresponding internal Cauchy stresses are then spherically symmetric in unconstrained equilibrium, as may be concluded from the expression (9.9, 1) for $\overline{q}_{ij}$. This is in accord with intuitive conceptions of 'viscous' stresses in a fluid. – A detailed version of Pfeifroth's work will be published elsewhere [11].

References

1. E. Becker, W. Bürger, *Kontinuumsmechanik*, Stuttgart, B.G. Teubner 1975.
2. W.T. Koiter, 'Thermodynamics of elastic stability', in: *Proceedings, Third Canadian*

Congress of Applied Mechanics, P.G. Glockner ed., Calgary 1971.
3. S. Chandrasekhar, *An introduction to the study of stellar structure*, Chicago, University of Chicago Press 1939.
4. H.A. Buchdahl, *The concepts of classical thermodynamics*, Cambridge University Press, 1966.
5. H.A. Buchdahl, *Twenty lectures on thermodynamics*, Rushcutters Bay NSW, Pergamon Press (Australia) 1975.
6. L.C. Woods, *The thermodynamics of fluid systems*, Oxford, Clarendon Press 1975.
7. B.T. Chu, 'Thermodynamics of deformation of solids', in: *A critical review of thermodynamics*, B. Stuart, B. Gal-Or, A.J. Brainard eds., Baltimore, Mono Book Corp., 1970.
8. E. Becker, 'Chemically reacting flows', in: *Annual Review of Fluid Mechanics*, M. Van Dyke, W.G. Vincenti, J.V. Wehausen eds., Palo Alto, Annual Reviews Inc. 1972.
9. E. Becker, *Gasdynamics*, New-York, London, Academic Press 1968.
10. K.C. Valanis, *Irreversible thermodynamics of continuous media, CISM Courses and Lectures no. 77*, Wien, New-York, Springer 1972.
11. L. Pfeifroth, 'Newtonsche Flüssigkeiten als elastische Materialien mit inneren Variablen', to be published in *Z. Angew. Math. Mech.*

Finite Element Methods

*J.F. Besseling**

Summary

Linear algebra offers itself as an ideal tool for the presentation of the principles of the classical linear theory of structures. Problems of ill-conditioning in solution procedures have been alleviated considerably by the so-called natural factor formulation, both in the displacement method and in the force method. An algebraic approximation of those differential equations of the continuous field theory, which are subject to boundary conditions, can be given in terms of finite element equations. They can be derived on the basis of the principle of virtual work, supplemented by a principle of virtual heat. A distinction is made between spatial finite elements for problems of flow, and material finite elements for problems of deformation. Aspects of stability of structures and of kinematic analysis of mechanisms are considered in terms of finite element concepts.

1. Introduction

The essence of the finite element methods lies in the observation that solutions of the field equations of continuum theories can always be approximated locally to any desired degree of accuracy by interpolation polynomials of low degree, except for a neighbourhood of singularities in the field. Since knowledge of the state of a body at any time will be considered adequate in a phenomenological sense if in a finite number of subregions this local approximation is available, finite element descriptions of continuous fields aim at providing these local approximations.

For the derivation of the finite element equivalent of the non-scalar equations of mechanics the variational formulation of these equations is most appropriate. The expression itself, that is being considered, represents a scalar, and hence is frame invarient, while it implies the boundary conditions as well as the differential equations of the field. Thus for purely mechanical problems the principle of virtual work is employed to formulate the laws of balance of momentum and of moment of momentum. For thermal problems a principle of virtual heat incorporates the heatflow equation with the proper boundary conditions. After a discussion of the theory of linear structural mechanics and associated solution procedures, it will be demonstrated that the variational statements of the two principles mentioned suffice for the derivation of the finite element equations in terms of the primary variables: stress, temperature, density, velocity and displacement.

* Laboratory of Engineering Mechanics, Mekelweg 2, Delft, The Netherlands

2. Linear structural mechanics

The direct sum of the vector spaces of the nodal displacements of each finite element is a finite dimensional vector space $\mathscr{U}^e$ with elements in $\mathscr{R}^{n^e}$. By means of the displacement interpolation functions on the boundary of the finite element the contact forces acting on the element can be represented by their virtual work as a linear function $\mathscr{L}(\mathscr{U}^e;\mathscr{R})$ from $\mathscr{U}^e$ into the real numbers $\mathscr{R}$. If in addition to contact forces there are also forces distributed over the domains of the finite elements, in these domains an interpolation between nodal displacements must be given in order to determine the contribution of these distributed forces to the virtual work.
$\mathscr{L}(\mathscr{U}^e;\mathscr{R})$ is the duel space of $\mathscr{U}^e$, and we denote it by the notation $\mathscr{F}^e$. Thus for each finite element we define

$$\int_{\partial B^e} t_i \delta u_i \, dA + \int_{B^e} \rho f_i \delta u_i \, dV = \langle f^e, \delta u^e \rangle, \; f^e \in \mathscr{F}^e \wedge \delta u^e \in \mathscr{U}^e. \tag{2.1}$$

On the left hand side we have the usual expression for the virtual work in terms of a surface integral and a volume integral with the cartesian components of the surface tractions, t_i, of the force density, ρf_i, and of the virtual displacements, δu_i. The bracket $\langle,\rangle$ operation on the right hand side is the scalar product, which can be viewed as a function

$$\langle,\rangle : \mathscr{F}^e \times \mathscr{U}^e \rightarrow \mathscr{R}. \tag{2.2}$$

The infinitesimal rigid body motions of a finite element form a subspace of $\mathscr{U}^e$, which will be called the vector space of degrees of freedom, $\mathscr{V}^e$. Virtual translations and rotations without deformation are defined by

$$\delta u^e \in \operatorname{Im} A = \{A \, \delta \nu^e \mid \delta \nu^e \in \mathscr{V}^e\}. \tag{2.3}$$

For the linear transformation A holds by virtue of the definition of $\mathscr{V}^e$

$$\operatorname{Dim} \operatorname{Im} A = \operatorname{Dim} \mathscr{V}^e \Leftrightarrow \operatorname{Ker} A = \{o\}. \tag{2.4}$$

Thus the linear transformation A is one-to-one (injective): for every vector $\delta u^e \in \operatorname{Im} A$ there exists only one element $\delta \nu^e \in \mathscr{V}^e$ such that $\delta u^e = A \, \delta \nu^e$.

A necessary condition for equilibrium of the finite element reads according to the principle of virtual work

$$\langle f^e, A \, \delta \nu^e \rangle = 0 \quad \forall \delta \nu^e \in \mathscr{V}^e, \tag{2.5}$$

or with the dual A^T of the linear transformation A

$$\langle A^T f^e, \delta \nu^e \rangle = 0 \quad \forall \delta \nu^e \in \mathscr{V}^e. \tag{2.6}$$

Hence in the case of equilibrium must hold

$$f^e \in \operatorname{Ker} A^T = \{f^e \mid A^T f^e = 0\}. \tag{2.7}$$

It is a property of the dual A^T that

$$\text{Dim Im } A^T = \text{Dim Im } A = \text{Dim } \mathscr{V}^e. \tag{2.8}$$

We shall call Im A^T the space of resultants, $\mathscr{P}^e$. Forces $f^e \in$ Ker A^T have zero resultants and satisfy the equations of equilibrium. Ker A^T may appropriately be called the space of generalized stresses, to be denoted by Σ^e. The stresses determine the state of loading of the element.

The stress space Σ^e has elements in $\mathscr{R}^{m^e}$, while this displacement space $\mathscr{U}^e$ had elements in $\mathscr{R}^{n^e}$. For a finite element in two-dimensional Euclidian space we have $m^e = n^e - 3$, and in three-dimensional Euclidean space $m^e = n^e - 6$.

A mapping of a natural basis for Σ^e = Ker A^T into the force space $\mathscr{F}^e$ can be determined by means of the row canonical form A^{KT} of the matrix A^T. Possibly after rearrangement of columns in A^T and of the corresponding elements in f^e, we can write

$$A^{KT} f^e = [I, B] \begin{vmatrix} f^1 \\ f^2 \end{vmatrix} = f^1 + Bf^2 = 0. \tag{2.9}$$

It follows that the columns of the matrix

$$C^{eT} = \begin{bmatrix} -\mathbf{B} \\ \mathbf{I} \end{bmatrix} \tag{2.10}$$

are linearly independent and from a basis for Ker A^{KT} = Ker $A^T \mathscr{F}^e$. Hence we have constructed the one-to-one linear transformation

$$C^{eT}: \Sigma^e \to \mathscr{F}^e. \tag{2.11}$$

The equilibrium condition for the finite element can now be formulated as follows

$$f^e \in \text{Im } C^{eT} = \{f^e \mid f^e = C^{eT} \sigma^e, \sigma^e \in \Sigma^e\}. \tag{2.12}$$

The principle of virtual work assumes the form

$$< f^e, \delta u^e > = < C^{eT} \sigma^e, \delta u^e > \; \forall \delta u^e \in \mathscr{U}^e, \tag{2.13}$$

or, with the dual C^e of the linear transformation (2.11),

$$< f^e, \delta u^e > = < \sigma^e, C^e \delta u^e > \; \forall \delta u^e \in \mathscr{U}^e. \tag{2.14}$$

The linear transormation

$$C^e : \mathscr{U}^e \to \mathscr{E}^e \tag{2.15}$$

determines a subspace $\mathscr{E}^e$ of $\mathscr{U}^e$, which can be called the space of the virtual deformations of the finite element. We have

$$\text{Dim Im } C^e = \text{Dim Im } C^{eT} = \text{Dim } \Sigma^e. \tag{2.16}$$

The deformation space $\mathscr{E}^e$ is dual of the stress space Σ^e and the scalar product $< \sigma^e, \delta \epsilon^e >$, $\sigma^e \in \Sigma^e \wedge \delta \epsilon^e \in \mathscr{E}^e$, is called the virtual work of deformation.

In the linearized theory of elasticity small, but finite deformations $\epsilon^e \in \mathscr{E}^e$ are brought into a one-to-one correspondence to generalized stresses $\sigma^e \in \Sigma^e$ by the bijective transformation

$$S^e : \mathscr{E}^e \rightarrow \Sigma^e. \tag{2.17}$$

As it was already observed the stresses determine the state of loading of a finite element. The limit of elasticity is then determined in terms of a yield function

$$\phi^e(\sigma^e) = 0. \tag{2.18}$$

Next we consider a structure as a composition of finite elements. The nodal displacements of the elements are in intersecting subspaces of the displacement space of the complete structure, $\mathscr{U}$. For each finite element we have the transformation

$$T^e : \mathscr{U} \rightarrow \mathscr{U}^e. \tag{2.19}$$

This linear transformation is onto (surjective): for each $\boldsymbol{u}^e \in \mathscr{U}^e$ there exists at least one element $\boldsymbol{u} \in \mathscr{U}$ such, that $\boldsymbol{u}^e = \boldsymbol{T}\boldsymbol{u}$. We now define the linear transformation

$$D^e = C^e T^e : \mathscr{U} \rightarrow \mathscr{E}^e. \tag{2.20}$$

The space of small defomations, $\mathscr{E}$, and the dual space of generalized stresses, Σ for the structure as a whole are introduced as the direct sum of the deformation spaces and stress spaces of the individual finite elements. Fig. 1 illustrates the various spaces and transformations which are being introduced. From (2.20) we deduce

$$D : \mathscr{U} \rightarrow \mathscr{E}. \tag{2.21}$$

For the virtual work of the forces acting on the structure as a whole we have with the continuous virtual displacement field of the finite elements

$$\int_{\partial B} t_i \delta u_i \, dA + \int_B f_i \delta u_i \, dV = \langle f, \delta \boldsymbol{u} \rangle, f \in \mathscr{F} \wedge \delta \boldsymbol{u} \in \mathscr{U}. \tag{2.22}$$

Forces on the structure are by (2.22) defined as elements in the force space $\mathscr{F}$, which is the dual spece of $\mathscr{U}$.

The virtual work of the forces acting on the structure is equal to the sum of the virtual work of the forces on all individual elements. According to (2.14) and (2.21) the necessary condition for equilibrium of a structure, composed of N elements, is therefore given by

$$\langle f, \delta \boldsymbol{u} \rangle = \sum_{e=1}^{N} \langle f^e, \delta \boldsymbol{u}^e \rangle = \sum_{e=1}^{N} \langle \sigma^e, \delta \epsilon^e \rangle = \langle \sigma, \delta \epsilon \rangle$$

$$\forall \delta \boldsymbol{u} \in \mathscr{U} \wedge \delta \epsilon = \boldsymbol{D} \delta \boldsymbol{u}. \tag{2.23}$$

With the dual of the transformation $\boldsymbol{D}$ we have

$$\langle f, \delta \boldsymbol{u} \rangle = \langle \boldsymbol{D}^T \sigma, \delta \boldsymbol{u} \rangle \; \forall \delta \boldsymbol{u} \in \mathscr{U}, \tag{2.24}$$

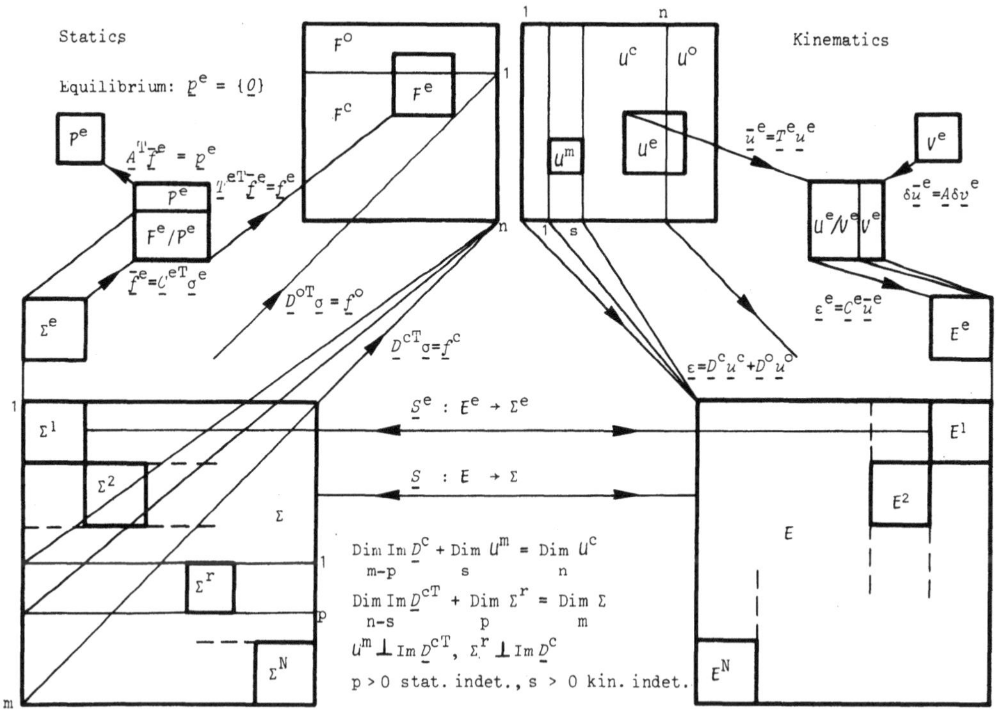

Fig. 1. Dual spaces and transformations in linear structural mechanics.

from which we obtain the equilibrium equations

$$D^T\sigma = f. \tag{2.25}$$

In the linearized theory we have furthermore the continuity relations

$$\epsilon = Du. \tag{2.26}$$

After splitting up the displacement space $\mathscr{U}$ into a subspace of prescribed displacements $\mathscr{U}^0$ and its direct complement $\mathscr{U}^c$ (see Fig. 1), the equilibrium and continuity equations for the structure can be rewritten as follows

$$D^{cT}\sigma = f^c, \quad \epsilon - D^0u^0 = D^cu^c. \tag{2.27}$$

$$D^{0T}\sigma = f^0,$$

The reactions f^0 form a subspace $\mathscr{F}^0$ of $\mathscr{F}$, which is the dual of the subspace of prescribed displacements, $\mathscr{U}^0$. $\mathscr{F}/\mathscr{F}^0$ is the space of prescribed forces, $\mathscr{F}^c$.

The static characteristics of the structure are contained in the linear transformations D^{cT} and D^{0T}. We observe that

$$\text{Dim Im } D^{cT} + \text{Dim Ker } D^{cT} = \text{Dim } \Sigma \; . \tag{2.28}$$

The subspace Ker D^{cT} of the stress space Σ is the space of the so-called redundant stresses, Σ^{r}. Equilibrium requires

$$f^c \in \text{Im} D^{cT}. \tag{2.29}$$

In case of equilibrium the complete preimage σ of f^c under D^{cT} is the equivalence set

$$\sigma = \{\sigma + \sigma^r \mid D^{cT}\sigma = f^c, \sigma^r \in \Sigma^{r}\}. \tag{2.30}$$

The duals of D^{cT} and D^{0T}, the linear transformations D^c and D^0, determine the kinematic characteristics of the structure. We have

$$\text{Dim Im } D^c + \text{Dim Ker } D^c = \text{Dim } \mathscr{U}^c. \tag{2.31}$$

The subspace Ker D^c of $\mathscr{U}^c$ is the space of kinematical degrees of freedom, $\mathscr{U}^m$, of the structure. All possible rigid motions of the structure can be expressed in a vector $u^m \in \mathscr{U}^m$. Continuity of the structure requires

$$\epsilon - D^0 u^0 \in \text{Im } D^c. \tag{2.32}$$

In case of continuity the complete preimage u^c of $\epsilon - D^0 u^0$ under D^c is the equivalence set

$$u^c = \{\bar{u}^c + u^m \mid D^c \bar{u}^c = \epsilon - D^0 u^0, u^m \in \mathscr{U}^m\}. \tag{2.33}$$

We notice that the image space of the linear transformation D^c is orthogonal to Ker D^{cT}, because for $\sigma \in$ Ker D^{cT} holds $f^c = D^{cT}\sigma = o$ and

$$< f^c, \delta u^c > = < \sigma, D^c \delta u^c > = 0 \;\; \forall \sigma \in \text{Ker } D^{cT}. \tag{2.34}$$

Hence the continuity conditions (2.32) can also be written in the form

$$< \sigma^r, \epsilon - D^0 u^0 > = 0 \;\; \forall \sigma^r \in \Sigma^r = \text{Ker } D^{cT}. \tag{2.35}$$

Likewise Ker D^c is orthogonal to the image space of the linear transformation D^{cT}, because for $\delta u^c \in$ Ker D^c holds $\delta\epsilon = D^c \delta u^c = o$ and

$$< \sigma, \delta\epsilon > = < D^{cT}\sigma, \delta u^c > = 0 \;\; \forall \delta u^c \in \text{Ker } D^c. \tag{2.36}$$

It follows that the equilibrium conditions (2.29) can also be written in the form

$$< f^e, \delta u^m > = 0 \;\; \forall \delta u^m \in \mathscr{U}^m = \text{Ker } D^c. \tag{2.37}$$

In (2.37) we recover the original equilibrium requirement according to the principle of virtual work: for all infinitesimal rigid body motions the work done by the forces acting on the body should be equal to zero.

The illustration of Fig. 1 summarizes the duality of the static and kinematic aspects of a discretized, finite element model of a structure. In addition to the static and kinematic relations in linear elasticity theory for the direct sum of the deformation spaces and stress spaces of the individual elements holds the bijective transformation

$$S : \mathscr{E} \to \Sigma \ , \tag{2.38}$$

if for the individual elements we have (2.17).

In the concise language of linear algebra [1] we have discussed the classical linear theory of structural mechanics, as it has evolved over more than hundred years. In view of the fact that a course in linear algebra is nowadays standard in any engineering curriculum it would seem advisable to develop the theory of structural mechanics in undergraduate teaching along the lines here indicated. Of course, as in the presentation with the older mathematical tools, many examples are needed to illustrate the physical content of the mathematical relations. The teaching by illustration ought to be facilitated by the fact that the analysis of a complex structure starts by the consideration of the rigid body motion of geometrically simple elements.

3. Solution procedures

The most straightforward solution procedure for the equations of linear structural mechanics is the matrix displacement method. It rests upon the equations obtained by the introduction of the linear transformations (2.38) and (2.21) in the equilibrium equations (2.25)

$$\mathbf{D}^T\sigma = \mathbf{D}^T\mathbf{S}\epsilon = \mathbf{D}^T\mathbf{S}\mathbf{D}\mathbf{u} = \mathbf{f}. \tag{3.1}$$

Provided Ker $D^c = \{o\}$ the solution for the nonprescribed quantities in a structural problem is uniquely determined by

$$\begin{aligned} \mathbf{u}^c &= [\mathbf{D}^{cT}\,\mathbf{S}\mathbf{D}^c]^{-1}\,(\mathbf{f}^c - \mathbf{D}^{cT}\,\mathbf{S}\mathbf{D}^0\mathbf{u}^0), \\ \epsilon &= \mathbf{D}^c\mathbf{u}^c + \mathbf{D}^0\mathbf{u}^0, \\ \sigma &= \mathbf{S}\epsilon, \\ \mathbf{f}^0 &= \mathbf{D}^{0T}\sigma. \end{aligned} \tag{3.2}$$

The displacement method of analysis may be contrasted with the matrix force method. The continuity conditions in the form (2.35) can be used to determine the redundant stresses in (2.30), once a linear relation between σ and ϵ has been established in the form of the inverse of the transformation (2.38). The equations then obtained are identical to the equations for the redundant stresses as they are derived in classical treatments with the aid of Castigliano's second theorem, or on the basis of the principle of minimum complementary energy for the discretized model of the structure [2].

In order to obtain a basis for the redundant stresses σ^r in the stress space Σ we can put the matrix $\mathbf{D}^{cT}$ by elementary row operations into row canonical form. Possibly after rearrangement of columns of the matrices and of the elements in the corresponding vectors we can replace by means of the Gauss-Jordan algorithm the matrix equations

$$\mathbf{D}^{cT}\sigma = \mathbf{I}\,\mathbf{f}^c \tag{3.3}$$

by

$$\begin{array}{c} \\ 1 \\ \vdots \\ n-s \\ \vdots \\ n \end{array}\begin{array}{c} 1\,..\,m-p\,..\,m \\ \left[\begin{array}{cc} \mathbf{I} & \mathbf{B} \\ \mathbf{O} & \mathbf{O} \end{array}\right] \end{array}\sigma = \begin{array}{c} \\ 1 \\ \vdots \\ n-s \\ \vdots \\ n \end{array}\begin{array}{c} 1\,...\,m-p\,...\,n \\ \left[\begin{array}{cc} [\mathbf{D}^{ccT}]^{-1} & \mathbf{O} \\ -\mathbf{B}^{*T} & \mathbf{I} \end{array}\right] \end{array}\mathbf{f}^c, \; n-s = m-p. \tag{3.4}$$

The equivalence set (2.30) is now defined by

$$\sigma = \Sigma^{0T}\,\mathbf{f}^c + \mathbf{Z}^T\sigma^r \tag{3.5}$$

with

$$\Sigma^{0T} = \begin{array}{c} \\ 1 \\ \vdots \\ m-p \\ \vdots \\ m \end{array}\begin{array}{c} 1\,..\,n-s\,..\,n \\ \left[\begin{array}{cc} [\mathbf{D}^{ccT}]^{-1} & \mathbf{O} \\ \mathbf{O} & \mathbf{O} \end{array}\right] \end{array}, \quad \mathbf{Z}^T = \left[\begin{array}{c} -B \\ \mathbf{I} \end{array}\right]. \tag{3.6}$$

In addition the equilibrium equations

$$\mathbf{Z}^{*T}\mathbf{f}^c = \mathbf{o} \tag{3.7}$$

must be satisfied, where according to (3.4)

$$\mathbf{Z}^{*T} = [-\mathbf{B}^{*T}\;\mathbf{I}]. \tag{3.8}$$

The Gauss-Jordan algorithm applied to the equations

$$\mathbf{D}^c\mathbf{u}^c = \mathbf{I}(\epsilon - \mathbf{D}^0\mathbf{u}^0), \tag{3.9}$$

possibly again after rearrangement of columns of the matrices and of the corresponding elements in the vectors, leads to

$$\begin{array}{c} \\ 1 \\ \vdots \\ m-p \\ \vdots \\ m \end{array}\begin{array}{c} 1\,..\,n-s\,..\,n \\ \left[\begin{array}{cc} \mathbf{I} & \mathbf{B}^* \\ \mathbf{O} & \mathbf{O} \end{array}\right] \end{array}\mathbf{u}^c = \begin{array}{c} \\ 1 \\ \vdots \\ m-p \\ \vdots \\ m \end{array}\begin{array}{c} 1\,..\,n-s\,..\,m \\ \left[\begin{array}{cc} \mathbf{D}^{cc^{-1}} & \mathbf{O} \\ -\mathbf{B}^T & \mathbf{I} \end{array}\right] \end{array}(\epsilon - \mathbf{D}^0\mathbf{u}^0), \; n-s = m-p. \tag{3.10}$$

The equivalence set (2.30) assumes the form

$$\mathbf{u}^c = \Sigma^0(\epsilon - \mathbf{D}^0\mathbf{u}^0) + \mathbf{Z}^*\mathbf{u}^m. \tag{3.11}$$

In addition the continuity equations

$$\mathbf{Z}(\epsilon - \mathbf{D}^0 \mathbf{u}^0) = \mathbf{O} \tag{3.12}$$

must be satisfied.

The fact that the matrix $\mathbf{Z}^*$ occurs both as a basis of the space $\mathscr{U}^m$ in the equivalence set (3.11) and as the matrix of coefficients in the equilibrium equations (3.7) is explained by the orthogonality of Ker D^c and Im D^{cT}, as expressed by (2.37). Likewise the orthogonality of Ker D^{cT} and Im D^c, expressed by (2.35), shows up in the simultaneous appearance of the matrix $\mathbf{Z}$ in the continuity equations (3.12) and as the basis of the space Σ^r in the equivalence set (3.5).

Introduction of the elasticity relation (2.38) and of the equivalence set (3.5) into the continuity equations (3.12) gives the equations for the redundant stresses

$$\mathbf{Z}\mathbf{S}^{-1}\mathbf{Z}^T\sigma^r = -\mathbf{Z}\mathbf{S}^{-1}\Sigma^{0T}\mathbf{f}^c + \mathbf{Z}\mathbf{D}^0\mathbf{u}^0. \tag{3.13}$$

Provided the equilibrium equations (3.7) are satisfied the force method determines the nonprescribed quantities in a structural problem by the following expressions

$$\begin{aligned}
\sigma &= [\mathbf{I} - \mathbf{Z}^T[\mathbf{Z}\mathbf{S}^{-1}\mathbf{Z}^T]^{-1}\mathbf{Z}\mathbf{S}^{-1}]\Sigma^{0T}\mathbf{f}^c + \mathbf{Z}^T[\mathbf{Z}\mathbf{S}^{-1}\mathbf{Z}^T]^{-1}\mathbf{Z}\mathbf{D}^0\mathbf{u}^0,\\
\mathbf{f}^0 &= \mathbf{D}^{0T}\sigma,\\
\epsilon &= \mathbf{S}^{-1}\sigma,\\
\mathbf{u}^c &= \Sigma^0(\epsilon - \mathbf{D}^0\mathbf{u}^0) + \mathbf{Z}^*\mathbf{u}^m.
\end{aligned} \tag{3.14}$$

The force method lost its one time prominence to the displacement method, largely because of the greater computational effort which is required in elastic stress analysis by the force method, except in some problems of frame structures and of shear panels. Still the force method offers some attractive features for limit analysis and optimization problems [2, 3].

Ill-conditioning of the matrix to be inverted may spoil the accuracy of the solution (3.2) by the displacement method and also of the solution (3.14) by the force method, but the ill-conditioning arises in the two methods from different causes.

If a very stiff part of the structure is fully imbedded in parts with a high flexibility, unequilibrated forces acting on the stiff part will cause it to move with relatively very small deformations of itself. Under these circumstances the unknown nodal displacements of the structure approach a linear dependence and this results in an ill-conditioned matrix $\mathbf{D}^{cT}\mathbf{S}\mathbf{D}^c$.

In the force method the stresses corresponding to two or more vectors of the basis of the space Σ^r may be linearly dependent in a large part of the structure, or in a part of the structure with a relatively very high flexibility (in the structure as a whole they are of course by definition linearly independent). Then the redundant stress systems are strongly coupled and the matrix $\mathbf{Z}\mathbf{S}^{-1}\mathbf{Z}^T$ will be ill-conditioned.

By the so-called natural factor formulation, introduced in [4] and elaborated in [5], problems due to ill-conditioned matrices can largely be avoided. The method rests upon the observation that after a simple Cholesky decomposition of the individual element stiffness matrices into upper and lower triangular matrices, $\mathbf{S}^{ut} = [\mathbf{S}^{lt}]^T$, the system matrix of the displacement method can be written in the form

$$\mathbf{K} = \mathbf{D}^{cT}\,\mathbf{S}\,\mathbf{D}^{c} = \mathbf{D}^{cT}\,\mathbf{S}^{lT}\,\mathbf{S}^{ut}\,\mathbf{D}^{c} = \mathbf{A}^{T}\mathbf{A}. \tag{3.15}$$

Now advantage is taken of the property of any $m \times n$ matrix of rank n ($m \geq n$), that it can be decomposed into an orthogonal matrix $\mathbf{Q}$ ($\mathbf{Q}^T\mathbf{Q} = \mathbf{I}$) and an upper triangular matrix $\mathbf{K}^{ut}$ by the following algorithm.

$$\mathbf{A} = \mathbf{Q}\,\mathbf{K}^{ut},$$

$$K^{ut}_{kk} = \left[\sum_{i=1}^{m} A_{ik}A_{ik} - \sum_{p=1}^{k-1} K^{ut}_{pk}K^{ut}_{pk}\right]^{1/2}, \qquad k = 1, 2 \dots n,$$

$$Q_{ik} = \left[A_{ik} - \sum_{p=1}^{k-1} Q_{ip}\,K^{ut}_{pk}\right]\left(K^{ut}_{kk}\right)^{-1}, \qquad i = 1, 2,\dots m;\ \ k = 1, 2,\dots n,$$

$$K^{ut}_{pk} = \sum_{i=1}^{m} Q_{ip}A_{ik},\ p = 1, 2, \dots, k-1;\ k = 2, \dots, n. \tag{3.16}$$

The matrix $\mathbf{K}^{ut}$ is the Choleski factor of the matrix $\mathbf{K}$ in (3.15). It has been shown [6] that the relative error in the calculation of $\mathbf{K}^{ut}$ can be considerably reduced when applying the algorithm (3.16) instead of the usual Choleski procedure. If necessary the accuracy can be further improved by the iterative procedure

$$\mathbf{Q}^{i} = \mathbf{Q}^{i+1}\,\mathbf{V}^{i},\ \ [\mathbf{K}^{ut}]^{i+1} = \mathbf{V}^{i}\,[\mathbf{K}^{ut}]^{i}. \tag{3.17}$$

In [5] it was derived that the $m \times n$ orthogonal matrix $\mathbf{Q}$ has a significance of its own. From

$$\mathbf{K}\mathbf{u}^{c} = \mathbf{D}^{cT}\mathbf{S}^{lt}\mathbf{S}^{ut}\mathbf{D}^{c}\mathbf{u}^{c} = \mathbf{f}^{c} - \mathbf{D}^{cT}\mathbf{S}^{lt}\mathbf{S}^{ut}\mathbf{D}^{0}\mathbf{u}^{0} \tag{3.18}$$

it folows that

$$\mathbf{u}^{c} = [\mathbf{K}^{ut}]^{-1}\left[[\mathbf{K}^{lt}]^{-1}\,\mathbf{f}^{c} - \mathbf{Q}^{T}\mathbf{S}^{ut}\mathbf{D}^{0}\mathbf{u}^{0}\right] \tag{3.19}$$

and

$$\sigma = \mathbf{S}^{lt}\mathbf{Q}[\mathbf{K}^{lt}]^{-1}\,\mathbf{f}^{c} + \mathbf{S}^{lt}\,[\mathbf{I} - \mathbf{Q}\mathbf{Q}^{T}]\,\mathbf{S}^{ut}\mathbf{D}^{0}\mathbf{u}^{0}. \tag{3.20}$$

Hence if the displacements are not explicitely demanded we can omit the backward substitution procedure of the $[\mathbf{K}^{ut}]^{-1}$ multiplication in (3.19). But the main advantage is that the accuracy of expression (3.20) is always better, and in critical cases far better than the expression for the stresses in terms of the calculated displacements $\mathbf{u}^{c}$. For details the reader is referred to [5].

Also in the force method the $\mathbf{Q}$ method relieves the conditioning problems. If we write the system matrix of the force method in the form (3.15), or

$$\mathbf{Z}\mathbf{S}^{-1lt}\mathbf{S}^{-1ut}\mathbf{Z}^{T}\sigma^{r} = -\mathbf{Z}\mathbf{S}^{-1lt}\mathbf{S}^{-1ut}\Sigma^{0T}\mathbf{f}^{c} + \mathbf{Z}\mathbf{D}^{0}\mathbf{u}^{0}, \tag{3.21}$$

then we obtain the following expressions

$$\mathbf{u}^c = \Sigma^0 \mathbf{S}^{-1lt} [\mathbf{I} - \mathbf{Q}\mathbf{Q}^T] \mathbf{S}^{-1ut} (\Sigma^{0T} \mathbf{f}^c - \mathbf{S}\mathbf{D}^0 \mathbf{u}^0) \tag{3.22}$$

and

$$\sigma = [\mathbf{S}^{-1ut}]^{-1} [\mathbf{I} - \mathbf{Q}\mathbf{Q}^T] \mathbf{S}^{-1ut} \Sigma^{0T} \mathbf{f}^c + [\mathbf{S}^{-1ut}]^{-1} \mathbf{Q}\mathbf{Q}^T \mathbf{S}^{-1ut} \mathbf{S}\mathbf{D}^0 \mathbf{u}^0, \tag{3.23}$$

where

$$\mathbf{S}^{-1ut} \mathbf{Z}^T = \mathbf{Q}\mathbf{F}^{ut}. \tag{3.24}$$

Here $\mathbf{F}^{ut}$ replaces $\mathbf{K}^{ut}$ in the algorithm (3.16).

In the statically determinate situation we have Dim $\Sigma^r = 0$ and in (3.22) and (3.23) $\mathbf{Q} = \mathbf{O}$. In that case Dim Σ = Dim $\mathscr{U}^c$ and in (3.20) holds $\mathbf{Q}\mathbf{Q}^T = \bar{\mathbf{I}}$ or prescribed displacements cannot cause stresses. $\mathbf{Q}\mathbf{Q}^T = \mathbf{I}$ in the expressions (3.22) and (3.23) only happens in the unlikely case that Dim Σ^r = Dim Σ, which implies that all displacements are prescribed.

It may be expected that the natural factor formulation will gain on the Cholesky procedure, especially for the relatively small computers, that now come into use on a large scale.

4. The principles of virtual work and of virtual heat

The preceding paragraphs have been restricted to the case of mechanical equilibrium, and for the elasticity problem to the linearized equations at that. The determination of the elasticity relation (2.17) for the finite elements so far has remained an open question.

Now we shall discuss the finite element method as an approximation theory for the thermo-mechanical continuum theory, be it that we are not able to go into the details of mathematical convergence and error analysis. In this realm great progress has been made for linearized finite element equations, derived on the basis of inequalities [7, 8].

For the derivation of the finite element equivalent of the general thermo-mechanical equations we turn to variational statements of these equations, even though the physics of the problem does not provide us with suitable inequalities. The expressions, which we consider, are scalars, and hence frame invariant. They can be approximated by an algebraic expression with the aid of interpolation polynomials.

In case of purely mechanical problems the principle of virtual work suffices. Like in par. 2, but now in the continuum description for the non-equilibrium case we shall employ this principle in the version which in [9] is attributed to Piola. The principle states as a necessary condition for any motion

$$\int_{\partial B} t_i \delta u_i \, dA + \int_B \rho f_i \delta u_i \, dV = \int_B \rho \ddot{u}_i \delta u_i \, dV, \tag{4.1}$$

for all rigid virtual displacements δu_i. A rigid virtual displacement is characterized by the condition

$$\frac{1}{2}\left(\frac{\partial \delta u_i}{\partial x_j} + \frac{\partial \delta u_j}{\partial x_i}\right) = \delta \epsilon_{ij} = 0, \tag{4.2}$$

in all points of B. Introduced by means of a symmetric tensor of multipliers t_{ij}, condition (4.2) transforms (4.1) into

$$\int_{\partial B} t_i \delta u_i \mathrm{d}A + \int_B \rho f_i \delta u_i \mathrm{d}V = \int_B (\rho \ddot{u}_i \delta u_i + t_{ij} \delta \epsilon_{ij}) \mathrm{d}V. \tag{4.3}$$

This must hold for all kinematically admissible virtual displacements δu_i. By applying Green's transformation, we derive

$$t_i = t_{ij} n_j, \tag{4.4}$$

and

$$\frac{\partial t_{ij}}{\partial x_j} + \rho f_i = \rho \ddot{u}_i, \tag{4.5}$$

which identifies t_{ij} as the stress tensor of Cauchy. Thus $t_{ij}\delta\epsilon_{ij}$ in the integrand in the right hand side of (4.3) may rightly be called the virtual work of deformation.

In cases, where a non-mechanical energy flux is significant the principle of virtual work can be supplemented by a principle of virtual heat, equivalent to the virtual dissipation principle of Biot [10]. By the introduction of the entropy flux $\dot{h}_i$ the heatflux can be written as $q_i = T\dot{h}_i$, where T represents the absolute temperature ($T \geq 0$). The principle of virtual heat then states as a necessary condition for any heat flow, with k as coefficient of heat conductivity,

$$\int_{\partial B} T \delta h_i n_i \mathrm{d}A = - \int_B \frac{T}{k} \dot{h}_i \delta h_i \mathrm{d}V, \tag{4.6}$$

for all divergence-free virtual entropy displacements δh_i. If

$$\frac{\partial \delta h_i}{\partial x_i} = 0 \tag{4.7}$$

is introduced in (4.6) by means of a multiplier θ we have

$$\int_{\partial B} T \delta h_i n_i \mathrm{d}A = \int_B \left(\theta \frac{\partial \delta h_i}{\partial x_i} - \frac{T}{k} \dot{h}_i \delta h_i \right) \mathrm{d}V. \tag{4.8}$$

This condition must hold for all thermally admissible virtual entropy displacements (perfect insulation of the boundary implies $\delta h_i n_i = 0$). By Green's transformation we obtain from (4.8)

$$\theta = T \tag{4.9}$$

and

$$\frac{\partial \theta}{\partial x_i} = - \frac{T}{k} \dot{h}_i. \tag{4.10}$$

Hence the multiplier θ is the absolute temperature and equations (4.10) state Fourier's law of heat conduction.

If conditions (4.3) and (4.8) are both satisfied at any moment in a thermo-mechanical process, it is easily established that the energy equation and the entropy production law reduce to one scalar condition, locally to be fulfilled by the constitutive equations for t_{ij} and T. After introduction of the internal energy per unit mass, e, the entropy and the entropy production per unit mass, s and σ, the energy equation and the entropy production law can be written as follows ($\frac{\mathrm{d}}{\mathrm{d}t}\int_B \rho \mathrm{d}V = 0$)

$$\int_{\partial B} (t_i\dot{u}_i - T\dot{h}_i n_i)\,\mathrm{d}A + \int_B \rho f_i \dot{u}_i\,\mathrm{d}V =$$
$$= \int_B [\rho(\dot{e} - T\dot{s}) + \rho\ddot{u}_i\dot{u}_i + T\rho\dot{s}]\,\mathrm{d}V, \tag{4.11}$$

$$\rho\dot{s} = -\frac{\partial \dot{h}_i}{\partial x_i} + \rho\sigma,\ \sigma \geq 0. \tag{4.12}$$

Substitution of (4.10) and (4.12) into (4.11), and application of the divergence theorem, leads to

$$\int_B [\rho(\dot{e} - T\dot{s}) - t_{ij}\,\frac{1}{2}\left(\frac{\partial \dot{u}_i}{\partial x_j} + \frac{\partial \dot{u}_j}{\partial x_i}\right) - \frac{k}{T}\,\frac{\partial T}{\partial x_i}\,\frac{\partial T}{\partial x_i} + T\rho\sigma]\,\mathrm{d}V = 0, \tag{4.13}$$

or

$$T\rho\sigma = \frac{k}{T}\,\frac{\partial T}{\partial x_i}\,\frac{\partial T}{\partial x_i} + t_{ij}\,\mathrm{d}_{ij} - \rho(\dot{e} - T\dot{s}) \geq 0. \tag{4.14}$$

Here $d_{ij} = \frac{1}{2}\left(\frac{\partial \dot{u}_i}{\partial x_j} + \frac{\partial \dot{u}_j}{\partial x_i}\right)$ is the rate of deformation tensor.

We conclude that the variational conditions (4.3) and (4.8) are not only equivalent both to the laws of balance of momentum and of moment of momentum and to the heatconduction equation, but that the energy equation and the law of entropy production are covered as well, provided the constitutive equations for the stresses and the temperature satisfy locally condition (4.14).

The law of conservation of mass only plays a role in the case of spatial finite elements. Spatial finite elements have a fixed geometry, while matter flows through their bounding surfaces. The finite elements fill together the volume within a so-called control surface. Because of

$$\frac{d}{dt}\int_B \rho \mathrm{d}V = \int_B \left(\frac{\partial \rho}{\partial t} + \dot{u}_i\,\frac{\partial \rho}{\partial x_i} + \rho\,\frac{\partial \dot{u}_i}{\partial x_i}\right)\mathrm{d}V = 0 \tag{4.15}$$

within each spatial finite element the mass density must satisfy locally the equation

$$\frac{\partial \rho}{\partial t} + \dot{u}_i\,\frac{\partial \rho}{\partial x_i} + \rho\,\frac{\partial \dot{u}_i}{\partial x_i} = 0. \tag{4.16}$$

The mass of material finite elements is by definition constant and hence no equation for conservation of mass has to be considered.

5. Spatial finite element equations

The derivation of the spatial finite element equations is most straightforward. In (4.3) and (4.8) all continuous, piecewize differentiable, virtual material and entropy displacements are admissible, that satisfy the constraints at the fixed control surface: $\delta u_i = 0$ on ∂B^u, $\delta h_i n_i = 0$ on ∂B^h. We introduce for each spatial finite element polynomial interpolation functions $\psi_N^e(\xi_k)$, such that in the domain of the finite element holds

$$x_i = \psi_N^e(\xi_k) x_{Ni}, \tag{5.1}$$

where x_{Ni} are the fixed spatial coordinates of node N, which must be a node on the element under consideration. Further the functions $\psi_N^e(\xi_k) x_{Ni}$ must uniquely define the boundary between adjacent elements, i.e. the function

$$x_i = \sum_e \psi_N^e(\xi_k) x_{Ni}$$

must be continuous and single-valued. It is convenient to take for the local coordinates ξ_k so-called natural coordinates as indicated in Fig. 2.

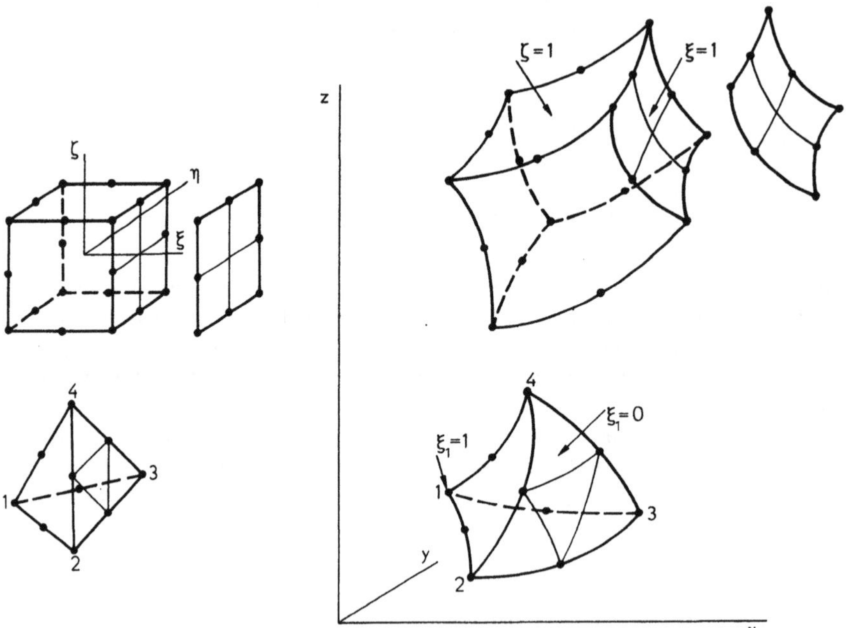

Fig. 2. Local coordinates and iso-parametric elements.

By approximating the velocity fields and the virtual displacements inside an element by the same interpolation functions as in (5.1),

$$\dot{u}_i = \psi_N^e(\xi_k)\dot{u}_{Ni}(t), \qquad \delta u_i = \psi_N^e(\xi_k)\delta u_{Ni},$$

$$\dot{h}_i = \psi_N^e(\xi_k)\dot{h}_{Ni}(t), \qquad \delta h_i = \psi_N^e(\xi_k)\delta h_{Ni}, \tag{5.2}$$

we get the so-called isoparametric finite elements [11]. They satisfy the necessary requirement, that the rigid and divergence-free virtual displacements are represented.

Each finite element is next assigned a stress distribution, a temperature distribution, and a mass density distribution by means of interpolation polynomials $\phi_p^e(\xi_k)$, such that

$$\begin{aligned} t_{ij} &= t_{ijp}^{(e)}\phi_p^{(e)}(\xi_k),\\ T &= T_p^{(e)}\phi_p^{(e)}(\xi_k),\\ \rho &= \rho_p^{(e)}\phi_p^{(e)}(\xi_k). \end{aligned} \tag{5.3}$$

Since t_{ij} and T can be looked upon as multipliers, taking conditions (4.2) and (4.7) into account, the number of parameters $t_{ijp}^{(e)}$, T_p^e, ρ_p^e should correspond to the number of deformation parameters and to the number of divergence parameters respectively, such as these numbers are determined by the virtual displacement fields.

If at the control surface appropriate boundary conditions,

$$\begin{aligned} &\dot{u}_i \text{ on } \partial B^u \text{ or } t_i \text{ on } \partial B^t,\\ &\dot{h}_i n_i \text{ on } \partial B^h \text{ or } T \text{ on } \partial B^T, \end{aligned} \tag{5.4}$$

are prescribed, then the integrals (4.3) and (4.8) can be evaluated. Taking into account the expression for the material acceleration (with $\dfrac{\partial \psi_M}{\partial \xi_k} = \psi_{M,k}$)

$$\ddot{u}_i = \frac{\partial \dot{u}_i}{\partial t} + \dot{u}_j \frac{\partial \dot{u}_i}{\partial x_j} = \ddot{u}_{Ni}\psi_N + \dot{u}_{Nj}\dot{u}_{Mi}\psi_N\psi_{M,k}\frac{\partial \xi_k}{\partial x_j}, \tag{5.5}$$

we obtain spatial finite element equations of the form

$$\delta u_k\,[t_i D_{ik}^\sigma + \rho_p(M_{pkl}\ddot{u}_l + N_{pklm}\dot{u}_l\dot{u}_m) - f_k^0] = 0 \;\; \forall \delta u_k, \tag{5.6a}$$

$$\delta h_k\,[T_p D_{pk}^\theta - \frac{T_p}{k} R_{pkl}\dot{h}_l - T_k^0] = 0 \;\; \forall \delta h_k. \tag{5.6b}$$

In (5.6) the vectors $\delta\mathbf{u}$, $\dot{\mathbf{u}}$, and $\ddot{\mathbf{u}}$, and the vectors $\delta\mathbf{h}$ and $\dot{\mathbf{h}}$ are vectors in finite dimensional isomorphic vector spaces, which are defined as the direct sum of the vector spaces at the spatial nodal points. The vectors $\mathbf{t}$, ρ, and $\mathbf{T}$ are vectors in the stress space, the mass density space, and the temperature space. These spaces are the direct sum of the corresponding finite element subspaces. The vectors $\mathbf{f}^0$ and $\mathbf{T}^0$ have prescribed elements and they are vectors in the dual space of the virtual displacement space.

It is clear that the equations (5.6) have to be supplemented by the constitutive equations for $\mathbf{t}$ and $\mathbf{T}$, and by the mass conservation law. The latter can be written in the form

$$\dot{\rho}_p + C_{pql}^\rho \rho_q \dot{u}_l + D_{pql}^\rho \rho_q \dot{u}_l = 0, \quad D_{pql}^\rho = 0 \text{ if } p \neq q, \tag{5.7}$$

after substitution of the interpolation polynomials for ρ and $\dot{u}_i$ from (5.2) and (5.3) into (4.16).

The constitutive equations can be of a quite complex nature. The elasto-viscoplastic model of material behaviour presents probably the most complicated example of constitutive equations, that have to be used in technical applications. Then only expressions for the stress rates can be given, since the deformation process is thermodynamically irreversible. Though usually coupling with thermal phenomena is not considered, an example of a complete set of equations for unlimited flow of an isotropic material with small elastic strains can be given by

$$\left(s_{ij} = t_{ij} - \frac{1}{3}\, t_{nn}\,\delta_{ij};\ \omega_{ij} = \frac{1}{2}\left(\frac{\partial \dot{u}_i}{\partial x_j} - \frac{\partial \dot{u}_j}{\partial x_i}\right)\right)$$

$$\dot{t}_{ij} = \frac{\partial t_{ij}}{\partial t} + \dot{u}_k \frac{\partial t_{ij}}{\partial x_k} = s_{kj}\,\omega_{ik} - s_{ik}\,\omega_{kj} + C(d_{nn} - \alpha\dot{T}) +$$

$$+ 2G\{\delta_{ik}\,\delta_{jl} - (1-h)\,Y_{ijkl}\}\left(d_{kl} - \frac{1}{3}\, d_{nn}\,\delta_{kl} - c_{kl}\right), \tag{5.8a}$$

$$\rho c_v \dot{T} = \rho c_v \left(\frac{\partial T}{\partial t} + \dot{u}_k \frac{\partial T}{\partial x_k}\right) = -C\alpha T_0 d_{nn} + T_0\left(-\frac{\partial \dot{h}_i}{\partial x_i} + \frac{k}{T^2}\frac{\partial T}{\partial x_i}\frac{\partial T}{\partial x_i}\right) +$$

$$+ \frac{T_0}{T}\{(1-h)\,s_{kl}(d_{kl} - c_{kl}) + s_{kl}c_{kl}\}, \tag{5.8b}$$

where

$$c_{kl} = \frac{3}{2}\,\frac{\gamma(T)}{(\sigma^y)^2}\left(\frac{3J_2}{(\sigma^y)^2}\right)^{n-1} s_{kl}$$

$$Y_{ijkl} = \frac{1}{2}\, J_2^{-1}\, s_{ij}s_{kl} \quad \text{and} \quad 0 \leqslant h < 1,$$

provided $s_{ij}(d_{ij} - c_{ij}) \geq 0,\ 3J_2 - (\sigma^y)^2 = 0,$

otherwise $Y_{ijkl} = 0$ and $h = 1$.

In the case of a hardening parameter $0 < h < 1$ the yield stress σ^y increases with the value of $J_2 = \frac{1}{2} s_{ij}s_{ij}$ in plastic deformation ($Y_{ijkl} \neq 0$), such that $(\sigma^y)^2 = (3J_2)$ max. Further the material is characterized by the material constants C, G, α, c_v, k and n, while $\gamma(T)$ represents the temperature dependence of the creep process. It can be shown [12, 13] that the equations (5.8) satisfy condition (4.14).

Variation of mass density need not be considered as long as only small elastic changes of volume occur. However anisotropic hardening and softening may be important. These effects can be incorporated in the description by the introduction of internal state variables. A simple form of such internal state variables are the subelement stresses of the so-called fraction model [14], in the finite-element technique also known as the sublayer model [11].

With the aid of the interpolation functions (5.2) and (5.3) equations analogous to (5.8) can be derived for the stress and temperature vectors of the spatial finite elements. An example without thermal coupling is given in [12]. Equations (5.6a) first have to be differentiated with respect to time before the equations for the stress rates can be substituted. In the quasi-static case thus equations of the following form are obtained.

$$[\mathbf{K} - \mathbf{P} + \mathbf{G}]\dot{\mathbf{u}} = \dot{\mathbf{f}} + \mathbf{q} \tag{5.9}$$

Here $\mathbf{K}$ is the stress-independent symmetric elasticity matrix, $\mathbf{P}$ is the stress-dependent symmetric plasticity matrix, and $\mathbf{G}$ is the unsymmetrical geometrical stiffness matrix. The rates of loading $\dot{\mathbf{f}}$ are prescribed and $\mathbf{q}$ is the stress-dependent creep vector. Equations (5.9) provide the predictor and equations (5.6) the corrector in a step-by-step integration procedure.

A considerable simplification is obtained if the material behaviour may be described by the incompressible, viscous model. Then the equations for the stresses read

$$t_{ij} = -p\delta_{ij} + \mu\left(\frac{\partial \dot{u}_i}{\partial x_j} + \frac{\partial \dot{u}_j}{\partial x_i}\right), \tag{5.10}$$

with a possible non-linear dependence of μ on s_{ij}. The coupling with thermal phenomena is now only through the temperature dependence of μ and it is usually neglected. In that case the flow problem is governed by the equations

$$-p_p D^{\rho}_{pk} + V_{kl}\dot{u}_l + (M_{kl}\ddot{u}_l + N_{klm}\dot{u}_l\dot{u}_m) = f^0_k, \tag{5.11a}$$

$$D^{\rho}_{ql}\dot{u}_l = 0. \tag{5.11b}$$

The second set of equations stems from (5.7) and expresses the incompressibility condition. The number of pressure parameters in a finite element should not exceed the number of divergence parameters, determined by the velocity field, otherwise equations (5.11b) will be linearly dependent.

Applications of equations (5.11) for steady flow ($\ddot{\mathbf{u}} = 0$) are discussed in [15], while in [16] also problems of unsteady flow are solved. In [11] examples are given of quasi-static flow of solids on the basis of the spatial finite element equations (5.11) (with $M_{kl} = N_{klm} = 0$), with a highly non-linear dependence of μ on the deviatoric components of stress, thus simulating visco-plastic flow.

Since free surfaces pose problems in the application of spatial finite elements, for the solution of most solid mechanics problems material finite elements seem to be more appropriate. Only in the pure heat conduction problem the distinction between spatial and material finite elements disappears, because the material remains at rest. The constitutive equation reduces to

$$\dot{T}_p = \frac{T_0}{\rho c_v}\left[-D^h_{pk}\dot{h}_k + \frac{k}{T^2(p)} G_{pqr} T_q T_r\right]. \tag{5.12}$$

As Biot suggests [10] usually sufficient accuracy is obtained with the linearized equations, in which the non-linear entropy production term is omitted. We then have from (5.6b) and

(5.12) for $\dot{h}_i n_i = 0$ on ∂B^h

$$\dot{T}_p = -\frac{T_0}{\rho c_v} D^h_{pk} \dot{h}_k = \frac{k}{\rho c_v} D^h_{pk} R^{-1}_{kl} (D^\theta_{ql} T_q - T^0_l). \tag{5.13}$$

The entropy velocity field has been eliminated and the resulting equations for the finite element temperatures are readily integrated. It should be observed that in the non-uniform state the temperature field will show discontinuities across finite element boundaries, which can however be made arbitrarily small by taking a finer mesh.

6. Material finite element equations

When we now turn to material finite elements we still have to differentiate between two types. In both cases the nodal velocities in (5.2) are taken to be not spatial velocities, but material velocities, which can be integrated to give displacements of the nodal points

$$u_i = \psi^e_N(\xi_k) u_{Ni}(t) = \psi^e_N(\xi_k)(x_{Ni}(t) - a_{Ni}). \tag{6.1}$$

The coordinates $a_i = \psi^e_N(\xi_k) a_{Ni}$ are the cartesian coordinates in the reference configuration.

In case we have to deal with large inelastic deformation it is again necessary to describe the flow of the material in the interior of the finite elements by the constitutive equations (5.8). Restricting ourselves to the quasi-static case ($\ddot{u}_i = 0$) we take as a starting point the virtual work equation in the form

$$\delta u_{Ni} \sum_e \left[\int_{\partial B^t} t_i \psi_N \, \mathrm{d}A + \int_{B^e} \rho f_i \psi_N \mathrm{d}V - \int_{B^e} t_{ij} \psi_{N,k} \frac{\partial \xi_k}{\partial x_j} \mathrm{d}V \right] = 0. \tag{6.2}$$

Now the domains of integration B^e and ∂B^t change and the geometry of the finite elements has to be updated continuously. Before the integrations are carried out the equations (6.2) are differentiated with respect to time and we get

$$\frac{d}{dt} \int_{B^e} t_{ij} \psi_{N,k} \frac{\partial \xi_k}{\partial x_j} \mathrm{d}V = \int_{B^e} \dot{t}_{ij} \psi_{N,k} \frac{\partial \xi_k}{\partial x_j} \mathrm{d}V +$$

$$+ \int_{B^e} t_{im} \psi_{N,k} \frac{\partial \xi_k}{\partial x_m} \psi_{M,l} \frac{\partial \xi_l}{\partial x_j} \dot{u}_{Mj} \mathrm{d}V - \int_{B^e} t_{im} \psi_{N,k} \frac{\partial \xi_k}{\partial x_j} \psi_{M,l} \frac{\partial \xi_l}{\partial x_m} \dot{u}_{Mj} \mathrm{d}V. \tag{6.3}$$

As it is shown for instance in [12] substitution of the constitutive equations (5.8a) again results in a system of equations of the form (5.9). All matrices, the matrix **K** included, have to be updated after each integration step, because not only the state of stress changes, but also the geometry of the finite elements. Properly, evenly dimensioned elements may in the course of the flow process be very much elongated in one direction, which will reduce their accuracy for subsequent analysis considerably. Of course with respect to accuracy nothing is gained by a transformation of (6.3) to the original geometry, as it is sometimes done. For the determination of the loading terms this transformation is advantagenous however, if the loading is by dead weight only.

The problem of mis-shapen finite elements, produced by a deformation process, can only be solved by a regeneration of a properly dimensioned mesh. Dealing with the free surface problem in the application of spatial finite elements may then be an alternative, worthy to be considered.

Material finite elements are particularly appropriate if the inelastic deformations remain small. The reference geometry then has physical relevance. Just as the deformation tensor in the continuum description may be considered as the sum of an elastic and an inelastic part, the generalized strains of the finite elements may be similarly decomposed, provided the condition of small inelastic deformation is satisfied. For large inelastic deformation, this decomposition loses its validity however [13, 17]. As it was pointed out above the problem is then a problem of flow rather than of deformation.

We consider the following transformation to the reference geometry with a_i as material coordinates and using the equality $\rho \mathrm{d}V = \rho_0 \mathrm{d}V_0$.

$$\int_B t_{ij}\delta\epsilon_{ij}\,\mathrm{d}V = \int_{B_0} \frac{\rho_0}{\rho}\,\frac{\partial x_k}{\partial x_i}\,t_{ij}\,\frac{\partial \delta u_k}{\partial x_j}\,\mathrm{d}V_0 =$$

$$= \int_{B_0} \frac{\rho_0}{\rho}\,\frac{\partial a_\alpha}{\partial x_i}\,t_{ij}\,\frac{\partial a_\beta}{\partial x_j}\,\frac{\partial x_k}{\partial a_\alpha}\,\frac{\partial \delta u_k}{\partial a_\beta}\,\mathrm{d}V_0. \tag{6.4}$$

After introduction of the pseudo stress tensor of Kirchhoff,

$$\sigma_{\alpha\beta} = \frac{\rho_0}{\rho}\,\frac{\partial a_\alpha}{\partial x_i}\,t_{ij}\,\frac{\partial a_\beta}{\partial x_j}, \tag{6.5}$$

and the Lagrangian strain tensor

$$E_{\alpha\beta} = \frac{1}{2}\left(\frac{\partial x_k}{\partial a_\alpha}\,\frac{\partial x_k}{\partial a_\beta} - \delta_{\alpha\beta}\right) = \frac{1}{2}\left(\frac{\partial u_\alpha}{\partial a_\beta} + \frac{\partial u_\beta}{\partial a_\alpha} + \frac{\partial u_k}{\partial a_\alpha}\,\frac{\partial u_k}{\partial a_\beta}\right), \tag{6.6}$$

we can write for (6.4)

$$\int_B t_{ij}\delta\epsilon_{ij}\,\mathrm{d}V = \int_{B_0} \sigma_{\alpha\beta}\,\delta E_{\alpha\beta}\,\mathrm{d}V_0. \tag{6.7}$$

The principal difference between $\delta\epsilon_{ij}$ and $\delta E_{\alpha\beta}$ lies in the fact that $\delta\epsilon_{ij}$ cannot, and $\delta E_{\alpha\beta}$ can be derived from a measure of finite strain as a first variation. These strains are by their elastic components $E'_{\alpha\beta} = E_{\alpha\beta} - E^0_{\alpha\beta}$ related to the stresses $\sigma_{\alpha\beta}$, that are the proper duals of the virtual deformations $\delta E_{\alpha\beta}$.

To allow of a maximum freedom in approximation methods we now write the virtual work condition (4.3) for an assembly of N finite elements with a number of multiplier functions, $\sigma^*_{\alpha\beta}$, t^*_i, and t^{**}_i.

$$-\int_{\partial B^t} \bar{t}_i\,\delta u^*_i\,\mathrm{d}A + \sum_{e=1}^{N} \int_{B_0^e} \Big[-\rho_0 (f_i - \ddot{u}^e_i)\,\delta u^e_i + \sigma^e_{\alpha\beta}\,\delta E^e_{\alpha\beta} +$$

$$+ \sigma^*_{\alpha\beta}\left\{\left(\delta_{k\alpha} + \frac{\partial u^e_k}{\partial a_\alpha}\right)\frac{\partial \delta u^e_k}{\partial a_\beta} - \delta E^e_{\alpha\beta}\right\} + \delta\sigma^*_{\alpha\beta}\left\{\frac{1}{2}\left(\frac{\partial u^e_\alpha}{\partial a_\beta} + \frac{\partial u^e_\beta}{\partial a_\beta} + \frac{\partial u^e_k}{\partial a_\alpha}\,\frac{\partial u^e_k}{\partial a_\beta}\right) - E^e_{\alpha\beta}\right\}\Big]\,\mathrm{d}V_0$$

$$+\sum_{e=1}^{N} \int_{\partial B_0^e} \left[(\delta u_i^* - \delta u_i^e)\, t_i^* + (u_i^* - u_i^e)\, \delta t_i^* \right] \mathrm{d}A_0 +$$

$$+\int_{\partial B^u} \left[-\delta u_i^* t_i^{**} + (\bar{u}_i - u_i^*)\, \delta t_i^{**} \right] \mathrm{d}A = 0. \tag{6.8}$$

The displacement fields u_i^* are only defined on the external surface and on the element interfaces, while $\sigma_{\alpha\beta}^e$ is related to $E_{\alpha\beta}^e$, for instance by the linear isotropic constitutive equations

$$\sigma_{\alpha\beta}^e = C\{E_{kk}^e - \alpha(T - T_0)\}\delta_{\alpha\beta} + 2G(E_{\alpha\beta}^e - E''_{\alpha\beta} - \frac{1}{3} E_{kk}^e \delta_{\alpha\beta}). \tag{6.9}$$

Here $E''_{\alpha\beta}$ represents the small inelastic strains. The prescribed quantities $\bar{t}_i$ and $\bar{u}_i$ will usually be given on the external boundary in the deformed state. By considering in (6.8) the variations $\delta E_{\alpha\beta}^e$, δu_k^e, and δu_i^* we derive the following equations.

$$\sigma_{\alpha\beta}^* = \sigma_{\alpha\beta}^e, \tag{6.10a}$$

$$\frac{\partial}{\partial a_\beta} \{\sigma_{\alpha\beta}^* \left(\delta_{k\alpha} + \frac{\partial u_k^e}{\partial a_\alpha}\right)\} + \rho_0 f_k = \rho_0 \ddot{u}_k^e \quad \text{in } B_0^e, \tag{6.10b}$$

$$\sigma_{\alpha\beta}^* \left(\delta_{k\alpha} + \frac{\partial u_k^e}{\partial a_\alpha}\right) n_\beta = t_k^* \quad \text{on } \partial B_0^e, \tag{6.10c}$$

$$t_{i(\mathbf{n})}^* = -t_{i(-\mathbf{n})}^* \quad \text{on inter element boundary surfaces,} \tag{6.10d}$$

$$t_i^{**}\, \mathrm{d}A = t_i^*\, \mathrm{d}A_0 \quad \text{on } \partial B^u \tag{6.10e}$$

$$t_i^*\, \mathrm{d}A_0 = \bar{t}_i\, \mathrm{d}A \quad \text{on } \partial B^t. \tag{6.10f}$$

From the variation of the multipliers we have an additional set of equations

$$E_{\alpha\beta}^e = \frac{1}{2}\left(\frac{\partial u_\alpha^e}{\partial a_\beta} + \frac{\partial u_\beta^e}{\partial a_\alpha} + \frac{\partial u_k^e}{\partial a_\alpha}\frac{\partial u_k^e}{\partial a_\beta}\right) \quad \text{in } B_0^e, \tag{6.11a}$$

$$u_i^e = u_i^* \quad \text{on } \partial B_0^e, \tag{6.11b}$$

$$u_i^* = \bar{u}_i \quad \text{on } \partial B^u. \tag{6.11c}$$

The most common type of finite elements is the so-called kinematically conforming type The equations (6.11) are satisfied apriori. With the aid of proper interpolation polynomials in (6.1) from the principle of virtual work a set of equations is obtained, which after differentiation with respect to time and after substitution of the constitutive equations for elasto-viscoplasticity is analogous to (5.9) [12].

$$[\mathbf{K}^0 + \mathbf{K}^u - \mathbf{P}^0 - \mathbf{P}^u + \mathbf{G}]\,\dot{\mathbf{u}} = \dot{\mathbf{f}} + \mathbf{q} \tag{6.12}$$

The geometrical non-linearities present themselves in the matrices $\mathbf{K}^u$, $\mathbf{P}^u$, and $\mathbf{G}$. Contrary to (5.9) now also the geometrical stiffness matrix $\mathbf{G}$ is symmetrical. A more manageable class of finite elements is derived on the basis of a polynomial interpolation for the strains $E^e_{\alpha\beta}$.

$$E^e_{\alpha\beta} = E^{(e)}_{\alpha\beta p}\, \phi^{(e)}_p\, (\xi_k). \tag{6.13}$$

We recall that in formulating the virtual work principle (4.1) rigid virtual displacements had to be characterized. Instead of (4.2) we now have $\delta E_{\alpha\beta} = 0$ as a condition to be satisfied for (4.1) to hold. Since in (6.13) there is only a finite number of strain parameters we can only approximate the condition $\delta E_{\alpha\beta} = 0$ in B^e_0 by requiring $\delta E^e_{\alpha\beta p} = 0$. In other words we may add $\int_{B^e_0} \sigma^{(e)}_{\alpha\beta} \delta E^{(e)}_{\alpha\beta} \mathrm{d} V_0$ for each finite element to expression (4.1), and require locally

$$E^e_{\alpha\beta_p} = \frac{1}{2} \left(\frac{\partial u^e_\alpha}{\partial a_\beta} + \frac{\partial u^e_\beta}{\partial a_\alpha} + \frac{\partial u^{(e)}_k}{\partial a_\alpha} \, \frac{\partial u^{(e)}_k}{\partial a_\beta} \right)_p , \tag{6.14}$$

in order to make sure that $\delta E^e_{\alpha\beta} = 0$ defines rigid virtual displacements. In order to avoid that the conditions $\delta E^e_{\alpha\beta p} = 0$ are linearly dependent, the number of strain parameters must not exceed the number of nodal displacements u^e_{Ni} minus the number of degrees of freedom of the finite elements as a rigid body. If further we consider continuous, piecewize differentiable displacements $u^e_i = u^*_i$, then we may replace (6.8) by

$$-\int_{\partial B^t} \bar{t}_i \delta u^*_i \, \mathrm{d}A + \sum_{e=1}^{N} \int_{B^e_0} \left[-\rho_0 \, (f_i - \ddot{u}^e_i) \, \delta u^{(e)}_i + \sigma^{(e)}_{\alpha\beta} \, \delta E^{(e)}_{\alpha\beta} \right] \mathrm{d} V_0 = 0, \tag{6.15}$$

leading to the equations

$$\delta u_k \, [-f_k + M_{kl} \ddot{u}_l + \sigma_i D_{i,k}] = 0 \; \forall \delta u_k. \tag{6.16}$$

The subsidiary conditions (6.14), to be taken into account in (6.15), define generalized strains

$$E_i = D_i(u_k), \; \delta E_i = D_{i,k} \, \delta u_k, \tag{6.17}$$

that appear as the parameters $E^e_{\alpha\beta p}$ in (6.13), while the stresses σ_i as the proper duals of δE_i are determined by relations of the form

$$\sigma_i = S_{ij} \, (E_j - E^0_j). \tag{6.18}$$

The stiffness matrix S_{ij} is calculated by substitution of (6.13) and (6.9) into (6.15). E^0_j represents the thermal and the inelastic strains.

The approach outlined above proved to be particularly fruitful in dealing with the geometrical non-linearities of the membrane deformations of plates and shells. It avoids the appearance of spurious membrane deformations, which are most detrimental to the accuracy of the solution, in particular if the plate or shell would permit inextensional bending [18]. The success of reduced integration techniques [11] for the improvement of the conditioning of the matrix, and of the accuracy of the solution, can also be understood in terms of the local enforcement of conditions (6.14), in number not exceeding the number of linearly independent virtual deformation parameters as determined by the number of nodal displacements. Early examples of this generalized stress-generalized strain technique or 'natural mode technique' in the finite element method are to be found in [19] and [20] respectively, apart from the much older applications in structural mechanics.

The description of the bending of plates and shells by finite elements meets with the wellknown difficulty, that not only the deflections, but also the rotations must be continuous at element interfaces for a kinematically conforming approach. Here a derivation of the element properties on the basis of a bending and twisting moment distribution provides an easy way-out for the problem of the continuity of the rotations.

We observe that for the static case ($\ddot{u}_i^e = 0$) in the absence of volume forces ($f_k = 0$) the displacement fields u_i^e disappear from expression (6.8), if the equations (6.10b) and (6.10c) are satisfied. By a rigid rotation of the finite element, provided the element is chosen sufficiently small, these equations can be reduced to the linear form

$$\frac{\partial \sigma^*_{\alpha\beta}}{\partial a_\beta} = 0, \quad \sigma^*_{\alpha\beta} n_\beta = t^*_\alpha. \tag{6.19}$$

It is then possible to find a stress distribution, and by analogy a moment distribution in the case of a plate or shell element, that satisfies the homogeneous equilibrium equations in the interior, and that determines edge forces and moments, for which continuity conditions (6.10d) are obtained by the variation of the interface displacements (and rotations) u_i^*. Expression (6.8) reduces to

$$-\int_{\partial B^t} \bar{t}_i \delta u_i^* \, \mathrm{d}A + \sum_{e=1}^{N} \left[\int_{B_0^e} -\delta \sigma^*_{\alpha\beta} E^*_{\alpha\beta} \, \mathrm{d}V + \int_{\partial B_0^e} (\delta u_i^* t_i^* + u_i^* \delta t_i^*) \, \mathrm{d}A_0 \right]$$

$$+ \int_{\partial B^u} \left[-\delta u_i^* t_i^{**} + (\bar{u}_i - u_i^*) \delta t_i^{**} \right] \mathrm{d}A = 0. \tag{6.20}$$

We should realize that t_i^* in (6.20) and t_α^* in (6.19) differ by the rigid rotation, that had to be applied to the finite element in order to free the equilibrium equations from the displacements. Since such rigid rotations have to be expressed in terms of nodal displacements the geometrical non-linearity enters expression (6.20) through this rigid rotation.

The strains $E^*_{\alpha\beta}$ are related to $\sigma^*_{\alpha\beta}$, for instance by the inverted relation (6.9)

$$E^*_{\alpha\beta} = \alpha(T - T_0)\delta_{\alpha\beta} + E''_{\alpha\beta} + \frac{1}{3C}\sigma^*_{kk}\delta_{\alpha\beta} + \frac{1}{2G}(\sigma^*_{\alpha\beta} - \frac{1}{3}\sigma^*_{kk}\delta_{\alpha\beta}) \tag{6.21}$$

For the stresses $\sigma^*_{\alpha\beta}$ we take a polynomial approximation,

$$\sigma^*_{\alpha\beta} = \sigma^{(e)}_{\alpha\beta_p} \phi_p^{(e)}(\xi_k), \tag{6.22}$$

such that the equations (6.19) are satisfied. Assuming that the surface displacement fields u_i^* are in agreement with the kinematical boundary conditions, we obtain from (6.20) equations of the following form

$$\delta u_k \left[-f_k + D_{i,k}(u_l)\sigma_i \right] + \delta\sigma_i \left[-S_{ij}^{-1}\sigma_j - E_i^0 + D_i(u_k) \right] = 0 \; \forall \delta u_k \wedge \forall \delta \sigma_i. \tag{6.23}$$

In case the number of stress parameters in (6.22) exceeds the number of linearly independent virtual deformation parameters of the element, as determined by the number of nodal displacements u^*_{Ni}, the kernel in the linear transformation $D_{i,k}\sigma^e_i$ determines a space of redundant stresses (see (2.28)), which for each element separately can be eliminated by the continuity equations. The finite elements thus created became known as hybrid stress elements [21]. Since the matrix S^{-1} in (6.23) can be inverted for each element separately, the equations of the system can however be set up directly in the form

$$D_{i,k}(u_l)\,\sigma_i = D_{i,k}(u_l)S_{ij}\,\{D_i(u_l) - E^0_i\} = f_k. \tag{6.24}$$

Furthermore a number of stress parameters in excess of the number of deformation modes generally gives the element a spurious stiffness.

The triangular element for plate bending with linearly varying bending and twisting moments, with nodal displacements in the three corner points and in the three midside points, and with two parameters for rotations about each side, is a noteworthy example of the success of the hybrid procedure with an equal number of stress and deformation modes. In this example, provided the plate surface carries concentrated loads only, the element is also a statically conforming element [22].

If the doubly curved triangular shell element is approximated by a flat element, that is given an initial curvature, element properties can be derived from quadratic displacement polynomials for the middle surface, linearly varying membrane deformations, and linearly varying bending and twisting moments. Thus a combination of the procedures (6.15) and (6.20) leads to a shell element, that is suitable for linear and non-linear problems [18, 23].

7. Stability of structures and kinematics of mechanisms

For material finite elements the equations of motion read according to (6.16)

$$\sigma_i D_{i,k} + M_{kl}\ddot{u}_l = f_k. \tag{7.1}$$

Here σ_i is in general determined by the history of the deformation and heating process, $D_{i,k}$ is a function of the nodal displacements, and M_{kl} is a constant massmatrix.

In case of small elastic disturbances u_k from an equilibrium state with external loads f^0_k we have

$$\sigma^0_i D_{i,k} = f^0_k, \tag{7.2}$$

and

$$D_{i,k}S_{ij}D_{j,l}u_l + \sigma^0_i D_{i,kl}u_l + M_{kl}\ddot{u}_l = 0. \tag{7.3}$$

If the eigenvalues of the matrix of coefficients of u_k are all positive, the structure will be stable. Instability occurs if one or more eigenvalues are negative. Thus a sufficient condition for stability is given by

$$\delta u_k\,[D_{i,k}S_{ij}D_{j,l} + \sigma^0_i D_{i,kl}]\,\delta u_l > 0 \;\; \forall\, \delta u_k \neq 0. \tag{7.4}$$

The matrix $\sigma^0_i D_{i,kl}$ is the so-called geometrical stiffness matrix. The effect of the initial stresses σ_i^0 can be stabilizing or destabilizing.

In case the equilibrium state is determined with sufficient accuracy by the linearized equations, we may express σ_i^0 by means of a load factor λ as $\sigma_i^0 = \lambda\sigma_i^I$, and we obtain from (7.4) the buckling criterion

$$\mathrm{Det}(D_{i,k}S_{ij}D_{j,l} + \lambda\sigma_i^I D_{i,kl}) = 0, \tag{7.5}$$

which is a linear eigenvalue problem. For buckling and postbuckling analysis with the aid of the deformation parameters $D_i(u_k)$ we refer to [18].

For a structure that can perform rigid motions, the nodal displacements according to (3.11) can be expressed by

$$u_k^c = \Sigma_{ki}^0(\epsilon_i - D_{i,k}^0 u_k^0) + Z_{kp}^* u_p^m, \tag{7.6}$$

where Z_{kp}^* is a basis for the kernel of the transformation $D_{i,k}(Z_{kp}^* D_{i,k} = 0)$. Hence the stability criterion (7.5) reads

$$\delta u_p^{(m)} Z_{kp}^* \sigma_i^0 D_{i,kl} Z_{lq}^* \delta u_q^{(m)} > 0 \; \forall \delta u_p^m \neq 0. \tag{7.7}$$

We conclude that also in the case, when the structure contains one or more mechanisms, the structure may be stable if the proper initial stresses are present. Thus a cable hangs in a stable form under the influence of the statically determinate normal forces in the cable due to gravity. The cable has a gravitational stiffness expressed by $\sigma_i^0 D_{i,kl}$.

An interesting possibility of stabilizing structures is discussed in [24]. If in (3.4) and (3.10) we have $p > 0$ and $s > 0$ the structure is simultaneously statically and kinematically indeterminate. Then it may be possible to adjust the redundant stresses such that the mechanisms are stabilized. With the aid of the matrix **Z** from (3.5) the stability condition is

$$\delta u_p^{(m)} Z_{kp}^* Z_{ir} \sigma_r^r D_{i,kl} Z_{lq}^* \delta u_q^{(m)} > 0 \; \forall \delta u_p^m \neq 0 \tag{7.8}$$

Simple examples are of course the strings of music instruments. Less trivial applications of this idea are termed tensegrity structures [24].

Once generalized strains as proper measures of deformation have been defined for the material finite elements, an analysis of the kinematics of mechanisms is readily carried out. We note that the first and second order kinematics for a mechanism with rigid links ($\epsilon - \mathbf{D}^0\mathbf{u}^0 = \mathbf{o}$) follow from

$$\dot{x}_k = Z_{kp}^* \dot{x}_p^m, \tag{7.9}$$

$$\ddot{x}_k = Z_{kp}^* \ddot{x}_p^m - \Sigma_{ki}^0 D_{i,rs} \dot{x}_r \dot{x}_s,$$

or

$$\ddot{x}_k = Z_{kp}^* \ddot{x}_p^m - \Sigma_{ki}^0 D_{i,rs} Z_{rp}^* Z_{sq}^* \dot{x}_p^{(m)} \dot{x}_q^{(m)}. \tag{7.10}$$

For each position of the mechanism the matrices $\mathbf{Z}^*$ and $\mathbf{\Sigma}^0$ can be calculated, once the functions $D_i(u_k)$ are known. The position vector $\mathbf{x}$ as a function of $\mathbf{x}^m$, or the so-called zero order kinematics, follows by integration, where the expressions (7.9) and (7.10) are to be used as predictors and the conditions $D_i(u_k) = 0$ as correctors in a step by step procedure. A detailed account of this development is given in [25], while also the brief outline in [18] discusses some examples of special elements, devised for the analysis of mechanisms, like sliders and gears.

References

1. R.M. Bowen and C.C. Wang, *Introduction to vectors and tensors*, Vol. 1. Plenum Press, New York and London (1976).
2. B. Fraeijs de Veubeke, 'The numerical analysis of structures', *Proc. 13th Int. Cong. of Theor. and Appl. Mech*, pp. 20-28. Springer Verlag, Berlin-Heidelberg-New York (1973).
3. J.F. Besseling, 'The force method and its application in plasticity problems', *Computers and Structures*, **8**, (1978) 323-330.
4. J.H. Argyris and O.E. Brönlund, 'The natural factor formulation of the stiffness for the matrix displacement method', *Comp. Meths. Appl. Mech. Eng.*, **5** (1975) 97-119.
5. D.W. Scharpf, 'A new method of stress calculation in the matrix displacement analysis', *Computers and Structures*, **8** (1978) 465-477.
6. Th.L. Johnsen and J.R. Roy, 'On systems of linear equations of the form $A^T Ax = b$; Error analysis and certain consequences for structural applications', *Comp. Meths Appl. Mech. Eng.*, **3** (1974) 357-374.
7. J.T. Oden and J.N. Reddy, *An introduction to the mathematical theory of finite elements*, John Wiley and Sons, New York, London, Sydney, Toronto (1976).
8. Ph.G. Ciarlet, *The finite element method for elliptic problems*, North-Holland Publ. Co., Amsterdam, New York, Oxford (1978).
9. C. Truesdell and R. Toupin, 'The classical field theories', p. 596, *Encyclopedia of Physics*, Vol. III/1, Springer Verlag, Berlin, Göttingen, Heidelberg (1960).
10. M.A. Biot, 'Variational Lagrangian-thermodynamics of nonisothermal finite strain mechanics of porous soldis and thermomolecular diffusion', *Int. J. Solids Structures*, **13** (1977) 579-597.
11. O.C. Zienkiewicz, *The finite element method*, third edition, McGraw-Hill Book Co., London (1977).
12. J.F. Besseling, 'Nonlinear stress and deformation analysis', *Proc. World Congr. Fin. Elem. Meths. Struct. Mech.*, Bournemouth (1975) L1-L47, Also in revised edition, Dept. Mech. Eng. TH-Delft, WTHD-111 (1978).
13. J.F. Besseling, 'A thermodynamic approach to rheology', *Proc. IUTAM Symp. on irreversible aspects of continuum mechanics*, Springer Verlag, Wien (1968) pp. 16-53.
14. J.F. Besseling, 'A theory of elastic, plastic and creep deformations of an initially isotropic material', *J. Appl. Mech.* **25** (1958) 529-536.
15. D.K. Gartling and E.B. Becker, 'Finite element analysis of viscous incompressible fluid flow', *Comp. Meths. Appl. Mech. Eng.*, I, **8**, 1 (1976) 51-60, II, **8**, 2 (1976) 127-138.
16. M. Kawahara, N. Yoshimura, K. Nakagawa, H. Ohsaka, 'Steady and unsteady finite element analysis of incompressible viscous fluid', *Int. J. Num. Meth. in Eng.*, **10**, 2(1976) 437-456.
17. E.H. Lee, 'Elastic-plastic deformation at finite strains', *J. Appl. Mech.*, **36**, 3(1969) 1-14.
18. J.F. Besseling, L.J. Ernst, A.U. de Koning, E. Riks, K. van der Werff, 'Geometrical and physical nonlinearities, some developments in the Netherlands', *Proc. Fenomech. 1978*, North Holland Publ. Co., Amsterdam; *Comp. Meths. Appl. Mech. Eng.* 17/18 (1979) 131-157.

19. J.H. Argyris, 'Continua and discontinua', *Proc. Conf. Matrix Meth. in Struct. Mech.*, Wright Patterson Air Force Base, AFFDL. TR.66.80 (1966) 11-189.
20. J.F. Besseling, 'The complete analogy between the matrix equations and the continuous field equations of structural analysis', *Int. Symp. on analogue and digital techniques applied to aeronautics*, Liège (1963) 223-242.
21. T.H.H. Pian, 'A historical note about "hybrid elements" ', *Int. J. Num. Meth. in Eng.* **12**, 5 (1978) 891-892.
22. G. Sander, 'Application of the dual analysis principle', *Proc. IUTAM Symp. on high speed computing of elastic structures*, University de Liège (1971) 167-208.
23. L.J. Ernst, *A finite element approach to shell problems*, Dept. Mech. Eng. TH. Delft, WTHD-114 (1979).
24 C.R. Calladine, 'Buckminster Fuller's "tensegrity" structures and Clark Maxwell's rules for the construction of stiff frames', *Int. J. Solids Structures*, **14** (1978) 161-172.
25 K. van der Werff, *Kinematic and dynamic analysis of mechanisms; a finite element approach*, thesis Delft University Press (1977).

Reliability of Structures

*V. V. Bolotin**

Summary

A survey of general concepts of the theory of the reliability of structures is presented. Stochastic models of failures of structures are discussed, and the methods for evaluation of reliability and longevity factors are developed. The ways to predict the service life of an individual structure using the current inspection procedure and the load history recording are discussed and related optimization problems are stated.

1. Introduction

Mechanics of solids develops the methods for calculating stresses, strains- and displacements-fields in structures. The final decision on safety and reliability of a structure in the design and development phase as well as during the operational phase is to be made using the theory of reliability. From the point of view of mechanics of solids, an appropriate version of the reliability theory was proposed in [1] and developed in [2-3]. According to this theory, a failure of a structure is interpreted as a result of stochastic variation of the state of the structure.

The main applications of the theory of reliability in mechanics of solids are in the structural design under loads of essential random nature. Seismic action [1, 4, 5, 6], ocean wave action [7], wind loads [8], etc. are examples of such loadings. The role of the theory is significant also when random fracture phenomena are involved such as damage accumulation, cracksinitation and propagation in stochastically inhomogeneous solids [9, 10]. Significant applications are expected in stability problems for thin plates and shells which are very sensitive to random imperfections of their middle surface [2, 11].

The reliability approach is practically the only way to logically based structural design in situations when various and just contradicting requirements to the structure and many loading conditions must be taken into account. As example, the optimization design of shock absorbers and vibration insulators under combined random loading are to be mentioned [3, 12, 13]. An engineering problem of the service life estimation of machines and structures as well as a problem of the residual life forcasting for an individual structure using the current inspection datas [14, 15] can be stated and solved properly using the reliability approach too.

The last but not the least application area of the reliability theory is substantiation of design codes in various fields of engineering such as civil engineering, shipbuilding, etc. Codified magnitudes of such significant design parameters as design loads, design ultimate stresses, safety factors etc. can be well-found using the reliability approach only [2].

* Academy of Sciences of the USSR, Moscow Energetics Institute

Further, almost without direct references to special mechanical applications, basic concepts and methods of the reliability theory are developed.

2. General concepts

Let us consider a structure under random loading [3]. The state of the structure is described by the vector **u** which is an element of the state space **U**. Evolution of the structure in time t is described by random vector process $\mathbf{u}(t)$. The basic equation to evaluate the process $\mathbf{u}(t)$ is the operator equation $\mathbf{Lu} = \mathbf{q}$ where $\mathbf{q}(t)$ is a random vector process in the load space Q. The choice of the spaces U and Q and of the operator **L** is conditioned mutually and presents a problem of mechanics of solids and of the structural design.

Operational and service requirments put certain restrictions on the parameters of the structure. Service properties of the structure are described by the vector **v** which is an element of the quality space V. The vectors **u** and **v** are connected by the operator equation $\mathbf{v} = \mathbf{Mu}$. A subset of states admissible from the point of view of the service requirements corresponds to the open region Ω in the quality space V. The boundary Γ of the region Ω corresponds to the limit states. The first crossing by the vector process $\mathbf{v}(t)$ of the limit surface Γ in the direction of the external normal corresponds to the failure of the structure. Such crossings (called also positive crossings, out-crossings, excursions, etc.) play an important part in the version of the reliability theory presented in this paper. Further, when it would cause no misunderstanding, we use a simple term 'crossing' meaning an exit of the quality vector from the admissible region.

Including into the reliability theory inspection procedures during the service life and life monitoring, an additional space is to be introduced. Let certain parameters of the structure be measured during the operation. Results of these measurements build up the diagnosis vector **w**. The corresponding vector space W is called the diagnosis space [14].

Let the load process $\mathbf{q}(t)$ and/or the operator **L** be stochastic. Then the vector processes $\mathbf{u}(t)$ and $\mathbf{v}(t)$ will be stochastic too, and failure will be a random event. The principal reliability and longevity measure is the reliability function $R(t)$. This function is introduced as the probability of the non-failure operation of the structure during the time-segment $[0, t]$, i.e. the probability that the vector $\mathbf{v}(t)$ is staying within the admissible region Ω during this time-segment:

$$R(t) = P\{\mathbf{v}(\tau) \in \Omega; \text{ for all } \tau \in [0, t]\}. \tag{2.1}$$

Many other reliability and longevity factors [16] are expressed by the function $R(t)$. For example, the probability distribution function of the life time T (more precisely, of the time till the first failure) is determined as $F(T) = 1 - R(T)$. Mathematical expectation $E(T)$ of the life time T and failure rate $\nu(t)$ are determined as

$$E(T) = \int_0^\infty R(t)\,\mathrm{d}t, \quad \nu(t) = -\frac{R'(t)}{R(t)}, \tag{2.2}$$

etc. The definition of the reliability function $R(t)$ can be generalized to cases when repeated failures, repairs and recoveries are included into the consideration. The parameter can be interpreted not only as the physical time but also as the operating time, the bulk of product, etc.

On the design phase, the structure is to be considered as a randomly chosen element from the general ensemble of similar structures. Let properties of a structure be characterized by finite number of random parameters which form the vector **r** interpreted usually as the strength vector. The operators **L** and **M** and the admissible region Ω generally depend on the vector **r** also. The conditional reliability function for a structure with a fixed value of the vector **r** is

$$R(t\,|\,\mathbf{r}) = P\{\mathbf{v}(\tau\,|\,\mathbf{r})\in\Omega(\mathbf{r});\ \tau\in[0, t]\} \tag{2.3}$$

where $\mathbf{v}(t|\mathbf{r})$ is the conditional vector process. The unconditional (complete) reliability function is calculated as

$$R(t) = \int R(t|\mathbf{r})\,p(r)\,\mathrm{d}r$$

where $p(\mathbf{r})$ is the probability density for the vector **r**.

The condition of the structural reliability is

$$R(t) \geqslant R_*(t),\ \ t\in[0, T_*] \tag{2.4}$$

where $R_*(t)$ is the required reliability function, and T_* is the designed life. Reliability condition is usually stated at $t = T_*$. Denoting $R_*(T_*) \equiv R_*$ we write Eq. (2.4) as $R(T_*) \geqslant R_*$. If a structure is an element of a set of many similar structures working in comparable loading conditions, the the statistical interpertation of probability is valid. For unique or small-scale production structures, probabilistic values may be interpreted in the meaning of the decision theory. In both cases, the reliability approach presents the more logical and appropriate way for structural design in situations when statistically given or incomplete information is involved [2].

3. Stochastic models of failures

Various properties of the quality process $\mathbf{v}(t)$ can be used for the classification of models of failures: stationarity and non-stationarity, dependence on the history, special properties of distributions, etc.

In earlier publications on the structural reliability problems, very simple models based on the elementary probability theory concepts were used. The time was excluded from consideration, and the reliability design was reduced to comparison of two random vectors: load vector **q** and strength vector **r**. The reliability factor is calculated as

$$R = \int_{\Omega} p(\mathbf{q}, \mathbf{r})\,\mathrm{d}\mathbf{q}\,\mathrm{d}\mathbf{r} \tag{3.1}$$

where $p(\mathbf{q}, \mathbf{r})$ is the joint probability density for the vectors **q** and **r**. The typical condition under which $\mathbf{v} \in \Omega$ is $\mathbf{q}-\mathbf{r} < 0$ where the both vectors are of the same dimension.

The other class of stochastic models of failures are Markov models. Let the evolution of the system in the space V to be a n-dimensional continuous Markov process. Let the system at time t_0 to be in the point $\mathbf{v}_0$. The transition probability density $p(\mathbf{v},t|\mathbf{v}_0,t_0)$ considered as the function of the variable $\mathbf{v}$, t satisfies the forward Kolmogorov equation, whereas the function of the variables $\mathbf{v}_0$, t_0 satisfies to the backward Kolmogorov equation. The conditional reliability function $R(t|\mathbf{v}_0, t_0)$ (equal to the probability that the system being at t_0 in the point $\mathbf{v}_0 \in \Omega$ is staying in Ω during the time segment $[t_0, t]$) is

$$R(t|\mathbf{v}_0, t_0) = \int_{\Omega} p_r(\mathbf{v}, t|\mathbf{v}_0, t_0)\,\mathrm{d}\mathbf{v}. \tag{3.2}$$

Here $p_r(\mathbf{v}, t|\mathbf{v}_0, t_0)$ is the transition probability density determined for the set of trajectories $\mathbf{v}(t)$ which are absorbed by the boundary Γ. If the process intensivities do not depend on time, the reliability function $R(t|\mathbf{v}_0, t_0)$ satisfies the backward Kolmogorov equation with the initial condition $R = 1$ $(t = t_0, \mathbf{v}_0 \in \Omega)$ and the boundary condition $R = 0$ $(t \geqslant t_0, \mathbf{v}_0 \in \Gamma)$. The time to the first crossing of the boundary Γ is interpreted as the life of the system under condition that at t_0 the quality vector was equal to $\mathbf{v}_0 \in \Omega$. This time is a random value with the moments

$$T_k(\mathbf{v}_0, t_0) = \int_{t_0}^{\infty} \left| \frac{\partial R(t|\mathbf{v}_0, t_0)}{\partial t} \right| t^k\,\mathrm{d}t \tag{3.3}$$

satisfying the generalized Pontriagin equation with the boundary conditions $T_k = 0$ ($\mathbf{v}_0 \in \Gamma$, $k = 1, 2, \ldots$). Using the listed facts, a number of approaches were proposed to evaluate structural reliability factors. Survey of these approaches and its applications to mechanical vibration and stability problems can be found in [3, 17].

Let the sequence of failures form a homogeneous Poisson stream. Then the reliability function is

$$R(t) = e^{-\nu t} \tag{3.4}$$

where ν is the failure rate. Eq. (3.4) is the well known 'exponentional law of reliability' [16]. A natural heuristic generalization of this law is based on the assumption that the sequence of failures is a singular stream of events and that the failures are seldom events. Instead of Eq. (3.4) we obtain te approximate formula

$$R(t) = \exp\{-E[N(t)]\}. \tag{3.5}$$

Here $E(\cdot)$ is the operator of mathematical expectation and $N(t)$ is a random function of t equal to the number of excursions of the process $\mathbf{v}(t)$ out of the region Ω occured at the time segment $[0, t]$. Under certain conditions Eq. (3.5) can be interpreted as a rigorous asymptotic approximation for exact distribution of random seldom crossings.

Further, all failure models using the moments of the numbers $N(t)$ will be called the models of the Poisson's type. One of the main area of application of these models is the earthquake engineering and seismic design of structures [1, 4, 5].

It is well known that the most failures of structures and machines are results of gradual damage accumulation such as development of plastic deformations, crack propagation, fatigue, wear, etc. The quality factors are becoming worse gradually and quasimonotonously during the life. The appropriate failure models for such life conditions are cumulative models [2]. Let the admissible region Ω a bounded convex set in the space V, and the point $\mathbf{v} = 0$ belongs to Ω. Then the condition $\mathbf{v} \in \Omega$ can be written in the form $\|\mathbf{v}\| < 1$, where $\|\mathbf{v}\|$ is a properly chosen norm in the space V. The process $\mathbf{v}(t)$ is called a cumulative one on the time segment T if for each $t_1, t_2 \in T$ the condition holds

$$\|\mathbf{v}(t_2)\| \geqslant \|\mathbf{v}(t_1)\| \quad (t_2 > t_1). \tag{3.6}$$

If $\mathbf{v}(t_1) \in \Omega$, the probability of staying the vector $\mathbf{v}(t)$ in Ω on the segment $[t_1, t_2]$ coincides with the probability of its presense in Ω at the time t_2. Hence

$$R(t) = P\{\mathbf{v}(t) \in \Omega\}. \tag{3.7}$$

This property of cumulative models facilitates the evaluation of the reliability factors essentially in comparison with the general case when the study of crossings distribution is necessary.

4. Approximate and two-side estimates for the reliability function

Following [2, 3, 18], we consider approximate and two-side estimates applicable to the models of Poisson type. Let at $t = 0$ a system with probability one be in the admissible region Ω, and the sequence of crossings of the surface be a singular stream of events. We denote the probabilities that during the time segment $[0, t]$ one, two, etc. positive crossings occure by $Q_1(t)$, $Q_2(t)$, etc. respectively. The reliability function $R(t)$ is connected with the probabilities $Q_N(t)$ as follows

$$R(t) = 1 - \sum_{N=1}^{\infty} Q_N(t). \tag{4.1}$$

Then we introduce the moments of positive crossings as

$$E[N^\alpha(t)] = \sum_{N=1}^{\infty} N^\alpha Q_N(t). \tag{4.2}$$

From Eqs. (4.1) and (4.2) a number of inequalities for the function $R(t)$ can be derived, for example,

$$1 - E[N(t)] \leqslant R(t) \leqslant 1 - \frac{3}{2} E[N(t)] + \frac{1}{2} E[N^2(t)]. \tag{4.3}$$

Moments $E[N^\alpha(t)]$ are expressed by means of joint probability density of the process $\mathbf{v}(t)$ and its time derivative $\dot{\mathbf{v}}(t)$ at various time moments. So, the mathematical expectation of the number of positive crossings related to time unit is

$$\nu(t) = \int_{\Gamma} \mathrm{d}\Gamma \int_{\dot{v}_n > 0} p(\mathbf{v}_\Gamma, \dot{\mathbf{v}}; t)\, \dot{v}_n \, \mathrm{d}\dot{\mathbf{v}}. \tag{4.4}$$

Here $p(\mathbf{v}, \dot{\mathbf{v}}; t)$ is the joint probability density for the process $\mathbf{v}(t)$ and its derivative at the same time t, $\dot{v}_n$ is a normal component of $\dot{\mathbf{v}}$ on the surface Γ, and $\mathbf{v}_\Gamma$ is taken on the same surface. The surface Γ is supposed to be closed and smooth. The quantity $\nu(t)$ can be interpreted as the failure rate if a failure is associated with the excursion out of the region Ω. Formulae for the higher moments have the same structure, but dimension of the involved integrals grows correspondingly. Eq. (4.4) and related formulae are valid for the piece-wise-smooth surfaces also if only the process $\mathbf{v}(t)$ is sufficiently well-mixed.

Two-side estimates such as given by Eq. (4.3) are in close relation with the exponential estimate (3.5). If the failure stream is strictly Poissonian one, then the estimates given by Eq. (4.3) correspond to the expansion of the right-hand side of Eq. (3.5) in a power series. Keeping two terms of the series we obtain the lower estimate, and keeping three terms we obtain the upper estimate. The most of structures and machines under normal service conditions are high-reliable systems. The values of the reliability functions of these systems have to be close to unity during the full design life. Common magnitudes of the design reliability factors are 0,95, 0,99, and just 0,999 [2]. Under such conditions, the lower estimate from Eq. (4.3) can be used for approximate calculation of the reliability function leading to the linear estimate

$$R(t) \approx 1 - E[N(t)]. \tag{4.5}$$

For high-reliable structures, Eqs. (3.5) and (4.5) give practically the same numerical results. Eq. (4.5) has two advantages. Firstly, it presents the rigorous lower estimate. Secondly, the linear approximation is a little more convenient when further analytical calculation using the function $R(t)$ is to be performed.

Estimates given by Eqs. (3.5), (4.3) and (4.5) were studied in [3, 18] where their systematic application for various structural reliability problems was performed. A number of examples treating regions of various shape in multidimensional quality spaces are presented in [3] also.

5. Reliability of continuous systems

Mechanical models of structures, machines and its elements are, as a rule, continuous systems. Corresponding spaces of the reliability theory are functional phase spaces which elements are scalar, vector and tensor fields.

Quality requirements can be usually described introducing a norm $\|\mathbf{v}\|$ in the functional quality space V. The norm $\|\mathbf{v}\|$ is a scalar time function not depending on geometric nature of the process $\mathbf{v}(t)$. Hence, the problem of evaluation of the reliability function leads to analysis of crossings of scalar fields. As example, a body with a random stress field $\sigma_{jk}(x, t)$ can be considered. Let the following restrictions to be put from the strength considerations on the tensor field $\sigma_{jk}(x, t)$: the stress intensity $\sigma(x, t)$ in every point $x \in G$ of the body volume G has not to exceed the values of the random field $\sigma_*(x)$ corresponding to the local yield limit. Introducing the quality space as

$$V = \{\mathbf{v}: \sigma_{jk}(x, t); x \in G\}$$

and the norm (more exactly, the semi-norm)

$$\|\mathbf{v}\| = \sup_{x \leftarrow G} \frac{\sigma(x_1 t)}{\sigma_*(x)}$$

the admissible region became

$$\Omega = \{\mathbf{v}: \|\mathbf{v}\| < 1\}. \tag{5.1}$$

The analogous result can be obtained when a continuous damage accumulation model is considered with the scalar damage rate $v(x, t)$ and its local critical value $v_*(x)$. Other examples of the problem statements for continuous systems can be found in [3, 18, 19, 20].

For simplification we consider later the following case. Elements of the space V are scalar fields $v(x, t)$ given in the bounded region $G \subset \mathbb{R}^n$, and the admissible region $\Omega \subset V$ is given by inequality $v(x, t) < v_*$. Here v_* is a deterministic constant. The reliability function $R(t)$ is introduced as

$$R(t) = P\{\sup_{x \in G} \nu(x, \tau) < \nu_*; \tau \in [0, t]\}. \tag{5.2}$$

A following interpretation of positive crossings out of the region Ω is introduced in [19]. We introduce $(n + 1)$-dimensional space of the variables $y = \{x, \tau)$ (further notation $m = n + 1$ is used). The region of the variation for y is a cylinder $G_t = G \times [0, t]$. Connected sets of points in G_t in which $\nu(y) > \nu_*$ correspond to the excursions of the field $\nu(x, \tau)$ out of Ω. For these sets, we use the term 'excursions' too. Direct calculations of characteristics relating to excursions of random fields seems to be too difficult. Assuming the field $\nu(y)$ to be a double differentiable with respect to all the variables we try to connect the reliability factors and the characteristics of distribution of critical points of the field $\nu(y)$ in G_t.

The number of excursions $N(t)$, and the number of maxima $N_{max}(t)$ exceeding the level ν_* satisfy the obvious inequality $N(t) \leqslant N_{max}(t)$. If the level ν_* is sufficiently high and the field $\nu(y)$ is sufficiently well-mixed, then $N(t) \approx N_{max}(t)$. Instead of Eq. (3.5) we obtain an estimate

$$R(t) \approx \exp\{-E[N_{max}(t)]\}. \tag{5.3}$$

If a system is high-reliable, the linear estimate is also valid

$$R(t) \approx 1 - E[N_{max}(t)]. \tag{5.4}$$

Let the probability of intersection between excursions and the boundary of G_t be negligibly small. Then $N_{max}(t)$ may be identified with the number of analytical maxima of the field $\nu(y)$ in G_t. A method to calculate the mathematical expectation of this number is presented in [21]. For this calculation, the joint probability density $p(\nu, \varphi, \kappa; y)$ for the field $\nu(y)$, its gradient $\varphi = \{\nabla_\alpha \nu\}$, and the elements of the Hesse matrix $H = \{\nabla_\alpha \nabla_\beta \nu\}$ ordered into the $[m(m + 1)/2]$ = -dimensional vector κ has to be known. Then

$$E[N_{max}(t)] = \int_{G_t} dy \int_{\nu_*}^{\infty} d\nu \int_{\kappa \in K_m} p(\nu, 0, \kappa; y) \, | \det H | \, d\kappa. \tag{5.5}$$

Here K_m is a set of values κ corresponding to the definitely negative matrix H.

A more sophisticated approach is based on the methods of differential topology [18]. Considering each excursion as a m-dimensional compact manifold with given on it the double differentiable function $\nu(y)$, and mapping this manifold on the m-dimensional sphere, we can apply the Morse theorem connecting numbers of critical points on the sphere. When all excursions are one-connected, we obtain

$$E[N(t)] = \sum_{j=0}^{m} (-1)^{m-j} E[N_j(t)]. \tag{5.6}$$

Here $N_j(t)$ is a complete number of critical points of j-th type being above the level ν_*. The type of a point is equal to the number j of negative eigenvalues of the matrix H. The mathematical expectations of the numbers $N_j(t)$ are calculated using formulae similar to Eq. (5.5) Further details and applications to elastic plates and shells with random initial imperfections are in [3, 11, 18, 19, 20, 22, 23].

If the process $\nu(t)$ is cumulative one in the sense of Eq. (3.6) it results to some simplifications. The reliability function is expressed by the mathematical expectation of number of excursions intersecting with the top of the cylinder $G_t = G \times [0, t]$. We come to further simplifications reducing functional quality spaces to the finite-dimensional spaces. It corresponds to common engineering approaches to the reliability design when one checks most stressed and reponsible elements, joints and single points of a structure.

6. Damage accumulation in structures

General definition of cumulative failure models was introduced earlier. Evolution of the quality vector $\nu(t)$ is governed by kinetic equation

$$\frac{d\mathbf{v}}{dt} = f(\mathbf{v}, q, \mathbf{r}). \tag{6.1}$$

Its right-hand side is a non-negative vector-function of the process $\nu(t)$, the load process $q(t)$, and the strength vector $\mathbf{r}$. A more general case is the equation of hereditary type

$$\frac{d\mathbf{v}}{dt} = \underset{\tau=0}{\overset{\tau=t}{H}} \{\mathbf{v}(\tau), \mathbf{q}(\tau), \mathbf{r}\} \tag{6.2}$$

and the right-hand part is a non-negative functional of the processes $\nu(\tau)$ and $\mathbf{q}(\tau)$ on the time segment $[0, t]$. Initial condition is $\mathbf{v}(0) = \nu_0$ where the vector $\mathbf{v}_0$ characterizes the initial damage rate. One usually assumes that $\mathbf{v}_0 = 0$. Life time T is to be determined from the condition

$$\| \mathbf{v}(T) \| = 1. \tag{6.3}$$

The cumulative damage rule which is most used in structural design [2, 15, 24] is a special case of the model (6.1). Let $\nu(\mathbf{t})$ be a scalar time function, and the righ-hand part in Eq. (6.1) is $f = f_1(\nu)\, f_2(\mathbf{q}, \mathbf{r})$, where $1/f_1(\nu)$ and $f_2(q, r)$ are integratable functions of their variables. Introducing the function

$$g(\nu) = \int_0^\nu \frac{d\nu}{f_1(\nu)}$$

under the condition that $g(1) = 1$, we represent the damage accumulation process by the inverse function as $\nu(t) = g^{-1}[\theta(t)]$,

$$\theta(t) = \int_0^t \frac{d\tau}{T_q[q(\tau), \mathbf{r}]} \tag{6.4}$$

where notation $T_q(q, r) = 1/f_2(q, \mathbf{r})$ is used. The function $T_q(q, \mathbf{r})$ has the meaning of the life time under constant load level, i.e. at q = const. The equation $t = T_q(q, \mathbf{r})$ can be interpreted as an equation of 'fatigue curve', 'delayed fracture curve', 'wear-and-tear curve', etc. depending on physical nature of the considered damage accumulation. Putting in Eq. (6.4) $\theta(T) = 1$ which means $\nu(T) = 1$, we obtain the so-called 'linear' cumulative damage rule:

$$\int_0^T \frac{d\tau}{T_q[q(\tau), \mathbf{r}]} = 1. \tag{6.5}$$

The aim of these rather trivial calculations is to show that the rule (6.5) does not mean necessary the linear growth of damage rate under stationary loading. Moreover, the function $g(\nu)$ is generally nonlinear and takes into account an influence of the load history on the intermediate stages of damage accumulation.

The damage accumulation process $\nu(t)$ has two origins of stochastic nature. Firstly, the loading $q(t)$ is a random process, and secondly, the strength vector $\mathbf{r}$ is to be considered in the design phase as a random one. In structural design practice one says about load and strength scatter correspondingly. Analysing the influence of the load scatter, we introduce the conditional process $\nu(t|\mathbf{r})$.

Assume the characteristic correlation time of the process $q(t)$ to be small in comparison with characteristic time of the process $\nu(t|\mathbf{r})$. Then the central limit theorem generalized to integrals of random processes can be applied to the process $\nu(t|\mathbf{r})$. It results into a wide class of life distributions which typical representative is [14]

$$F(T|\mathbf{r}) = 1 - \Phi\{\frac{1 - E[\nu(t|\mathbf{r})]}{\sigma_\nu(t|\mathbf{r})}\}.$$

Here $\sigma_\nu^2(t|\mathbf{r})$ is the dispersion of the conditional process $\nu(t|\mathbf{r})$, and $\Phi(z)$ is the function of the normalized Gaussian distribution. When the life scatter is sufficiently small, the corresponding probability density can be approximated by the delta-function. Hence

$$p(T|\mathbf{r}) = \frac{\partial F(T|\mathbf{r})}{\partial T} \approx \delta[T - T_*(\mathbf{r})]$$

where $T_*(\mathbf{r})$ is the root of the equation $E[\nu(T|\mathbf{r})] = 1$.

Consideration of the strength scatter leads to significant difficulties. Firstly, a moderate scatter of strength parameters is associated with considerable scatter of life. Secondly, direct evaluation of functions containing the strength vector often seems to be impossible because of nonreproducibility of fracture tests relating to an individual specimen. For example, fracturing a specimen at q = constant we can find only one point of the curve $t = T_q(q, \mathbf{r})$. As another example, a common two-step procedure to verify the cumulative damage rule (6.5) can be indicated. The equation connecting the duration t_1 of the first step under the load q_1 = constant and the duration t_2 under the q_2 = constant till the fracture of specimen is:

$$\frac{t_1}{T_q(q_1, \mathbf{r})} + \frac{t_2(\mathbf{r})}{T_q(q_2, \mathbf{r})} = 1. \tag{6.6}$$

Eq. (6.6) contains three unknown values $T_q(q_1, \mathbf{r})$, $T_q(q_2, \mathbf{r})$ and $t_2(\mathbf{r})$ relating to the same specimen. To determine this values one must fracture the same specimen three times. Thus Eq. (6.6) cannot be verified directly.

The main conclusion from the study of probabilistic and statistic aspects connected with the damage accumulation rule is the following [24]: at this moment there are no weighty arguments to reject this rule as a basic engineering tool in structural design under random loading.

7. Prediction of individual reliability and life factors

Prediction of the service of an individual structure using the current inspection procedure and the load history records is of special interest being applied to large and expensive structures. From the point of view of the reliability theory, the prediction procedure consists of estimation of the state of the system and its parameters, and later stochastic extrapolation of the random vector processes upon the next time segments. Using the extrapolated processes, the reliability factors of the structure can be predicted and the value of the optimal residual service life can be recommended. This approach was developed in [14] where a way was suggested to the common use of the apriori information relating to the general ensemble of structures and the diagnostic information relating to a certain structure.

Let a structure to be inspected at $t_1, t_2, ..., t_k$. The results of inspections are presented by values of the diagnosis vector $\mathbf{w}$, equal to $\mathbf{w}_1, \mathbf{w}_2, ..., \mathbf{w}_k$. We assume that the vector $\mathbf{w}$ includes the information about the loading history also. For abbreviation we notate the set of time moments $T_k = \{t_1, t_2, ..., t_k\}$ and the set of corresponding inspection datas $\mathbf{w}(T_k) = \{\mathbf{w}_1, \mathbf{w}_2, ..., \mathbf{w}_k\}$.

The simpliest approach to individual life prediction is based on the extrapolation of the process $\mathbf{w}(t)$ in the diagnosis space W. The admissible region Ω is to be introduced in the same space. Denoting the aposteriori process $\mathbf{w}(t|\mathbf{r}, T_k)$ the following decision rule is to be used: a structure having been under normal operational conditions at $t \leqslant t_k$ is allowed to continue its operation till the time moment $t_{k+1} > t_k$ if $\mathbf{w}(t|\mathbf{r}, T_k) \in \Omega$ at all $t \in (t_k, t_{k+1})$.

A substantional volume of information is generally available on the structural design phase. It is expedient to use this information together with diagnostic datas relating to the considered structure. This approach requires extrapolation in the state space U as well as in the quality space V. Denoting the extrapolated processes $\mathbf{u}(t|\mathbf{r}, T_k)$ and $\mathbf{v}(t|\mathbf{r}, T_k)$ correspondingly, the following definition of the aposteriori reliability function is introduced

$$R(t|\mathbf{r}, T_k) = P\{\mathbf{v}(\tau|\mathbf{r}, T_k) \in \Omega; \tau \in (t_k, t)\}.$$

Using this function we can calculate the probability density $p(T|\mathbf{r}, T_k)$ for the individual life, the mathematical expectation of the residual life of the structure, etc. It is to be emphasized that $\mathbf{u}(t|\mathbf{r}, T_k)$ and $\mathbf{v}(t|\mathbf{r}, T_k)$ are random processes. Firstly, the future loading on the structure is given only stochastically. Secondly, although the aposteriori reliability function $R(t|\mathbf{r}, T_k)$ corresponds to a certain structure with a fixed value of the strength vector $\mathbf{r}$, this value remains in fact unknown and therefore ought to be considered as a random one. For example, if no failure occurs till the time t_k, the apriori probability density $p(\mathbf{r})$ is to be substituted by the truncated probability density $p(\mathbf{r}|T_k)$. Nondestructive methods for estimation of mechanical properties of the structure present only indirect information about these properties and thus present only probabilistic datas on the vector **r**. All these considerations seem to be essential for the proper theoretical substantiation of the various practical approaches to individual life prediction [15] using load counters, damage indicators, etc.

In [14] an optimization approach was proposed to estimate the 'best' moment for the next inspection, repair or stopping of service. Let the profit related to the time unit to be $\pi(t)$. Several types of failures of the system are possible, and the loss associated with each type is equal to L_α. The mathematical expectation of the winnings under condition that a system inspected last time at t_k remains in service till $t > t_k$ is

$$E[C(t|\mathbf{r}, T_k)] = \int_{t_k}^{t} R(\tau|\mathbf{r}, T_k)\,\pi(\tau)\,d\tau - \sum_\alpha Q_\alpha(t|\mathbf{r}, T_k) L_\alpha.$$

Here $Q_\alpha(t|\mathbf{r}, T_k)$ are the aposteriori probabilities of failures of certain types. The first term on the right-hand side grows with t, and the second term is a decreasing function of t. The 'best' time moment t_* is determined from the condition

$$E[C(t|\mathbf{r}, T_k)] \to \max.$$

This approach can be generalized on the case when the profit function $\pi(t)$ and losses L_α depend on the quality vector **v**. In this case the mathematical expectation of the winnings is expressed by the aposteriori probability density $p(\mathbf{v}; t|\mathbf{r}, T_k)$ of the aposteriori process $\mathbf{v}(t|\mathbf{r}, T_k)$.

8. Conclusion

The theory of structural reliability strictly does not belong to the mechanics of solids. At the same time this theory provides a way to efficient applications of achievements of the mechanics of solids to engineering practice. The principal aim of this survey is to develop the basic concepts and methods of the theory of structural reliability in the form matching with the concepts and methods of the mechanics of solids. It seems that the theory of structural reliability should became an origin for many new problems in the mechanics of solids.

9. References

1. Bolotin V.V., 'Statistical theory of the aseismic design of structures'. *Proc. of the Second World Cong. on Earthquake Engineering*, Tokyo, 1960.
2. Bolotin V.V., *Statističeskie methody v stroitel'noj mehanike*, Strojizdat, Moskva, 1961, 1965. Engl. transl.: *Statistical methods in structural mechanics*. Holden-Day, San Francisco, 1969.
3. Bolotin V.V., *Primenenie metodov teorii verojatnostej i teorii nadežnosti v rasčetah sooruženij*, Strojizdat, Moskva, 1971 (*Application of the methods of the probability theory and the reliability theory in calculations of constructions*).
4. Freudenthal A.M., Shinozuka M., 'Probability of structural failure under earthquake acceleration'. *Trans. Japan Soc. Civil Engrs*, 1965, No. 118.
5. Bogdanoff J.L., Shiff, A., 'Earthquake effects in the safety and reliability analysis of

engineering structures', *Proc. Internat. Conf. on Structural Safety and Reliability*. Pergamon Press, Oxford, 1972.
6. Kaul M.K., Penzien J., 'Stochastic seismic analysis of yielding offshore towers', *J. Engng Mech. Div.*, Proc. ASCE, 1974, **Vol. 100**, No. EM5.
7. Bea R.G., 'Selection of environmental criteria for offshore platform design', *J. Pet. Tech.*, 1974, No. 10.
8. Konishi I., 'Safety and reliability of suspension bridges', *Internat. Conf. on Structural Safety and Reliability*, Pergamon Press, Oxford, 1972.
9. Freudenthal A.M., *Statistic approach to brittle fracture. Fracture – an Advanced Treatise*, Vol. 2. Academic Press, New York, 1968.
10. Bolotin V.V., *Stochastic models of cumulative damage in composite materials. Progress in Fatigue and Fracture*, Pergamon Press, Oxford, 1976.
11. Bolotin V.V., Volohovskij V. Ju., *O deformacii uprugoj sferičeskoj oboločki so slučajnymi načal'nymi nepravil'nostjami*, Izv. AN SSSR, Mehanika tverdogo tela, 1972, No. 5. (*On the deformation of an elastic spherical shell with random initial imperfections*. Proc. USSR Ac. of Sci., Mechanics of Solids (1972) 5).
12. Radin, V.P., 'Ob optimizacii linejnyh vibrozaščitnyh sistem po nadežnosti', *Prikladnaja Mehanika*, 1972, **tom 8**, vyp 9. ('On the optimization of linear vibration absorbing systems from the point of view of reliability').
13. Volohovskij V. Ju., Radin V.P., 'O vybore optimal'nyh parametrov nelinejnyh vibrozaščitnyh sistem pri slučajnyh vozdejstvijah', *Izv. AN SSSR, Mehanika tverdogo tela*, 1972, No. 2 ('On the choice of optimal parameters of non-linear vibration absorbing systems in the presence of random excitations', Proc. USSR Ac of Sc., Mechanics of Solids (1972) 2).
14. Bolotin V.V., 'O prognozirovanii nadežnosti i dolgovečnosti mašin', *Mašinovedenie*, 1977, No. 5 (On the prediction of reliability and endurance of machines. Mechanical Engineering (1977) 5).
15. Payne A.O., *The fatigue of aircraft structures. Progress in Fatigue and Fracture*, Pergamon Press, Oxford, 1976.
16. Gnedenko V.V., Beljaev Ju. K., Solov'ev A.D., *Matematičeskie metody v teorii nadeznosti*, Nauka, Moskva 1965. (Mathematical methods in the theory of reliability, Science (ed.), Moscow (1965)).
17. Bolotin V.V., 'Statistical aspects in the theory of structural stability', *Proc. Internat. Conf. on Dynamic Stability of Structures*, Pergamon Press, Oxford, 1967.
18. Bolotin V.V., 'Zadači teorii nadežnosti v mehanike deformiruemyh tverdyh tel', *Mehanika splošnoj sredy i rodstvennye problemy analiza*, Nauka, Moskva 1972. ('Problems of the theory of reliability in the mechanics of deformable solid bodies', in *The mechanics of continuous media and related methods of analysis*, Science (ed.) Moscow (1972)).
19. Bolotin V.V., 'Teorija naadežnosti raspredelennyh mehaničeskih sistem', *Izv. AN SSSR, Mehanika tverdogo tela*, 1969 No. 6. ('The theory of reliability of distributed mechanical systems'). *Proc. USSR Ac. of Science, Mechanics of Solids* (1969) 6).
20. Bolotin V.V., *Reliability theory and stochastic stability. Study on Stability*, University of Waterloo, Waterloo, Canada, 1973, No. 6.
21. Longuet-Higgins M.S., 'The statistical analysis of a random moving surface', *Phil. Trans. Roy. Soc. London*, 1957, Vol. 249 A, No. 966.
22. Beljaev Ju. K., 'Raspredelenie maksimuma slučajnogo polja i ego priloženie k zadačam nadežnosti', *Izv. AN SSSR, Tehničeskaja kibernetika*, 1970, No. 2. ('The distribution of the maximum of a random field and its application on problems of reliability', *Proc.*

USSR Ac. of Science, Technological Cybernetics (1970) No. 2).
23. Volohovskij V. Ju., 'Issledovanie očenok dlja funkcii nadežnosti raspredelennyh mehaničeskih sistem metodom statističeskogo modelirovanija'. *Izv. AN SSSR, Mehanika tverdogo tela*, 1973, No. 5. ('The investigation of estimations for reliability functions of distributed mechanical systems by means of the method of statistical modelling', *Proc. USSR Ac. of Sc., Mechanics of Solids* (1973) 5.
24. Bolotin V.V., Ermolenko A.F., 'Summirovanie ustalostnyh povreždenij i statističeskij razbros pročnosti', *Mašinovedenie*, 1979, No. 1 ('Summation of fatigue damages and statistical scatter of strength'. *Mechanical Engineering* (1979) 1).

Buckling: Progress and Challenge

*B. Budiansky and J.W. Hutchinson**

Summary

The general theories of elastic and plastic buckling and post-buckling behavior of structures are summarized briefly, and several special topics of current interest are discussed.

1. Introduction

It is tempting to begin this survey of the subject of buckling with a long backward look at its history and to muse on the special fascination it has held for so many engineers and scientists. Everybody loves a buckling problem! Without sentimentality, however, we will limit ourselves to the simple but confident introductory assertion that great progress in the understanding of buckling phenomena has been achieved in the last four decades. It is symptomatic of the vigor of the subject that surveys, assessments, and recapitulations of this progress have been appearing with unusual frequency [e.g. 1, 2, 3]. Nevertheless, we welcome with pleasure this opportunity to present our own overview at this symposium in honor of the central contemporary figure in the field of buckling and stability of structures, Warner Koiter.

With deference to the fact that we face a mixed audience of specialists and non-specialists we will try not to get too technical as we summarize progress, outline solutions, and pose problems. To set the stage, we shall give a succinct précis of some of the main concepts and results of general elastic buckling theory, and then describe several of the mode-interaction problems that have recently been under scrutiny by various investigators. We will turn next to a description of the far less well appreciated results of the general theory of plastic buckling and post-buckling behavior, focus on some central unsolved basic problems, and indicate current attempts to reach acceptable solutions. Finally, we will conclude with some observations, necessarily brief, on several related topics that have engaged the attention of the buckling community, including optimum design, stochastic buckling, and the unresolved, nagging problem of the quest for a basic, general stability theorem.

2. General Theory of Elastic Buckling

The general theory of elastic buckling and post-buckling behavior was presented by Koiter in his 1945 Ph.D. thesis [4]. After a dormant period of over fifteen years the basic ideas of the

* Division of Applied Sciences, Harvard University, Cambridge, Massachusetts 02138, U.S.A.

theory started to become widely known, and by now have been the subject of numerous alternative (but essentially equivalent) expositions [3, 5-7]. The theory has found applications to many specific problems and, indeed, informs all current understanding of buckling phenomena. We shall summarize briefly some essential features of the general theory, in a functional notation [3] that, while not in common use, is succinct, general and convenient for applications.

In the study of the equilibrium displacements u produced in an elastic body by prescribed loads proportional to a scalar measure λ, it is convenient to introduce an inner-product functional $\langle u,v\rangle$ and a norm $\|u\| = \langle u,u\rangle^{1/2}$. We suppose the existence of a *fundamental* displacement $u_0(\lambda)$ that varies continuously from $\lambda = 0$ (see Fig. 1(a)), and contemplate the possibility that a distinct equilibrium path

$$u = u_0(\lambda) + v(\lambda) \tag{2.1}$$

sprouts from the fundamental path at the critical load $\lambda = \lambda_c$. The normalized *bifurcation* mode may be defined by

$$u_1 = \lim_{\lambda\to\lambda_c} v/\|v\| \tag{2.2}$$

and then, with the introduction of the scalar parameter

$$\xi = \langle v,u_1\rangle \tag{2.3}$$

the bifurcated path can be described via the dependence of v and λ on ξ. The expansions

$$u = u_0(\lambda) + \xi u_1 + \xi^2 u_2 + \ldots \quad (\langle u_n,u_1\rangle = 0, n \geq 2) \tag{2.4}$$

and

$$\lambda = \lambda_c + \lambda_1\xi + \lambda_2\xi^2 + \ldots \tag{2.5}$$

may then be used in a standard perturbation procedure for solving the governing equations of equilibrium and thereby deducing $u_1, u_2, \ldots$ and $\lambda_c, \lambda_1, \lambda_2, \ldots$.
In a conservative elastic system, equilibrium may be enforced by the variational assertion

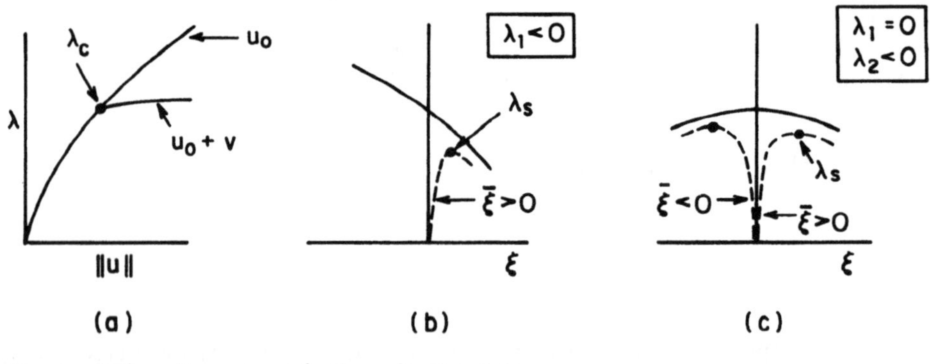

Fig. 1. Bifurcations, imperfections, limit points.

$$\delta\phi = \phi'[u;\lambda]\delta u = 0 \tag{2.6}$$

where ϕ is the potential energy functional of the system, the prime denotes Fréchet differentiation, and δu is an arbitrary admissible variation of displacement. Either (2.6), or its Euler-equation consequences, may serve as the vehicle for the generation of a hierarchy of governing equations of various orders in ξ that follow from the expansions (2.4) and (2.5). If (2.6) is used, the result is:

$$\begin{aligned}
&\xi[\phi_c'' u_1]\delta u && = 0\\
&+\xi^2[\phi_c'' u_2 + \lambda_1 \dot{\phi}_c'' u_1 + \tfrac{1}{2}\phi_c''' u_1^2]\delta u && = 0\\
&+\xi^3[\phi_c'' u_3 + \lambda_1(\dot{\phi}_c'' u_2 + \dot{\phi}_c''' u_1^2) + \lambda_1^2(\tfrac{1}{2}\ddot{\phi}_c'' u_1)\\
&\qquad + \lambda_2\dot{\phi}_c'' u_1 + \phi_c''' u_1 u_2 + \tfrac{1}{6}\phi_c^{iv} u_1^3]\delta u && = 0\\
&+\ldots && = 0
\end{aligned} \tag{2.7}$$

where $\phi_c^{(n)} \equiv \phi^{(n)}[u_0(\lambda_c), \lambda_c]$ and $(\dot{\ }) \equiv \dfrac{\partial}{\partial\lambda}(\ )$.

The perturbation equilibrium equation of order ξ provides a homogeneous problem for the lowest eigenvalue λ_c and the corresponding eigenfunction u_1, and if the solution for u_1 is unique (except for sign) the perturbation procedure is unambiguous. In this single-mode case, setting $\delta u = u_1$ in the equilibrium equation of order ξ^2 gives

$$\lambda_1 = \tfrac{1}{2}\phi_c''' u_1^3 / (-\dot{\phi}_c'' u_1^2) \tag{2.8}$$

and if $\lambda_1 = 0$, a similar calculation with the third-order equation yields

$$\begin{aligned}
\lambda_2 &= (\tfrac{1}{6}\phi_c^{iv} u_1^4 + \phi_c''' u_1^2 u_2)/(-\dot{\phi}_c'' u_1^2)\\
&= (\tfrac{1}{6}\phi_c^{iv} u_1^4 - \phi_c'' u_2^2)/(-\dot{\phi}_c'' u_1^2)
\end{aligned} \tag{2.9}$$

(Stability considerations imply positive values for the denominators of (2.8) and (2.9).)

If N linearly independent eigenfunctions u_{1i} are found at λ_c, they may be orthonormalized ($\langle u_{1i}, u_{1j}\rangle = \delta_{ij}$) and the expansion (4) may be modified to

$$u = u_0(\lambda) + \sum_{i=1}^{N} \xi_i u_{1i} + \sum_{i=1}^{N}\sum_{j=1}^{N} \xi_i \xi_j u_{2ij} + \ldots \qquad \langle u_{2ij}, u_{1k}\rangle = 0 \tag{2.10}$$

where $\xi_i = \langle v, u_{1i}\rangle$ and $\xi^2 = \sum_{i=1}^{N} \xi_i \xi_i$. In turn, each ξ_i is expanded in powers of ξ, which is now intrinsically positive. Then perturbation equations of successive order provide information for the determination of admissible bifurcation modes specified by $\nu_i = \lim_{\lambda\to\lambda_c} \xi_i/\xi$, the corresponding values of λ_1, and, if necessary, the u_{2ij} and λ_2.

The major significance of the quantities λ_1 and λ_2 is their implication concerning *imperfection sensitivity*. If, in the single-mode case, $\lambda_1 \neq 0$, or if $\lambda_1 = 0$ and $\lambda_2 < 0$, small initial geometrical imperfections will generally induce snap buckling at loads λ_s smaller than λ_c (see Fig. 1(b, c)). (The same is true for the multi-mode case, except that, since ξ is then intrinsically positive, $\lambda_1 > 0$ does not necessarily imply imperfection sensitivity.) Explicit asymptotic formulas for λ_s can be found on the basis of a modified potential $\overline{\phi}[u, \overline{u}]$ valid when an initial displacement $\overline{u}$ is present. Thus, for the special case of a *linear fundamental state*, in which $u_0(\lambda)$ is proportional to λ, the *knockdown-factors* λ_s/λ_c are:

$$\lambda_s/\lambda_c \approx 1 - 2 \left(-\frac{\lambda_1 \overline{\xi}}{\lambda_c}\right)^{1/2} \qquad (\lambda_1 \overline{\xi} < 0) \tag{2.11}$$

$$\lambda_s/\lambda_c \approx 1 - 3 \left(-\frac{\lambda_2 \overline{\xi}^2}{4\lambda_c}\right)^{1/3} \qquad (\lambda_1 = 0, \lambda_2 < 0) \tag{2.12}$$

where

$$\overline{\xi} = \langle \overline{u}, u_1 \rangle \tag{2.13}$$

is the 'amount' of bifurcation mode u_1 contained in an initial imperfection $\overline{u}$. The same formulas apply in the multi-mode case if $\overline{u} = \overline{\xi} \sum_{i=1}^{N} \nu_i u_{1i}$ is in the shape of one of the bifurcation modes. A method for deducing the influence of initial imperfections in the shapes of a few *nearly* coincident eigenmodes u_{1i} is to use the expression (2.10) directly in the variational equation

$$\delta\overline{\phi}\,[u, \Sigma\overline{\xi}_i u_{1i}] = 0 \tag{2.14}$$

and thereby deduce approximate relations among λ, the displacement measures ξ_i, and the imperfection parameters $\overline{\xi}_i$. In some problems only the terms of order ξ_i will be needed; in others the quadratic terms are necessary, and then the u_{2ij} can be chosen in the forms they *would* have had if the modes were actually coincident [21]. More about this later.

The importance of the results (2.11) and (2.12) is that they offer a rational explanation for the inability of many structures (particularly shells) to achieve their ideal buckling strengths: they demonstrate the sensitivity of their actual strengths to initial imperfections, and they provide useful quantitative estimates for the magnitude of this sensitivity. We should also mention the significant fact, emphasized by Koiter, that the equilibrium along the descending portions of the curves in Fig. 1(b, c) is *unstable*. Numerous references to specific problems that have been solved by application of the general theory are contained in the surveys [1], [8], and [9].

3. Mode Interaction Problems

Despite the apparent special character, from a mathematical viewpoint, of multiple-eigenvalue problems, practical situations in which several bifurcation modes interact must be considered frequently. There is a fundamental reason for this: optimum design tends to produce structures having nearly equal resistances to more than one mode of failure. There are two well-known primitive structural systems – the spherical shell under external pressure and the circular cylindrical shell under axial stress – that are already, in a sense, optimum, and have a large number of independent bifurcation modes at, or very close to, the lowest bifurcation load. The notorious imperfection-sensitivity of these shells has been studied extensively, and although unresolved questions remain, these problems will not be addressed here. Rather, we shall direct our attention to recent attempts to understand mode-interaction problems of column, plate, and shell assemblies that are designer-induced by the quest for optimality.

The model structure shown in Fig. 2(a) captures many of the characteristic features of mode-interaction phenomena. It consists of an assembly of two elementary spring-supported structures, with the little component providing support to the big one. The strain-energy functions for the restraining springs are taken as

$$\mathscr{E}_z = l\left[\frac{z^2}{2} + \frac{Bz^4}{4}\right] \tag{3.1a}$$

$$\mathscr{E}_y = \lambda_E\left[\frac{y^2}{2} + \frac{Ay^4}{4}\right] \tag{3.1b}$$

respectively. The special choices $B = \pm 1$, $A = \pm 1$ lead to the familiar initial post-buckling behaviors sketched in Fig. 2(b, c) for the two components considered as separate structures.

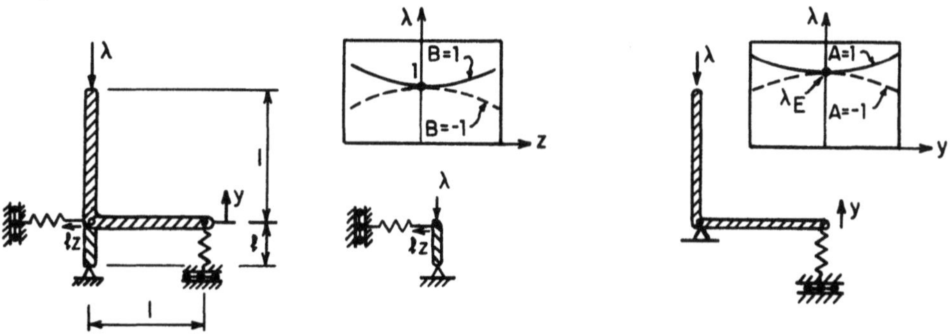

Fig. 2. Mode-coupling model.

For the combined structure of Fig. 2(a) the governing potential energy functional (to fourth-degree terms) is

$$\phi = \mathscr{E}_z + \mathscr{E}_y - \lambda\left\{l\left[\frac{z^2}{2}(1+y) + \frac{z^4}{8}(1+l)\right] + \frac{y^2}{2} + \frac{y^4}{8}\right\} \tag{3.2}$$

and from the equilibrium conditions $\frac{\partial\phi}{\partial z} = \frac{\partial\phi}{\partial y} = 0$ a rich variety of behaviors emerge.

For the sake of future analogy, we dub z and y the displacements corresponding to *local* and *overall* modes, respectively. The associated buckling loads of separate structures are 1 and λ_E. These remain the two critical loads for the combined structure, with the post-buckling behaviors sketched in Fig. 3 for the case $A = B = 1$ and $l = 1/4$. For $\lambda_E < 1$ the lowest bifurcation is stable, but a secondary bifurcation at $\lambda = \lambda_s$ lurks in its vicinity. Furthermore,

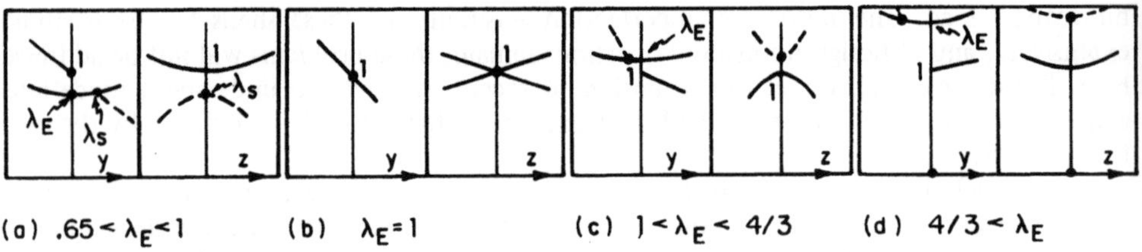

Fig. 3 Mode-interaction behavior of model ($A = B = 1$).

this secondary bifurcation is unstable for $\lambda_E > .65$. When $\lambda_E = 1$, so that the lowest eigenvalue is a multiple one, an unstable mode reveals itself right at bifurcation, and instability of the fundamental mode persists up to $\lambda_E = 4/3$. For $\lambda_E > 4/3$, the critical mode is once more stable, but now no secondary bifurcation, unstable or not, exists in the vicinity of the buckling load $\lambda = 1$. (In this range determination of the maximum load of the perfect model structure would depend on higher order terms in the potential energy, not taken into account in the present analysis.)

Analogous results could be sketched for other combinations of A and B corresponding to unstable post-buckling behavior of one or both of the individual structures. Further, the effects of initial imperfections in y or z or both can be assessed. A small sampling of the kinds of results obtainable is exhibited in Fig. 4. In this figure stable bifurcations are denoted by solid lines, unstable ones by dashed lines, and limit-point buckling by dotted lines. The various behaviors of Fig. 3 for the perfect structure, with $A = B = 1$, correspond to Fig. 4(a), which also shows results for an imperfection $\bar{y} = .01$ in the overall mode. The imperfection-sensitivity in the neighborhood of $\lambda_E = 1$ is evident. The results in Fig. 4(b) (imperfection-sensitive overall mode) and Fig. 4(c) (both modes imperfection-sensitive) display enlarged domains of imperfections-sensitivity, but it is notable that near $\lambda_E = 1$ the magnitudes of the knockdown factors are not much dependent on A and B, at least for the very small imperfection considered. In other words, even if the decoupled structures are individually imperfection-insensitive, they can interact unstably, and near $\lambda_E = 1$ they may do so in a way that is not changed very much by the introduction of imperfection-sensitivity into the decoupled structures.

Two underlying mechanisms of unstable mode interaction are evident from our model; they may conveniently be characterized as *load amplification* and *support degradation*. Thus

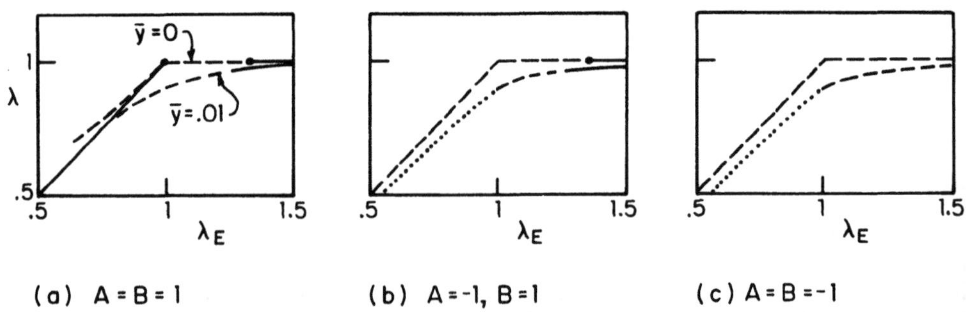

Fig. 4. Stable bifurcations (——), unstable bifurcations (---), limit points (...).

(see Fig. 2(a)) a positive displacement y in the overall mode makes the load applied to the local supporting structure greater than λ. On the other hand, a displacement z in the local mode reduces the effective support supplied to the primary structure at the juncture of the two components. These two basic mechanisms of mode interaction can be recognized in the more realistic structures to which we turn next.

Each of the structures shown in Fig. 5 has received detailed analysis during the last decade, and each involves the interaction of recognizable local and overall modes. The lattice column [7, 10] displays interaction between overall Euler buckling and the local buckling of the vertical members as columns over many supports. Here there is negligible post-buckling increase of load associated with both the individual modes, corresponding to essentially neutral stability of each. The two-flange column [11-14] and the wide, stiffened-plate column [15-20] both permit local, multi-wave, plate-element buckling (stable) to interact with overall Euler buckling (neutral). Finally, the axially loaded, longitudinally stiffened cylinder [21, 22] permits interaction between local, longitudinally multi-wave, interstiffener skin buckling (either stable or unstable) and an overall, circumferentially multi-wave, skin-stiffener mode (usually unstable). In each problem-type, unstable mode interaction occurs, with results that are ofen remarkably similar to those displayed by our simple two-degree-of-freedom model. We will not exhibit these special results, but a qualitative discussion of various lines of attack on the problem of Fig. 5 may be of some interest.

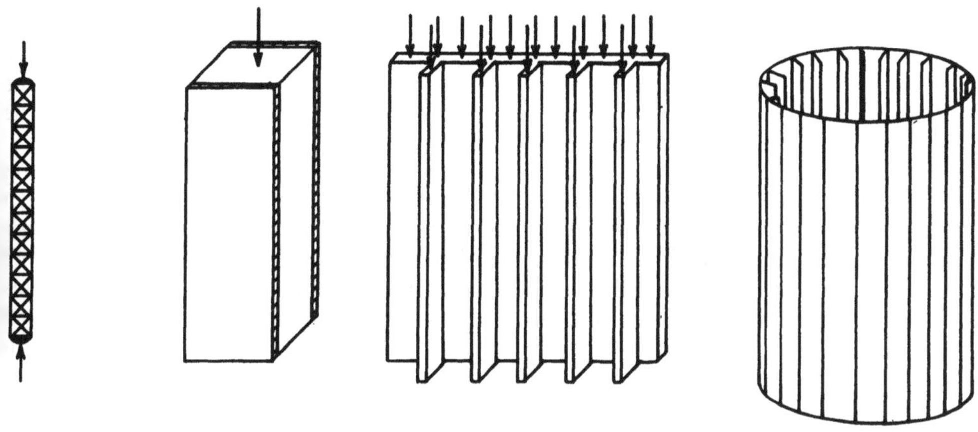

Fig. 5. Mode-interacting structures.

Three different approaches have been made. The first (and oldest) analyzes the decoupled local buckling behavior, and then presents to the overall structure an associated degraded local support stiffness. The coupling mechanism of load amplification is thereby also taken into account, because overall deformations will automatically induce forces that drive the local buckling. This approach goes back, at least in concept if not execution, to the early thirties and effective-width ideas. Thus, in the lattice column, the axial stiffnesses of the vertical members are computed as functions of local load and initial local imperfection, and used for the estimation of reduced bending stiffness in an overall column buckling analysis [7, 10]. Similar use has been made of the effective overall bending stiffness supplied by locally buckled flanges (Fig. 5(b)) [11-13] and buckled plate assemblies (Fig. 5(c)) [17, 20]. (In [19] Koiter and Pignataro show how to make rapid engineering estimates of the effective post-buckling stiffnesses of various kinds of plate assemblies.) But this approach has at least two apparent deficiencies. First, it does not seem to fit comfortably into the framework of the general theory, and so appropriate measures for improvements on its results are not obvious. Second, there is a conceptual flaw in the use of an effective local stiffness as a truly local property in the overall buckling phenomenon, because the local buckling deformations along the length of the structure are not decoupled from each other, as is implicitly assumed when the local stiffnesses are considered to depend only on the local loads. Both objections are met by the approaches taken in [15, 16] for the stiffened panel problem, and in [21] for the stiffened cylinder problem, which are based essentially on the use of the expansion (2.10) in the variational Eq. (2.14) of the general theory, and take into account two, nearly coincident buckling loads, with their associated buckling modes. But the story can not end there. In the problems illustrated in Fig. 5 there may be *many* local buckling modes with nearly equal buckling loads, and local buckling deformations in various parts of the overall structure may be nearly decoupled from each other. The consequence is that the first approach may, after all, be more nearly correct, since the more formal second method embodies only a single local buckling mode. This dilemma has been resolved by Koiter et al [14, 18, 19, 22] in a series of papers that presents the third approach, containing the technique of *amplitude modulation*. Here the apparatus of the general theory is invoked, but now the amplitude of the local buckling mode is permitted to vary from point to point in the overall structure. This is not rigorous, but provides an effective recognition of the existence of many local modes having nearly the same critical loads. Not only does the approach recapture the results of the earlier local-effective-stiffness methods, it points the way toward useful extensions to other problems.

In recent work Koiter and van der Neut [23] come to grips with mode interaction problems wherein several local modes that do *not* share nearly equal critical loads play an essential role. This occurs, for example, in the intriguing problem of the interactive buckling of a long square tube of uniform thickness. An elementary explanation of the mode-coupling involved can be seen with the help of Fig. 6. Plate bending w produces longitudinal strains proportional to w^2. Accordingly, combining the strains due to overall bending (proportional to z) with those due to plate buckling leads to a strain energy contribution per unit column length given approximately by

$$\oint [c_1 z + (c_2 w_1 + c_3 w_2)^2]^2 \, ds \tag{3.3}$$

where the coefficients c_1, c_2, c_3 are proportional to the amplitudes of the modes in overall (Euler) buckling and the two plate modes shown in Fig. 6(a) and 6(b). The consequence is that the only overall-local coupling term that survives the symmetry of the cross-section is

$$4 c_1 c_2 c_3 \oint (z w_1 w_2) \, ds \tag{3.4}$$

and so the mode interaction phenomenon would be completely missed if the mode types in either Fig. 6(a) or Fig. 6(b) were omitted from the analysis.

What of the future? While the studies of the last ten years have produced much understanding, and clever approximate analyses have swept away some seemingly complictated computational barriers, our feeling (shared with Tvergaard [1]) is that the subject of mode interaction in practical structures may be ripe for brutal computerization. Given the nearly

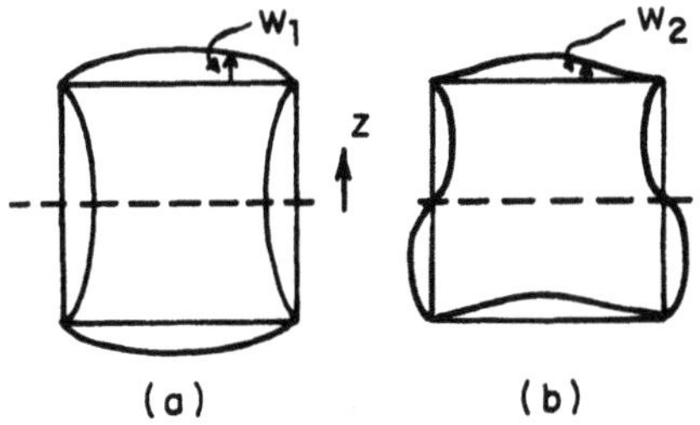

Fig. 6. Plate modes in square tube.

routine but sophisticated computer capabilities that now exist for the calculation of bifurcation loads and modes (e.g. [24-26]) and the theoretical foundations provided by the general theory of post-buckling behavior, it does not seem too optimistic to expect that a concerted effort would produce powerful, multi-mode computer procedures for the assessment of mode interaction in column, plate, and shell assemblies.

4. Plastic Buckling

Compared to elastic buckling, plastic buckling is an underdeveloped subject. Conditions for bifurcation are well established, and some fairly general results characterizing behavior immediately following bifurcation are known. But a general understanding of post-bifurcation behavior and imperfection-sensitivity in the plastic range does not yet exist. Here an attempt will be made to survey broadly various pieces of the subject which are understood and to point to aspects on which progress may be possible. We recognize that numerical methods are indispensable for solving practical plastic buckling problems, even more so than for elastic buckling. Nevertheless, the absence of a theory of plastic buckling and post-buckling behavior leaves the subject at the present as little more than a collection of rather unrelated example problems.

4.1. Bifurcation in the plastic range

Shanley's [27] rationalization of the tangent modulus load for compressed columns has been generalized for elastic-plastic continua by Hill [28]. Some of the unresolved issues related to choice of constitutive law will be touched upon below. Irrespective of these issues, Hill's formulation embraces most plasticity laws of conceivable interest in buckling. It is fair to say that the mathematical aspects of the bifurcation problem for the lowest bifurcation load are well founded, apart from a difficulty analogous to that in elastic buckling related to sufficiency conditions for stability. Hill's theory has now gained wide-spread acceptance. Sewell [29] has reviewed Hill's theory and has related it to earlier buckling work. Sewell's article contains an organized list of over 600 references on plastic buckling. More recent survey reports of limited circulation have been prepared by Sewell [30] and Storåkers [31]. Tvergaard's survey [1] covers aspects of plastic buckling and post-buckling.

As in the elastic case the fundamental solution whose uniqueness is in question, $u_0(\lambda)$, is assumed to be a function of a single load or deformation parameter λ which increases from zero. Bifurcation from the fundamental solution is assumed to occur prior to the occurrence of a limit point in λ. On the bifurcated branch we consider monotonic growth of the bifurcation mode.

Hill [28] introduces a quadratic functional of the displacement-rate for testing for bifurcation. The instantaneous moduli which enter into this functional are called comparison moduli. Loosely speaking, at each point in the body the comparison moduli are the softest branch of available moduli at that state. With a smooth yield surface, the comparison moduli are simply the plastic loading moduli wherever the stress is at yield, and the instantaneous elastic moduli elsewhere. The situation is more complicated when the stress is at the corner of a yield surface. If a total loading regime exists (i.e., a range of stress-rates in which all potential active yield systems are actually activated), then the comparison moduli are the total loading moduli. Bifurcation is first possible at the load λ_c where the quadratic functional vanishes for a non-zero admissible displacement-rate.

4.2 Post-bifurcation behavior

The search for the lowest bifurcation load has been reduced to a standard eigenvalue problem subject to the constraint that bifurcation does occur in such a way that the comparison moduli *are* activated by the bifurcation mode. It is this constraint which gives rise to bifurcation under increasing load and which determines the linear combination of the eigenmode and the fundamental solution increment comprising the bifurcation mode. We will use an axially compressed, simply supported column to illustrate initial post-bifurcation behavior.

The slender, solid cylindrical column shown in Fig. 7 has been analyzed within the framework of the approximations of simple column theory (Hutchinson [32]) wherein strain-rates are assumed to be linearly distributed through the thickness and the stress is taken to be uniaxial at each point. Later we will comment on the relation of these results to a more accurate 3-dimensional analysis.

The lowest bifurcation load (Shanley's tangent modulus load) is

$$\lambda_c = \pi^2 E_t^c I/L^2 \tag{4.1}$$

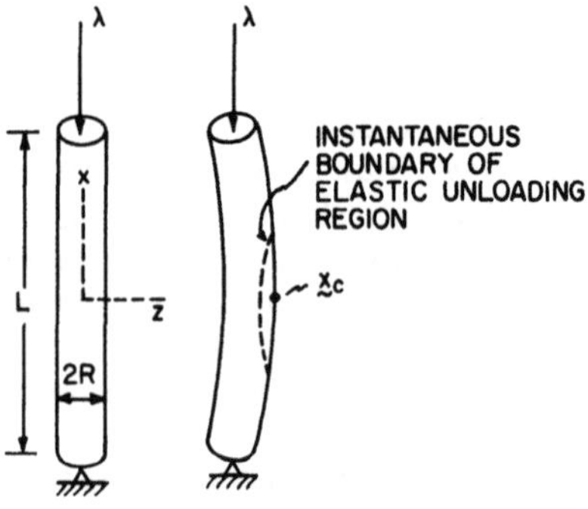

Fig. 7. Simply supported column with solid circular cross-section.

where I is the moment of inertia of the cross-section and E_t^c is the tangent modulus of the uniaxial stress-strain curve at the bifurcation stress. The lateral deflection component of the eigenmode is

$$w_1 = R \cos \left(\frac{\pi x}{L}\right) . \tag{4.2}$$

The amplitude of the eigenmodal contribution to the total lateral deflection w is denoted by ξ so that, by definition,

$$w = \xi w_1 + \widetilde{w} \tag{4.3}$$

where $\widetilde{w}$ is orthogonal to w_1.

Before discussing the behavior of the elastic-plastic column, it is revealing to discuss the behavior of a fictional comparison column with elastic unloading suppressed. This makes the column, effectively, nonlinearly elastic, and it can then be analyzed by the methods discussed earlier. The lowest bifurcation load is still given by (4.1) with the same eigenmode (4.2). The initial post-bifurcation response is necessarily symmetric in ξ, and the expansion (2.5) has the form

$$\lambda = \lambda_c + \lambda_2^e \xi^2 + \ldots \tag{4.4}$$

where the superscript e labels quantities limited to the comparison nonlinear elastic column. The post-bifurcation coefficient λ_2^e depends on the first and second derivatives of the tangent modulus at λ_c; it may be positive or negative. Regardless of the sign of λ_2^e, strain-rate reversal occurs over the entire half of the nonlinear elastic column towards which the lateral deflection takes place. As is well known, a nonlinear elastic column ceases to behave like an elastic-plastic column as soon as bifurcation occurs.

Bifurcation of the actual elastic-plastic column must take place under increasing load in such a way that the tangent modulus E_t governs the response everywhere in the column. With

$$\lambda = \lambda_c + \lambda_1 \xi + \ldots ,$$

the stress immediately following bifurcation is

$$\sigma = \sigma_0^c + \xi(\lambda_1 \dot{\sigma}_c + \sigma_1) + \ldots \tag{4.5}$$

where $\dot{\sigma}_c = (\partial \sigma_0 / \partial \lambda)_c$ and σ_1 is the eigenmodal contribution. The constraint that plastic loading occurs everywhere in the column as ξ increases from zero requires

$$\lambda_1 / \lambda_c \geq 4. \tag{4.6}$$

On the other hand, if $\lambda_1/\lambda_c > 4$ then, by continuity, no elastic unloading occurs in some finite neighborhood of λ_c. But this is not possible since then, for sufficiently small ξ, (4.4) would hold giving rise to immediate unloading over half the column. Elastic unloading must *start* at bifurcation with $\lambda_1/\lambda_c = 4$. In other words, the bifurcation mode, which is the combination of the fundamental solution-rate and the eigenmode of order ξ in (4.5), must satisfy neutral loading somewhere in the body. For the column problem this translates into

$$\max(\lambda_1 \dot{\sigma}_c + \sigma_1) = 0 \tag{4.7}$$

which leads to $\lambda_1/\lambda_c = 4$. For the column with the solid circular cross-section there is exactly one point $\mathbf{x}_c = (R, 0, 0)$ where (4.7) is attained, assuming without loss in generality $\xi > 0$. This point is indicated in Fig. 7. Elastic unloading spreads from $\mathbf{x}_c$ as ξ increases. The next term in the initial post-bifurcation expansion reflects the growth of the elastic unloading region. The general form of the initial post-bifurcation expansion, for the column and more general situations, is [32, 33]

$$\lambda = \lambda_c + \lambda_1 \xi + \lambda_2 \xi^{1+\beta} + \ldots \tag{4.8}$$

where β depends on the geometry in the neighborhood of $\mathbf{x}_c$. The general formula for λ_2 depends only on the bifurcation mode. For the column of Fig. 7,

$$\beta = \frac{1}{3} \quad \text{and} \quad \frac{\lambda_2}{\lambda_c} = -3 \left\{ \frac{3\pi E_t^c}{E - E_t^c} \left[1 + \left(\frac{\pi R}{2L}\right)^2 \left(\frac{\mathrm{d}E_t}{\mathrm{d}\sigma}\right)_c \right] \right\}. \tag{4.9}$$

The lowest order equation for the surface separating the spreading lens-shaped zone of elastic unloading from the plastically loading region is

$$(R - z)/R + \frac{1}{2}(\pi x/L)^2 = \frac{1}{3}(-\lambda_2/\lambda_c)\xi^{1/3}. \tag{4.10}$$

The above results were also obtained from an approximate 3-dimensional analysis [33], except that λ_2 is then a weak function of Poisson's ratio which agrees with (27) in the limit of incompressibility. In the 3-dimensional treatment a boundary layer analysis is used to obtain the lowest order equation (4.10) governing the growth of the unloading region. Because of the presence of the negative term $\lambda_2 \xi^{4/3}$ in (4.8), the rate of increase of the load following bifurcation is rapidly eroded. This explains why the maximum support load is often only slightly above λ_c in spite of a large initial rate of load increase. From (4.10) it can be seen that the unloading region rapidly encroaches on the column. In the general case the rate at which elastic unloading makes its presence felt in the expansion (4.8) through the term $\lambda_2 \xi^{1+\beta}$ depends on the geometry in the neighborhood of the point, or set of points, at which elastic unloading starts. Several other examples are shown in Fig. 8.

Tvergaard and Needleman [34, 35] have studied several examples where structural asymmetry results in post-buckling behavior in the plastic range which is entirely different from that in the elastic range. A good example is the simply supported column with a solid, equilateral triangular cross-section shown in Fig. 8 and again in Fig. 9.

Within the framework of column theory a linearly elastic column has a symmetric buckling behavior even though the cross-section is asymmetric. This is not the case for a comparison nonlinearly elastic column [34]. For buckling about the y-axis,

$$\lambda = \lambda_c + \lambda_1^e \xi + \ldots \tag{4.11}$$

where

$$\frac{\lambda_1^e}{\lambda_c} = \frac{8\sqrt{2}}{15\pi} \frac{q}{1+q} \quad \text{with} \quad q = \frac{1}{18}\left(\frac{\pi h}{L}\right)^2 \left(\frac{\mathrm{d}E_t}{\mathrm{d}\sigma}\right)_c. \tag{4.12}$$

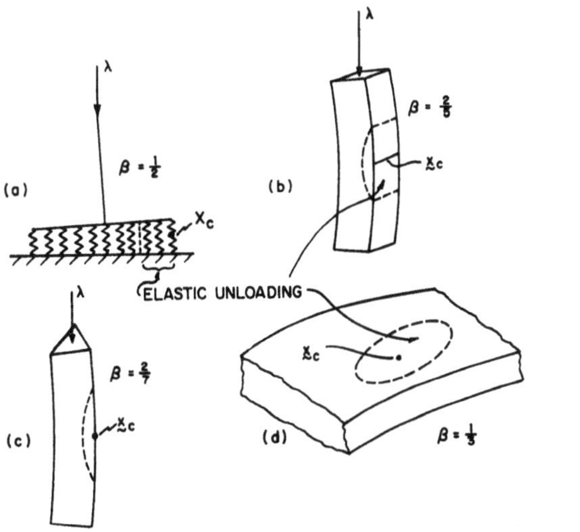

Fig. 8. Onset and spread of elastic unloading region.

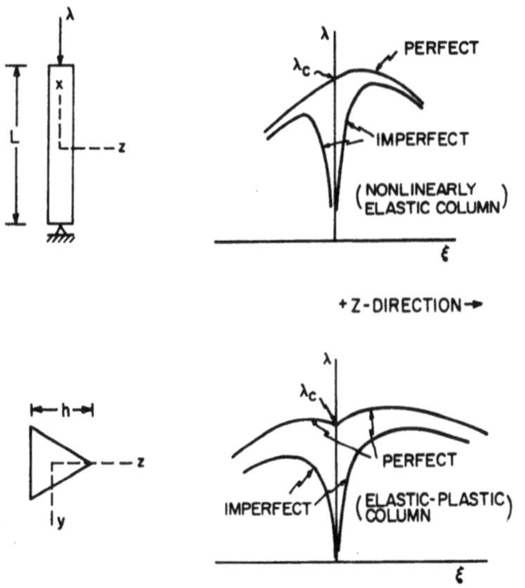

Fig. 9. Buckling of a simply-supported column with solid triangular cross-section about y-axis.

Here λ_c is given by (4.1) and, instead of (4.2), $w_1 = (h/2\sqrt{3}) \cos(\pi x/L)$ with ξ defined by (4.3). For a typical stress-strain curve, with E_t decreasing as $|\sigma|$ increases, $dE_t/d\sigma$ is positive for compression. Then (4.11) implies that the load falls for buckling deflections towards the flat side of the cross section ($\xi < 0$) and increases when buckling occurs towards the point of the triangle ($\xi > 0$) as depicted in Fig. 9.

For the elastic-plastic column

$$\lambda/\lambda_c = 1 + 2\sqrt{2}\xi + (\lambda_2/\lambda_c)\xi^{9/7} + \dots, \qquad \xi \geq 0 \tag{4.13}$$

$$\lambda/\lambda_c = 1 + \sqrt{2}(-\xi) + (\lambda_2/\lambda_c)(-\xi)^{7/5} + \dots, \qquad \xi \leq 0 \tag{4.14}$$

where expressions for λ_2 are given in [34]. The initial rate of increase of load is larger by a factor of 2 for buckling towards the point of the triangular cross-section ($\xi > 0$). Thus the initial post-bifurcation behaviors of both the nonlinearly elastic and the elastic-plastic column indicate a greater reserve of post-buckling strength for buckling with $\xi > 0$. A numerical analysis of the column (including effects of initial imperfections) was also carried out by Tvergaard and Needleman. The greater post-buckling resistance for $\xi > 0$ was indeed verified with findings such as those depicted in Fig. 9.

Using the same set of techniques, Tvergaard and Needleman [35] also investigated the important problem of the buckling of an eccentrically stiffened wide-column such as that shown in Fig. 5(c). The post-buckling strength of the wide column in the plastic range was found to be the greatest for buckling towards the side of the column on which the stiffeners are attached. The column is significantly less sensitive to imperfections which promote buckling towards the stiffeners compared to those which cause deflections in the opposite direction.

While the initial post-bifurcation expansion (4.8) gives insight into how the rate of load increase diminishes so rapidly following bifurcation, it does not usually provide a means of accurately predicting the maximum load and the associated buckling deflection. Van der Heijden [36] has proposed an alternative approximate analysis aimed at predicting the maximum load of a structure with or without imperfections. The simple continuous spring model of Fig., 8 was used to try out the approach.

With reference to Fig. 10, let λ_c denote the lowest bifurcation load and let λ_{rm} denote the reduced modulus load of von Karman [37] where bifurcation takes place under constant load to lowest order in ξ. Using perturbation methods it is possible to determine the expansion of the solution bifurcating at λ_{rm}, i.e.

$$\lambda = \lambda_{rm} + \lambda_2^{rm}\xi^2 + \dots . \tag{4.15}$$

Central to van der Heijden's method is the determination of the locus of maxima $\lambda^*(\xi^*)$ associated with solutions bifurcating in the range $\lambda_c \leq \lambda \leq \lambda_{rm}$. An expansion about λ_{rm} gives

$$\lambda^* = \lambda_{rm} + A\xi^* + B\xi^{*2} + \dots . \tag{4.16}$$

Van der Heijden approximates the solution associated with bifurcation at the lowest load by

$$\lambda = \lambda_c + \lambda_1\xi + b\xi^2 + c\xi^3 \tag{4.17}$$

where λ_1 is the known initial slope. The coefficients b and c are chosen such that $\lambda(\xi)$ in (4.17) crosses the locus of maxima (4.16) with zero slope and with the same curvature as in (4.15) at maximum load. That is,

$$\frac{d\lambda}{d\xi} = 0, \quad \frac{d^2\lambda}{d\xi^2} = 2\lambda_2^{rm} \quad \text{where} \quad \lambda = \lambda^* \quad \text{at} \quad \xi = \xi^*. \tag{4.18}$$

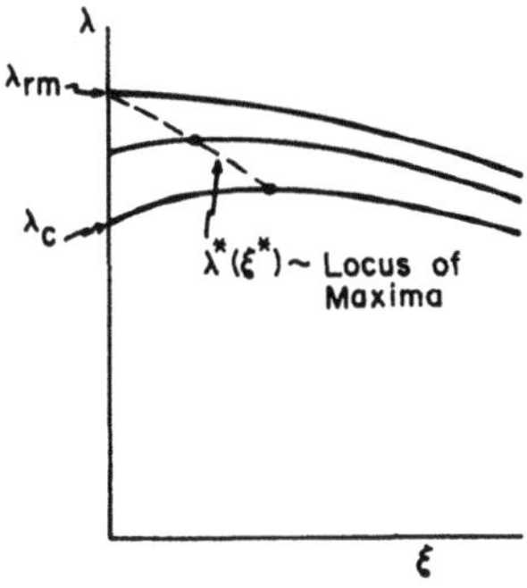

Fig. 10. Bifurcation responses.

In principle, an expansion of the curvature of the solutions, where they cross the locus of maxima, could be developed about λ_{rm} and used to improve the estimate of $d^2\lambda/d\xi^2$ in (4.18). But this expansion is quite complicated and, by comparison with numerical solutions, van der Heijden found that the curvature at λ_{rm} served as a reasonable approximation. The method was also extended to estimate the effects of small imperfections on the maximum support load.

There is clearly nothing rigorously asymptotic about this method. Indeed, it is unlikely that any purely analytical method can be found to predict the maximum load since it occurs at a finite deflection beyond bifurcation. The method does make use of information which can be determined from expansions about λ_c and λ_{rm}.

Given the history of column buckling, it is interesting that van der Heijden's method brings in Shanley's tangent modulus load and von Karman's reduced modulus load on roughly an equal footing.

4.3. Imperfection-sensitivity

For linearly elastic bodies instability and imperfection-sensitivity derive from geometric nonlinearity. In the plastic range, material nonlinearity in the form of decreasing moduli with increasing deformation is an additional destabilizing influence. In addition, under certain conditions there is an anomalous material-based sensitivity to extremely small imperfections associated with the assumption of a smooth yield surface.

Very few general results are available for the effect of imperfections on the maximum support load in the plastic range. Here we will confine ourselves to the effects of small geometric imperfections of the form $\bar{\xi}u_1$. In many problems of interest the fundamental solution of the perfect structure undergoes no elastic unloading as λ is increased. Elastic unloading starts at bifurcation at λ_c in the manner discussed above. One general result which then holds asymptotically for sufficiently small imperfections is that elastic unloading starts when the slope of the relation between λ and ξ is reduced to

$$\frac{d\lambda}{d\xi} = \lambda_1 \qquad (4.19)$$

where, as in (4.4), λ_1 is the initial slope for the perfect structure [38]. The onset of elastic unloading is a pivotal point in any buckling analysis; it is the beginning of inherently irreversible behavior whose effects seem to persist for all larger ξ.

For problems such as that of the column or the continuous spring model, where the stress history at each point is uniaxial, the value of λ and of ξ at which elastic unloading starts in the imperfect structure is given by the asymptotic formula for small $\bar{\xi}$

$$\hat{\lambda} = \lambda_c - C_1 \sqrt{|\bar{\xi}|}, \quad \hat{\xi} = C_2 \sqrt{|\bar{\xi}|}. \tag{4.20}$$

These formulas also hold, in general, for other problems if it is assumed that, prior to elastic unloading, the material is characterized by a deformation theory of plasticity. Formulas for C_1 and C_2, which depend on the sign of $\bar{\xi}$, are given in [32] and [38].

Studies of discrete spring models and of the continuous spring model indicate that the maximum load is also reduced by an amount proportional to the square root of the imperfection for sufficiently small $\bar{\xi}$, i.e.

$$\lambda_{max} = \lambda^0_{max} - C_3 \sqrt{|\bar{\xi}|}. \tag{4.21}$$

The strong sensitivity to small imperfections implied by (4.21) is closely connected to the pivotal role of $\hat{\lambda}$. Model studies and numerical studies of more complicated examples show that the maximum load is often attained shortly after the onset of elastic unloading. It has not been shown that (4.21) is precisely valid for other than column-like problems but it seems likely that the asymptotic dependence on the square root of $\bar{\xi}$ may be a general feature, assuming a smooth yield surface is not used. As will be discussed below, a *much stronger* asymptotic imperfection-sensitivity occurs if bifurcation involves non-proportional plastic loading and if a plasticity theory based on a smooth yield surface is used. We discuss this next. The well-known example of the cruciform column shown in Fig. 11 will be used to illustrate what is involved [39, 40]. For simplicity an incompressible material characterized by J_2 flow (incremental) theory is assumed. The uniaxial stress-strain curve is assumed to be monotonically increasing. The compressive buckling stress of the perfect cruciform associated with torsional buckling is

$$\sigma_c = G(t/b)^2. \tag{4.22}$$

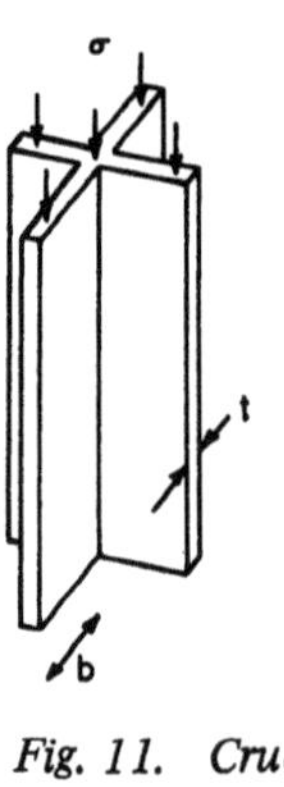

Fig. 11. Cruciform column.

The elastic shear modulus G governs the incremental shear response in the state of uniaxial compression and consequently controls bifurcation.

The cruciform column is unusual in that bifurcation occurs at constant load, with $\lambda_1 = 0$. From (4.19) it follows that elastic unloading in the slightly imperfect column will occur at essentially the maximum load. This observation was exploited in [40] where imperfection-sensitivity was studied with elastic unloading suppressed. The J_2 flow theory with elastic unloading suppressed is an inherently path-dependent hypo-elastic constitutive law. It was found that elastic unloading started just after attainment of the maximum load in the slightly imperfect column if the tangent modulus at bifurcation satisfied $E_t^c < E/3$. Thus in this range the maximum load could be calculated using the hypo-elastic characterization.

An asymptotically exact analysis of the equations governing the cruciform was carried out and the asymptotic expression for the maximum load was found to be

$$\lambda_{\max}/\lambda_c = 1 - \mu\bar{\xi}^s. \tag{4.23}$$

Here μ is a positive number of order unity which depends on details of the stress-strain curve and

$$s = \frac{E_t^c}{E + E_t^c/2}. \tag{4.24}$$

In the unloaded state the initial imperfection is a twist per unit length $\bar{\theta}$; the nondimensional imperfection amplitude is $\bar{\xi} = b^2\bar{\theta}/t$.

Values of s less than 1/10 are rather typical for buckling in the plastic range. Therefore, from (4.23) one concludes that an imperfection as small as $\bar{\xi} = 10^{-10}$ will have a non-negligible influence on the buckling load. This extraordinary sensitivity is due to the hypo-elastic character of the loading branch of J_2 flow theory. It is not restricted to the cruciform column, but can be expected in any problem where bifurcation involves a strongly non-proportional plastic response [28, 41]. The bifurcation load predicted using a flow theory with a smooth yield surface has little meaning when the presence of truly unavoidable imperfections renders it unattainable.

There has been some progress in recent years in developing phenomenological flow theories of plasticity which model incremental behavior at the corner of a yield surface and which are sufficiently simple that they can be used in numerical calculations [42, 43]. These theories can accommodate a description of a nearly-proportional loading response which is the same as that from the J_2 deformation theory of plasticity. Thirty years ago, Batdorf [44] rationalized the use of J_2 deformation theory in bifurcation analyses by appealing to a cornered yield surface. Thus the bifurcation load from one of these theories is the same as that from J_2 deformation theory, which is generally accepted as giving reasonably good estimates of buckling loads when compared with experimental loads. Furthermore, the corner theories will not give rise to a huge sensitivity to unavoidably small imperfections but are likely to display a dependence such as that in (4.21).

Even if sharp corners do not develop at the loading point on a yield surface, a corner theory may be a better model of behavior at a region of high curvature on the yield surface for the purposes of buckling analysis than one of the conventional smooth yield surface theories. If the conventional theory underestimates the local curvature of the yield surface, then it will overestimate the stiffness of the material response for non-proportional loading increments. A corner theory will tend to underestimate this stiffness. Judging from the success of J_2 deformation theory in predicting buckling loads, the corner theory should not be unduly conservative in its buckling predictions.

5. Optimum Design

We limit ourselves to three observations:

(1) The alarm was sounded early [45] with respect to the dangers of imperfection-sensitivity in structures having nearly-coincident local and overall buckling loads, with the further speculation that this might vitiate the requirement of critical load coincidence for optimality. Current overviews of the situation tend to confirm the first warning, but not necessarily the second. Optima tend to be smooth, and even if the best design, in the presence of imperfections, is not right at the equal-critical-load configuration, the consequent weight-saving may often be small. But when nearly-coincident critical loads are permitted, adequate knockdown factors must be invoked, and it should be well understood, for whatever inference one might wish to draw concerning the design, that should failure occur at such a design, it is likely to be catastrophic.

(2) An unusually thorough and rigorous optimization study was recently executed by Libai [46], in which an optimum design against elastic buckling (but not failure) was sought for the configuration shown in Fig. 12(a), consisting of a simply-supported square plate containing a central, one-sided, blade stiffener. For a given total end thrust P and plate dimension L, the dimensions h, t_w, and t_s for minimum weight under the constraint of no buckling were sought. Libai found a *local* minimum (in configuration space) at the design for which (a) symmetrical plate-stiffener buckling, (b) uncoupled torsional stiffener buckling, and (c) uncoupled, transversely antisymmetrical plate buckling have very nearly (but not exactly!) the same critical loads. In the practical range of structural index P/L^2, the stiffener is 'thin' at this configuration, with t_w/h usually less than 1/10. But a different local minimum, requiring a 'thick' stiffener having $t_w/h \approx 1/2$, was also found, providing a design with a very slightly lower weight. The results for (σ_{cr}/E) at the two design minima, shown in Fig. 12(b) as one curve, are indistinguishable on the scale used. The stiffener-to-plate area ratios

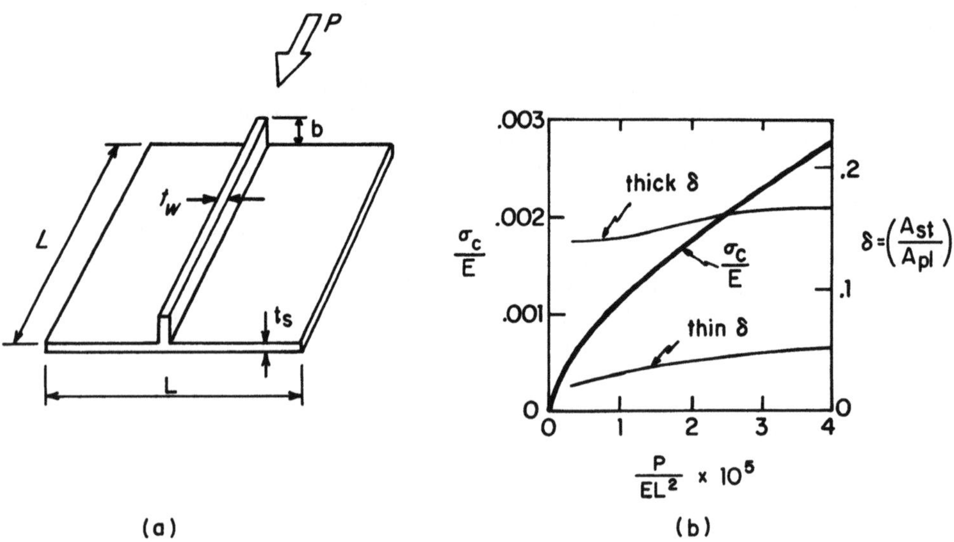

Fig. 12. Centrally-stiffened-plate optima.

are also shown in Fig. 12(b), with very different values at the two optima. At both the thick and thin minima the symmetric mode and the coupled transversely antisymmetric mode had equal buckling loads, but in the case of the thick minimum the stiffener was nowhere near its own torsional instability load. It therefore seems likely that the thick minimum is not imperfection sensitive, but that the thin minimum is. The possibility of finding similar results in the optimum design of other, more realistic configurations should not be overlooked.

(3) It has to be acknowledged that, despite theoretical progress, few practical design optimization problems have, as yet, been 'solved' analytically. In this respect, theory lags experiment, and a glance at a remarkable series of tests conducted over thirty years ago may be instructive [47, 48]. Over 150 2024-ST aluminum alloy zee- and hat-stiffened panels having systematically varied configurations were tested for ultimate compressive strength, with the global results summarized in Fig. 13. Here the appropriate structural index is $P_i/L\sqrt{c}$, where P_i is the load-per-unit-width of panel, L is the panel length, and c is an effective end-fixity coefficient ($c = 1$ for simple support; $c = 4$ for clamped ends). The curves show the compressive strengths of minimum weight designs. (The actual configurations providing these designs are obtainable from auxiliary curves not shown here.) The most striking feature of these old results is that (at least for equal sheet and stiffener thicknesses) they announce the superiority of zees over hats as stiffeners. These experimental results automatically incorporate the effects of certain representative initial imperfections, not to mention plasticity, discrete rivet attachments, finite corner radii – and they stand as a challenge to theoreticians to confirm or refute them, and deduce analogous results for other configurations and materials.

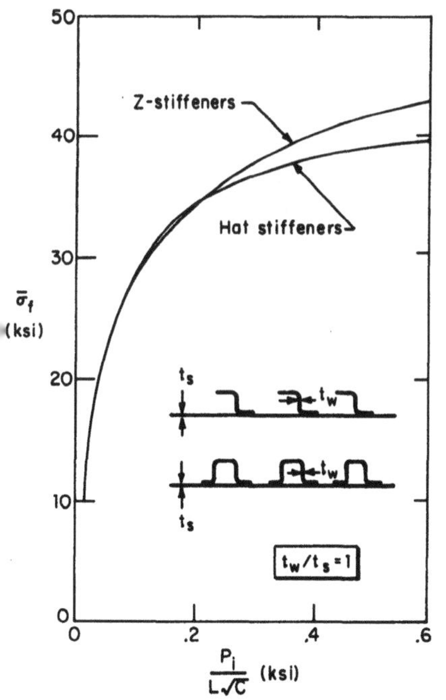

Fig. 13. Optimization of zee- and hat-stiffened panels.

6. Stochastic Buckling Problems

Given the importance of small imperfections in imperfection-sensitive structures, and the uncertainty of their magnitudes and shapes, it has long been believed that a rational approach to the design of such structures should relate failure probability under a given loading to appropriate statistical information concerning imperfections. Such correlations are straightforward, in principle, when single-mode buckling is pertinent, for then the probability of failure may be related directly to the probability distribution function of the amplitude of the relevant imperfection. This procedure may be extended when a few modes interact, but becomes cumbersome for many modes. On the other hand, if interacting modes can be idealized into a continuous spectrum, the techniques of generalized harmonic analysis become useful. Examples of the two approaches are contrasted in [49].

The intermediate situation, in which not a few but also not very many modes must be considered to have stochastic imperfections, has been addressed in [50], where the problem of a lattice column with random imperfections in the individual members is tackled. Through the use of effective approximations results for the means and variances of buckling loads are estimated, with accuracies confirmed by Monte Carlo calculations.

7. Stability

For some years Koiter has been concerned with a fundamental dilemma in the theory of elastic stability, and has tried to provoke interest in its resolution [51]. Briefly, the situation is this:

Stability is guaranteed at an equilibrium state u_0 if, for some constant c,

$$\frac{\Delta\phi}{||\delta u||^2} \equiv \frac{\text{£}[u_0+\delta u]-\phi[u_0]}{||\delta u||^2} > c > 0 \tag{7.1}$$

for sufficiently small $||\delta u||$. This implies that with respect to the assumed norm, ϕ has a strong minimum at u_0. Subsequent to an initial displacement $\delta u(0)$ at time $t=0$ (with velocity $\delta\dot{u}(0) = 0$) energy considerations dictate that with or without damping $\Delta\phi(t) \leq \Delta\phi(0)$ for all $t > 0$. Therefore $||\delta u(t)||$ will never exceed a sufficiently small assigned ϵ if the initial restrictions $||\delta u(0)|| \leq \epsilon$ and $\Delta\phi(0) \leq c\epsilon^2$ are imposed. If the choice of norm does not violate our engineering sensibilities, the definition of stability implicit in these facts is acceptable. Trouble arises, however, when, like all good engineers, we propose to replace the sufficient condition (7.1) by the requirement of a positive-definite second variation of the potential energy:

$$\frac{\delta^2\phi}{||\delta u||^2} > d > 0 \tag{7.2}$$

where

$$\delta^2\phi \equiv \frac{\partial^2}{\partial\eta^2}\,[u_0+\eta\delta u]_{\eta=0} \tag{7.3}$$

(Here the same norm used in (7.1) and (7.2) is invoked in the limit operation involved in the second derivative). The existence of a continuous second Fréchet derivative ϕ'' (which was blithely assumed in [3] and in the first part of this paper) implies the identity

$$\delta^2\phi = \phi''(\delta u)^2 \tag{7.4}$$

and then, as Como and Grimaldi [51] have indicated, all is well. For with $\phi' = 0$, $\Delta\phi = \frac{1}{2}\phi''(\delta u)^2 + R$, where $R/||\delta u||^2 \to 0$ for $||\delta u|| \to 0$, and then (7.4) and (7.2) imply (7.1), which implies stability. Indeed, by establishing the existence of Fréchet derivatives with respect to suitable norms Como and Grimaldi [53] verify Koiter's proof [54] that (7.2) is acceptable as a sufficient stability condition for shallow shells. But it turns out [55] that in non-linear 3-D elasticity theory the required Fréchet derivatives generally do not exist for reasonable choices of norm and strain energy.

This means that $\Delta\phi$ and $\delta^2\phi$ need not share the same sign, no matter how small $||\delta u||$ is made. A striking example given by Koiter [56] displays a very simple regular functional $\phi[u(x)]$ for which $\delta^2\phi$ is positive in the sense of (7.2), but for which $\Delta\phi$ can be made negative for arbitrarily small norm $||u|| = [\int_0^1 (u^2 + u_x^2)\,dx]^{1/2}$.

What is to be done about this? Koiter's counterexample involved very tortured deformation states with huge second derivatives, and this evidently gave him the successful idea of salvaging the second-variation criterion by introducing a modified strain-energy functional that depends on strain gradients as well as strains [56]. Despite Koiter's later [2] lack of enthusiasm for this device, it remains fairly persuasive as an argument for the *practical* reliability of the second-variation criterion. Another pragmatic viewpoint that one might adopt is to note that regular potentials of finite-dimensional systems have Fréchet derivatives of all order, and invoke a presumption of the accuracy of finite-element approximations.

But there is no doubt that at least an esthetic problem remains, and that a new, congenial definition of stability is desirable. Perhaps some statistical concepts may be fruitful. When the second-variation method fails, it appears that the minimum in ϕ is destroyed only by the presence of obscure secret passages in function space into which no self-respecting structure would venture except by wildly improbable accident. Accordingly, an appropriately defined *probability* of failure should, under these circumstances, be absurdly low. But we do not have any helpful suggestions concerning such a definition, which, in order to be useful in the assessment of the *practical* stability of a structure, should permit easy evaluation of the desired probability.

8. Concluding Remarks

Many very interesting recent buckling papers simply could not be mentioned in this survey, and we regret this; the subject is just too big. In particular, we have not mentioned current progress in experimentation, dynamic buckling, creep buckling, and numerical methods. Some wise comments concerning catastrophe theory might have been appropriate. But we happily leave these, and other topics, to the next surveyors, who, if past experience is any guide, are already preparing their reviews.

Acknowledgement

This work was supported in part by the National Science Foundation under Grant NSF-MCS78-07598, and by the Division of Applied Sciences, Harvard University.

References

1. Tvergaard, V., 'Buckling Behavior of Plate and Shell Structures', *Proc. 14th IUTAM Congress*, Delft, North-Holland, 233-247 (1976).
2. Koiter, W.T., 'Current Trends in the Theory of Buckling', *Proc. IUTAM Symp. on Buckling of Structures*, 1974 (Ed. B. Budiansky), Springer-Verlag, 1-16 (1976).
3. Budiansky, B., 'Theory of Buckling and Post-Buckling Behavior of Elastic Structures', in *Advances in Applied Mechanics*, Vol. 14, Academic Press, 1-65 (1974).
4. Koiter, W.T., *Over de stabiliteit van het elastisch evenwicht*, Thesis, Delft, H.J. Paris, Amsterdam (1945). English transl. (a) NASA TT-F10, 833 (1967) (b) AFFDL-TR-70-25 (1970).
5. Budiansky, B. and Hutchinson, J.W., 'Dynamic Buckling of Imperfection-Sensitive Structures', *Proc. 11th IUTAM Congress*, Munich, 636-651 (1964).
6. Sewell, M.J., 'A General Theory of Equilibrium Paths Through Critical Points', I, II, *Proc. Roy. Soc. A306*, 201-223, 225-238 (1968).
7. Thompson, J.M.T. and Hunt, G.W., *A General Theory of Elastic Stability*, Wiley, New York (1973).
8. Budiansky, B. and Hutchinson, J.W., 'A Survey of Some Buckling Problems', *AIAA J.*, **4**, 1505-1510 (1966).
9. Hutchinson, J.W. and Koiter, W.T., 'Postbuckling Theory', *App. Mech. Rev.*, **23**, 1353-1366 (1970).
10. Crawford, R.F. and Hedgepeth, J.M., 'Effects of Initial Waviness on the Strength and Design of Built-up Structures', *AIAA J.*, **13**, 672-675 (1975).
11. van der Neut, A., 'The Interaction of Local Buckling and Column Failure of Thin-Walled Compression Members', *Proc. 12th IUTAM Congress*, Springer-Verlag, 389-399 (1969).
12. Meijer, J.J. and van der Neut, A., *The Interaction of Local Buckling and Column Failure of Imperfect Thin-Walled Compression Members*, Delft University of Technology, Rep. VTH-160 (1970).
13. van der Neut, A., 'The Sensitivity of Thin-Walled Compression Members to Column Axis Imperfection', *Int. J. Solids Structures*, **9**, 999-1011 (1973).
14. Koiter, W.T. and Kuiken, G.D.C., *The Interaction Between Local Buckling and Overall Buckling on the Behavior of Built-up Columns*, Delft University of Technology, Rep. WTHD 23 (1971).
15. Tvergaard, V., 'Imperfection Sensitivity of a Wide Integrally Stiffened Panel Under Compression', *Int. J. Solids Structures*, **9**, 177-192 (1973).
16. Tvergaard, V., 'Influence of Post-Buckling Behavior on Optimum Design of Stiffened Panels', *Int. J. Solids Structures*, **9**, 1519-1534 (1973).
17. van der Neut, A., 'Mode Interaction with Stiffened Panels', *Proc. IUTAM Symp. on Buckling of Structures*, 1974 (Ed. B. Budiansky), Springer-Verlag, 117-132 (1976).
18. Koiter, W.T. and Pignataro, M., 'An Alternative Approach to the Interaction Between Local and Overall Buckling in Stiffened Panels', *Proc. IUTAM Symp. on Buckling of Structures*, 1974 (Ed. B. Budiansky), Springer-Verlag, 133-148 (1976).

19. Koiter, W.T. and Pignataro, M., *A General Theory for the Interaction Between Local and Overall Buckling of Stiffened Panels*, Delft University of Technology, Rep. WTHD 83 (1976).
20. van der Neut, A., *The Interaction of Local and Overall Buckling in Stiffened Panels*, Delft University of Technology, Rep. LR-255 (1977).
21. Byskov, E. and Hutchinson, J.W., 'Mode Interaction in Axially Stiffened Cylindrical Shells', *AIAA J.*, **15**, 941-948 (1977).
22. Koiter, W.T., *General Theory of Mode Interaction in Stiffened Plate and Shell Structures*, Delft University of Technology, Rep. WTHD 91 (1976).
23. Koiter, W.T. and van der Neut, A., *Interaction Between Local and Overall Buckling of Stiffened Compression Panels*, paper presented at Int'l. Conf. on Thin-Walled Structures, University of Strathclyde, April 3-6 (1979).
24. Williams, F.W., Wittrick, W.H. and Plank, R.J., 'Critical Buckling Loads of Some Prismatic Plate Assemblies', *Proc. IUTAM Symp. on Buckling of Structures*, 1974 (Ed. B. Budiansky), Springer-Verlag, 17-26, (1976).
25. Bushnell, D., 'Buckling of Elastic-Plastic Shells of Revolution with Discrete Elastic-Plastic Ring Stiffeners', *Int. J. Solids Structures*, **12**, 51-66 (1976).
26. Almroth, B.O., Meller, E. and Brogan, F.A., 'Computer Solutions for Static and Dynamic Buckling of Shells'. *Proc. IUTAM Symp. on Buckling of Structures*, 1974 (Ed. B. Budiansky), Springer-Verlag, 52-65 (1976).
27. Shanley, F.R., 'Inelastic Column Theory', *J. Aero. Sci.*, **14**, 261-267 (1947).
28. Hill R., 'Bifurcation and Uniqueness in Nonlinear Mechanics of Continua', in Muskhelishvili Volume, *Soc. Ind. Appl. Math.*, Philadelphia, Pennsylvania, 95-102 (1961).
29. Sewell, M.J., 'A Survey of Plastic Buckling', in *Stability* (Ed. H. Leipholz), Univ. Waterloo Press, Ontario, 85-197 (1972).
30. Sewell, M.J., *Elastic and Plastic Bifurcation Theory*, Report of Dept. of Mathematics, University of Reading, England (July 1975).
31. Storåkers, B., *Uniqueness and Stability at Finite Deformation of Inelastic Solids*, Publication No. 186, Dept. of Strength of Materials and Solid Mechanics, Royal Inst. of Tech., Stockholm (April 1973).
32. Hutchinson, J.W., 'Plastic Buckling', in *Advances in Applied Mechanics*, Vol. 14, Academic Press, 67-144 (1974).
33. Hutchinson, J.W., 'Post-Bifurcation Behavior in the Plastic Range', *J. Mech. Phys. Solids*, **21**, 163-190 (1973).
34. Tvergaard, V. and Needleman, A., 'On the Buckling of Elastic-Plastic Columns with Asymmetric Cross-Sections', *Int. J. Mech. Sci.*, **17**, 419-424, (1975).
35. Tvergaard, V. and Needleman, A., 'Buckling of Eccentrically Stiffened Elastic-Plastic Panels on Two Simple Supports or Multiply Supported', *Int. J. Solids Structures*, **11**, 647-663 (1975).
36. van der Heijden, A.M.A., *A Study of Hutchinson's Plastic Buckling Model* (January 1979), to be published.
37. von Karman, Th., 'Untersuchungen über Knickfestigkeit, Mitteilungen über Forschungsarbeiten', *Ver. Deut. Ing. Forschungsh.*, **81** (1910).
38. Hutchinson, J.W., 'Imperfection-Sensitivity in the Plastic Range', *J. Mech. Phys. Solids*, **21**, 191-204 (1973).
39. Onat, E.T. and Drucker, D.C., 'Inelastic Instability and Incremental Theories of Plasticity', *J. Aero. Sci.*, **20**, 181-186 (1953).
40. Hutchinson, J.W. and Budiansky, B., 'Analytical and Numerical Study of the Effects of

Initial Imperfections on the Inelastic Buckling of a Cruciform Column', *Proc. IUTAM Symp. on Buckling of Structures*, 1974 (Ed. B. Budiansky), Springer-Verlag, 95-105 (1976).
41. Needleman, A. and Tvergaard, V., 'An Analysis of the Imperfection-Sensitivity of Square Elastic-Plastic Plates under Axial Compression', *Int. J. Solids Structures*, **12**, 185-201 (1976).
42. Sewell, M.J., 'A Plastic Flow Rule at a Yield Vertex', *J. Mech. Phys. Solids*, **22**, 469-490 (1974).
43. Christoffersen, J. and Hutchinson, J.W., *A Class of Phenomenological Corner Theories of Plasticity*, Harvard University Report MECH-8 (January 1979).
44. Batdorf, S.B., 'Theories of Plastic Buckling', *J. Aero. Sci.*, **16**, 405-408 (1949).
45. Koiter, W.T. and Skaloud, M., comments in: Colloque sur le comportement postcritique des plaques utilisées en construction métallique, *Mem. Soc. Roy. d. Sciènces de Liège*, **5**, VIII, 64-68, 103, 104 (1963).
46. Libai, A., *On the Optimization of a Stiffened Square Panel under Buckling Constraints*, Harvard University Report DAS M-6 (December 1977).
47. Schuette, E.H., *Charts for the Minimum-Weight Design of 245-T Aluminum-Alloy Flat Compression Panels with Longitudinal Z-Section Stiffeners*, NACA Report No. 827 (1945).
48. Hickman,, W.A. and Dow, N.F., *Compressive Strength of 245-T Aluminum-Alloy Flat Panels with Longitudinal Formed Hat-Section Stiffeners Having a Ratio of Stiffener Thickness to Skin Thickness Equal to 1.00*, NACA Techn. Note 1439 (1947).
49. Amazigo, J.C., 'Buckling of Stochastically Imperfect Structures', *Proc. IUTAM Symp. on Buckling of Structures*, 1974 (Ed. B. Budiansky), Springer-Verlag, 172-182 (1976).
50. Miller, R.K. and Hedgepeth, J.M., 'The Buckling of Lattice Columns with Stochastic Imperfections', *Int. J. Solids Structures*, **15**, 73-84 (1979).
51. Koiter, W.T., 'A Basic Open Problem in the Theory of Elastic Stability', *Proc. IUTAM/IMU Symp. on Application of Methods of Functional Analysis to Problems of Mechanics*, Marseilles (1975).
52. Como, M. and Grimaldi, A., 'Stability, Buckling and Post-Buckling of Elastic Structures', *Meccanica*, **4**, 254-268 (1975).
53. Como, M. and Grimaldi, A., 'Stability Analysis of Structural Elastic Systems', *Proc. IUTAM Symp. on Variational Methods in the Mechanics of Solids*, 1978, Pergamon Press (to be published 1979).
54. Koiter, W.T., 'A Sufficient Condition for the Stability of Shallow Shells', *Proc. Kon. Ned. Ak. Wet., Series B*, **70**, 367-375 (1967).
55. Martini, R., 'On the Fréchet Differentiability of Certain Energy Functionals', *Proc. Kon. Ned. Ak. Wet., Series A*, **79**, 326-330 (1976).
56. Koiter, W.T., 'The Energy Criterion of Stability for Continuous Elastic Bodies', I, II, *Proc. Kon. Ned. Ak. Wet., Series B*, **68**, 178-202 (1965).

Nonlinear Vibrations

*C. Hayashi**

Summary

This paper is devoted largely to the exposition of forced oscillations in nonlinear systems. Using the theory of Hill's equation, we derived a criterion for the stability of periodic oscillations. This criterion is useful for testing the stability of higher-harmonic and subharmonic oscillations. The topological method of analysis is made applicable to the study of forced oscillations under transient conditions. The relationship between the initial condition and the resulting periodic oscillation is investigated. The mapping procedure is also used for the study of transient oscillations. Appearance of the homoclinic structure of invariant curves in the phase plane is worth noting. Almost periodic oscillations are presented by limit cycles or invariant closed curves in the phase plane.

1. Introduction

Nonlinear vibrations or nonlinear phenomena in general are not only associated with mechanical or electrical engineering, but widely with various fields of natural sciences. Furthermore, in recent years, those problems are also studied in connection with social sciences. The nonlinear vibrations, however, have a great variety and often exhibit anomalous phenomena. Therefore, before going into their details, it may be advisable to make a historical survey with reference to their simple examples.

The nonlinear vibrations were already discussed in a number of classical treatises including those of Lord Rayleigh [1] and H. Poincaré [2]. G. Duffing has studied the behavior of forced mechanical systems with nonlinear restoring force [3]. The system is described by the so-called Duffing's equation

$$\ddot{x} + k\dot{x} + f(x) = B \cos t \tag{1.1}$$

where $f(x)$ is a nonlinear function of x and dots indicate differentiation with respect to t. This equation may have a higher-harmonic or subharmonic solution in addition to the harmonic solution with period 2π. Another typical example of second-order nonlinear differential equations was given by van der Pol [4]. It is known as van der Pol's equation

$$\ddot{x} - \epsilon(1 - x^2)\dot{x} + x = 0, \quad \epsilon > 0 \tag{1.2}$$

When $|x|$ is small, this equation has a negative damping. It describes the self-excitation of a vacuum-tube oscillator.

* Kyoto University; also Department of Electrical Engineering, Osaka Institute of Technology

These investigations pioneered by Duffing and van der Pol have stimulated the interest of engineers and scientists in the field of nonlinear vibrations. A great volume of subsequent investigation was carried out in the Soviet Union. In the main, two groups of workers are to be distinguished. One comprises L. Mandelstam, N. Papalexi, A. Andronov, A. Vitt and S. Khaikin, and the other comprises N. Kryloff, N. Bogoliuboff and Yu. Mitropolsky. Among their contributions, Refs. [5] to [7] are of importance. N. Minorsky's book [8] includes those works of Russian scholars as well as his own contributions.

Besides in the Soviet Union, the nonlinear mechanics has been extensively studied in Poland, Czechoslovakia, Germany and other countries. For instance, the mathematical analysis of nonlinear vibration problems has been studied by S. Ziemba and his group (Poland), nonlinear vibrations pertaining to the rotor dynamics by A. Tondl (Czechoslovakia), nonlinear circuit analysis and synthesis by E. Philippow (GDR), hunting movements of railway vehicles by A.D.de Pater (Netherlands) and random vibrations by S. Crandall (USA). Some of their publications are listed in the References. There are, of course, a considerable number of works which are unable to mention here but are of high caliber. Finally, it is mentioned that the International Conferences on Nonlinear Oscillations are held every three years in the Eastern Europe.

We shall now describe in the following sections some important results of our study on nonlinear vibrations.

2. Periodic solutions and their stability [11], [12]

When a periodic force is applied to a linear system, the resulting motion is obtained by a superposition of the transient- and the steady-state components of the oscillation. The former is due to the free oscillation of the system, and the latter is related to the forced oscillation which arises from the action of the external force. When the system is (asymptotically) stable, the free oscillation is damped out after a sufficiently long period of time. Hence only the forced oscillation having the same frequency as the external force would be observed. Thus, in linear systems, the forced oscillation is uniquely determined, and it is affected in no way by the initial condition with which the oscillation was started.

In nonlinear systems, however, the circumstances are quite different in this respect. The principle of superposition will no longer be applicable. A non-linear system may possess a wide variety of periodic oscillations in addition to those which have the same frequency as the external force.

Contrary to the case of linear systems, it is hardly possible to obtain the general solution of a nonlinear differential equation for a given initial condition. In so far as we deal with a periodic solution, our conventional method of solution is to assume for it a Fourier series development with undetermined coefficients and then to fix them by the method of harmonic balance. By this method of solution, however, we can merely find the periodic states of equilibrium, which are not always realized, but are able to exist only when they are stable. The circumstances under which this condition obtains are determined by the stability investigation of periodic solutions.

As a typical but relatively general example, we consider the following second-order equation

$$\ddot{x} + f(x, \dot{x}) = e(t), \qquad e(t + T) = e(t). \tag{2.1}$$

Equations of this type, including Duffing's equation, are often encountered in physical systems. By the use of Floquet's theory, we investigate the stability of a periodic solution for Eq. (2.1). Let the solution be expressed by

$$x(t) = x_0(t) \tag{2.2}$$

where $x_0(t + nT) = x(t)$, n being a positive integer. A small variation from this periodic solution be denoted by ξ, then substitution of $x_0(t) + \xi$ for $x(t)$ in Eq. (2.1) leads to the variational equation

$$\ddot{\xi} + \left(\frac{\partial f}{\partial \dot{x}}\right)_0 \dot{\xi} + \left(\frac{\partial f}{\partial x}\right)_0 \xi = 0 \tag{2.3}$$

in which the symbol $(\ )_0$ denotes the insertion of $x_0(t)$ and $\dot{x}_0(t)$ after differentiation. If the variation ξ tends to zero with increasing t, then the corresponding periodic solution $x_0(t)$ is asymptotically stable. Since the coefficients of $\dot{\xi}$ and ξ are periodic functions of t, Eq. (2.3) is written as

$$\ddot{\xi} + F(t)\dot{\xi} + G(t)\xi = 0. \tag{2.4}$$

By introducing a new variable η defined by

$$\xi = \{\exp[-\frac{1}{2}\int F(t)\,\mathrm{d}t]\}\eta \tag{2.5}$$

and thus eliminating the first-derivative term in Eq. (2.4), we obtain

$$\ddot{\eta} + \{G(t) - \frac{1}{2}\frac{\mathrm{d}F}{\mathrm{d}t} - \frac{1}{4}[F(t)]^2\}\eta = 0. \tag{2.6}$$

This is a linear equation in which the coefficient of η is a periodic function of t and may be developed into a Fourier series. Equation (2.6) takes the form of Mathieu's or Hill's equation, and by Floquet's theorem the general solution is written as

$$\eta(t) = c_1 e^{\mu t}\phi(t) + c_2 e^{-\mu t}\psi(t) \tag{2.7}$$

where μ is called the characteristic exponent, and $\phi(t)$ and $\psi(t)$ are periodic functions of t having the same period as that of the coefficient of η in Eq. (2.6). As one sees from Eqs. (2.5) and (2.7), the variation ξ tends to zero with increasing t if both the real parts of $(-\delta \pm \mu)$ are negative, where 2δ is the constant term in the series of $F(t)$. Hence the stability condition for the periodic solution $x_0(t)$ is given by

$$\text{Real parts of } (-\delta \pm \mu) < 0. \tag{2.8}$$

As a concrete example, we consider an electric circuit of Fig. 1 in which the nonlinear oscillation takes place owing to the presence of a saturable inductor $L(\phi)$. An experimental result displayed in Fig. 2 shows the waveforms of the applied voltage and the resulting responses.

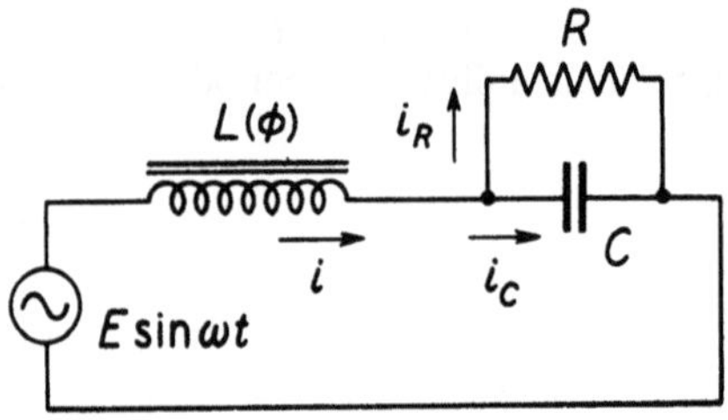

Fig. 1. Oscillatory circuit with nonlinear inductor.

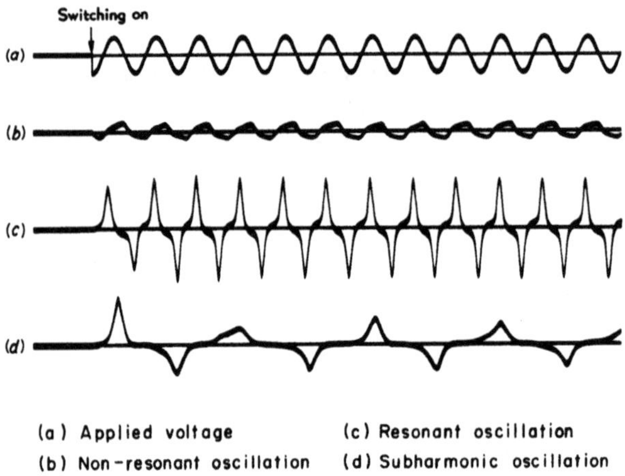

Fig. 2. Different types of oscillation in a nonlinear system.

Upon application of a sinusoidal voltage (*a*), the responses of (*b*) and (*c*) in the steady state have the same frequency as the applied voltage, and hence these are harmonic oscillations. However, the response (*d*) is a subharmonic oscillation of order 1/3, since the steady state frequency is one-third the driving frequency. These wave-forms are obtained by starting the oscillation with different initial charges on the capacitor. It is a salient feature of nonlinear systems that various responses may be obtained in a same system depending on the initial condition.*

2.1. Harmonic oscillations

To begin with, we deal with the harmonic oscillation in which the fundamental component having the same frequency as that of the external force is predominant, so that the higher harmonics may be neglected. We again take the electric circuit of Fig. 1. Its behavior is described by Duffing's equation (1.1). Then, the solution has the same frequency component as the external force $B \cos t$. Substitution of the solution into Eq. (1.1) gives the

* The relationship between the initial condition and the resulting response will be discussed in Sections 3 and 4

response characteristic as shown in Fig. 3, where r_0 is the amplitude of the oscillation and k is related to the damping of the circuit. After some algebra, the condition (2.8) for stability of the periodic solution leads to an inequality $dr_0/dB > 0$. The unstable portion of the characteristic curves is shown by shaded area. It corresponds to the first unstable region of the stability chart for the variational equation of the Mathieu type.†

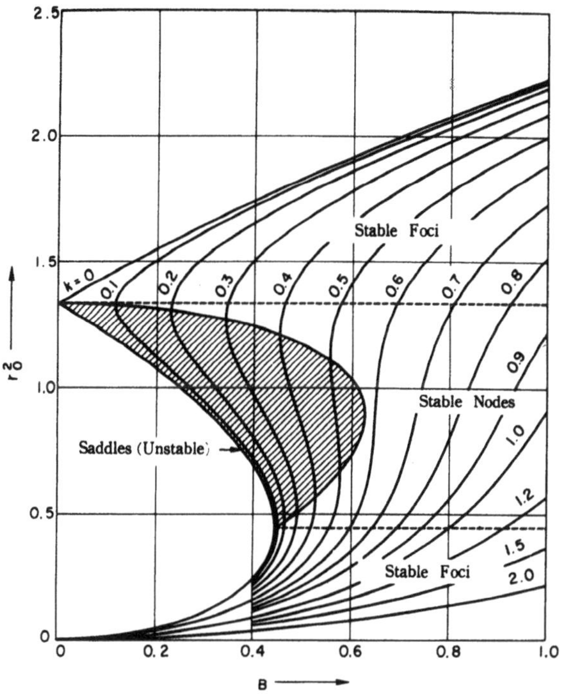

Fig. 3. Amplitude characteristic of the harmonic oscillation.

The discontinuous jump in r_0 occurs when the characeristic curve comes to the boundary of the shaded region. The lower values of r_0 correspond to the nonresonant state, and the higher values to the resonant state [cf. Fig. 2(b) and (c)]. They are both stable, but the intermediate one lying in the shaded region is unstable and transferred to one of the stable states as time increases.

2.2. Higher-harmonic and subharmonic oscillations

The stability condition (2.8) is particularly useful when we deal with the higher-harmonic or subharmonic oscillation. Higher-harmonic components become appreciable when the applied voltage is increased. An experiment is made for the series-resonance circuit of Fig. 1, in which the applied voltage is raised exceedingly. The result is shown in Fig. 4, where the oscillating current is plotted in heavy line for a wide range of the applied voltage. This

† A stability chart of Mathieu's equation is shown in Fig. 6

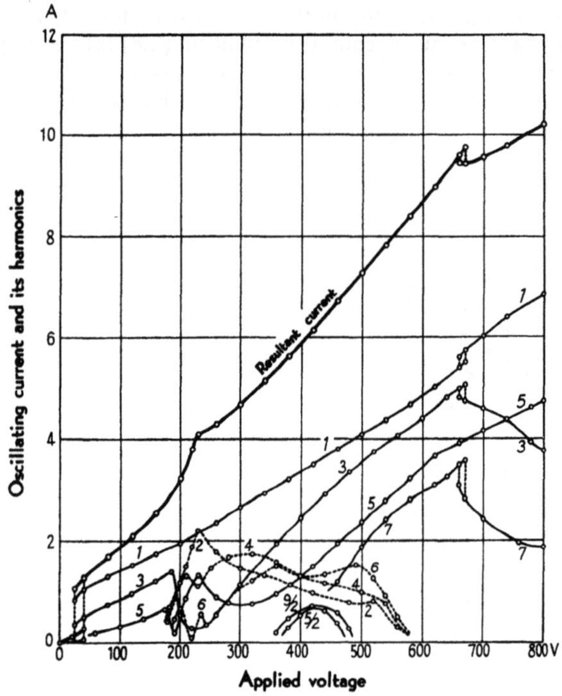

Fig. 4. Oscillating current and its harmonic components when the iron core is highly saturated.

current is analyzed into harmonic components; and they are shown in fine lines, the numbers on which indicate the order of harmonics. Instability of the characteristic curves occurs corresponding to the unstable regions of the stability chart of the variational equation. It is to be noted that, in the second unstable region extending from 180 to 580 V of the applied voltage, even-order harmics are excited as one may expect from the theory of Hill's equation.

Higher-harmonic oscillations occur frequently in a parallel-resonance circuit of Fig. 5. Since there is no series impedance, the inductor cores are easily saturated and higher-harmonic oscillations occur successively with the increase of the applied voltage. These oscillations are excited in opposite direction in the two resonant circuits, resulting in the anomalous neutral voltage V_n as illustrated in the figure. Those self-excited oscillations correspond to the unbounded solutions in the unstable regions of the stability chart for Mathieu's equation. See Fig. 6. As the applied voltage increases, the locus of the parameters θ_0 and θ_1 moves along the straight line 0, a_1, b_1, ... When the point (θ_0, θ_1) enters these unstable regions I, II, ..., higher-harmonic oscillations of the corresponding orders I, II, ... build up in the resonant circuits of Fig. 5.

The stability criterion is also applicable to the analysis of the subharmonic oscillation. As mentioned before, the 1/3-harmonic oscillation occurs in the series-resonance circuit of Fig. 1. Figure 7 shows an experimental result. In Fig. 7(a) the nonlinearity is given by an iron core available on the market, i.e. by a quintic or even higher-order polynomial of the

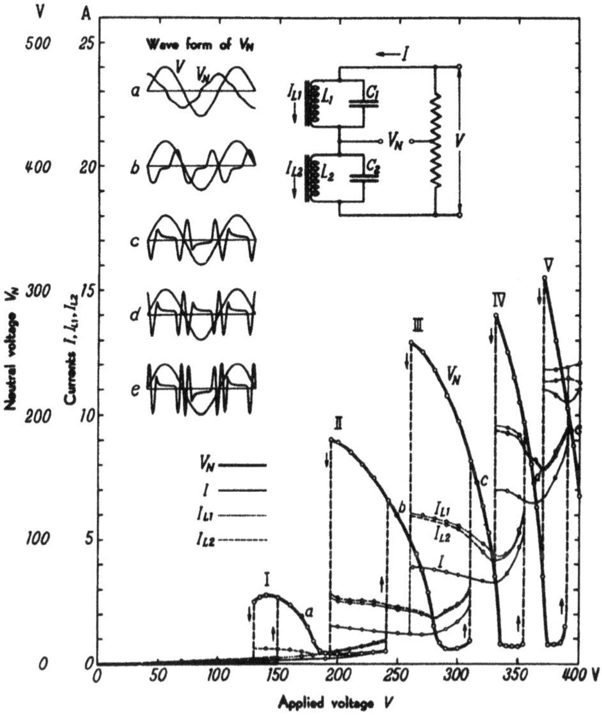

Fig. 5. Neutral instability caused by higher-harmonic excitation.

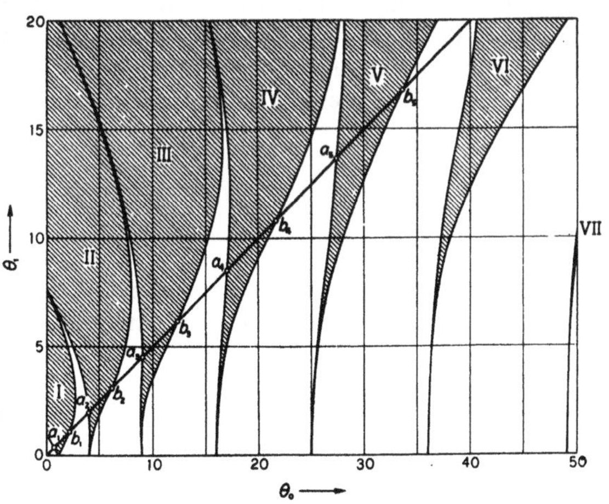

Fig. 6. Locus of point (θ_0, θ_1) on the stability chart for Mathieu's equation: $\ddot{\eta} + (\theta_0 + 2\theta_1 \cos 2t)\eta = 0$.

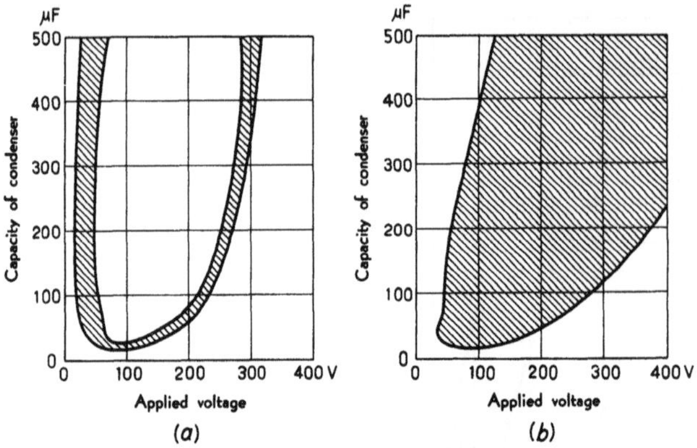

Fig. 7. Regions in which the 1/3-harmonic oscillation is sustained.

magnetization curve. Whereas in Fig. 7(b), the nonlinearity is given by a cubic function. In the former case the instability associated with the second unstable region of the stability chart gives rise to the excitation of an unbounded solution of the 2/3-harmonic and interrupts the original 1/3-harmonic oscillation. This kind of instability has not been detected hitherto by the stability criterion for linear systems.

3. Solution curves in the state plan [12]

As mentioned in the foregoing section, a number of responses may be obtained in a same system depending on different values of the initial condition. The relationship between the initial condition and the resulting oscillation must be discussed by considering the transient state of the oscillation before it settles down in the steady state. Two methods of analysis are proposed. The first is the graphical solution of integral curves in the state plane, and the second method makes use of the mapping procedure based on the transformation theory of differential equations. In this section we explain the first method with reference to some illustrative examples.

3.1 Harmonic oscillations

We consider an oscillation described by Duffing's equation (1.1). It is assumed, for the time being, that the fundamental component is predominant over the higher harmonics, and we write

$$v(t) = x(t)\sin t + y(t)\cos t \tag{3.1}$$

where $v(t)$ stands for x in Eq. (1.1). It is assumed that the amplitudes $x(t)$ and $y(t)$ are slowly varying functions of time t. Under this assumption we make use of the averaging method to obtain an autonomous system of the form

$$\dot{x} = X(x, y), \quad \dot{y} = Y(x, y). \tag{3.2}$$

By using the isocline method, we plot the solution curves, i.e. the integral curves of Eqs. (3.2) in the xy plane. The result is shown in Fig. 8. Periodic solutions are correlated with $x(t)$ = const. and $y(t)$ = const., i.e. with the singular points of Eqs. (3.2) where both $X(x,y)$ and $Y(x,y)$ vanish. As observed in Fig. 8, there are three singularities; points 2 and 3 are stable, and the corresponding periodic states are realized. The remaining singularity, point 1, is a saddle point, which is intrinsically unstable. The corresponding periodic state cannot be sustained, because any slight deviation from point 1 will result in passage to the stable state represented either by point 2 or point 3.

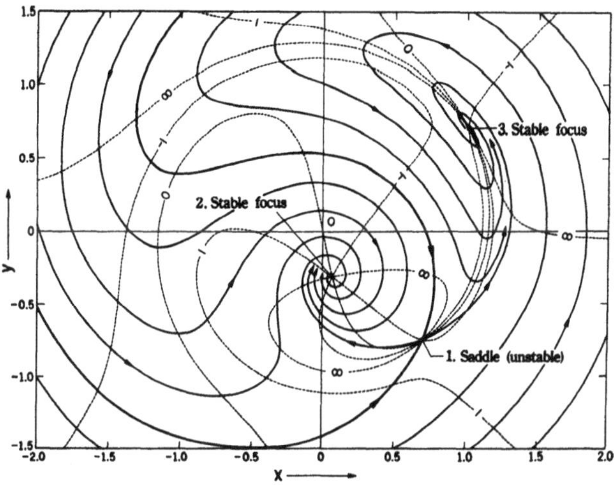

Fig. 8. Solution curves for the harmonic oscillation with three singular points.

Now suppose that we choose a point $[x(0), y(0)]$ in the xy plane as an initial condition. Then the representative point $[x(t), y(t)]$ moves, with the increase of time t, along the integral curve which passes through the initial point $[x(0), y(0)]$ and leads ultimately to the singular point 2 or 3. Thus the transient state is correlated with the integral curves of Eqs. (3.2) and the steady-state solution with the singular points in the xy plane. When the initial condition is given by a point far from the origin, the transient solution has a higher frequency than the external driving frequency, because the point $[x(t), y(t)]$ associated with the transient oscillation moves along the integral curve in the counterclockwise direction around the origin. It is worth noting that one of the integral curves (drawn in heavy line) that contains the saddle point divides the whole plane into two regions, in one of which all integral curves tend to the singular point 2 and in the other to the point 3. This particular integral curve is referred to as a separatrix.

3.2. Subharmonic oscillations

The preceding method of analysis is also applicable to the study of the subharmonic oscillation. We first consider the subharmonic oscillation of order 1/3. Let the sine and cosine components of the oscillation be $x(t)$ and $y(t)$, respectively. Then as before, we can derive an autonomous system analogous to Eqs. (3.2). The solution curves are shown in Fig. 9. In the figure we see that the singular points 1 to 3 represent the stable states of subharmonic oscillation and that they are equidistant from and equiangular about the origin. The angular distance between any two of these singular points corresponds to one complete cycle of the external force. This is a plausible result for the subharmonic oscillation of order 1/3. Any oscillation started from a point in the shaded regions leads to one of the singularities of points 1 to 3, resulting in the subharmonic response; whereas any oscillation started from the unshaded region leads ultimately to the origin, resulting in no subharmonic response.

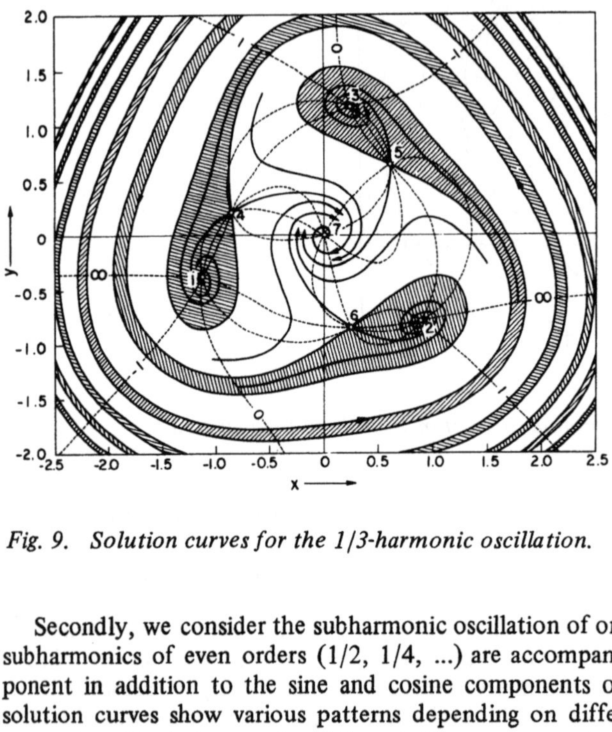

Fig. 9. Solution curves for the 1/3-harmonic oscillation.

Secondly, we consider the subharmonic oscillation of order 1/2. It is to be noted that the subharmonics of even orders (1/2, 1/4, ...) are accompanied with the nonoscillatory component in addition to the sine and cosine components of the oscillation. In this case the solution curves show various patterns depending on different combinations of the system parameters. Typical examples of the solution curves are shown in Fig. 10.

3.3. Almost periodic oscillations

Thus far we have dealt with the transient state of oscillations which ultimately tends to the periodic steady state correlated with the singular points in the state plane. There are also certain special cases in which the response is not periodic even when subjected to a periodic

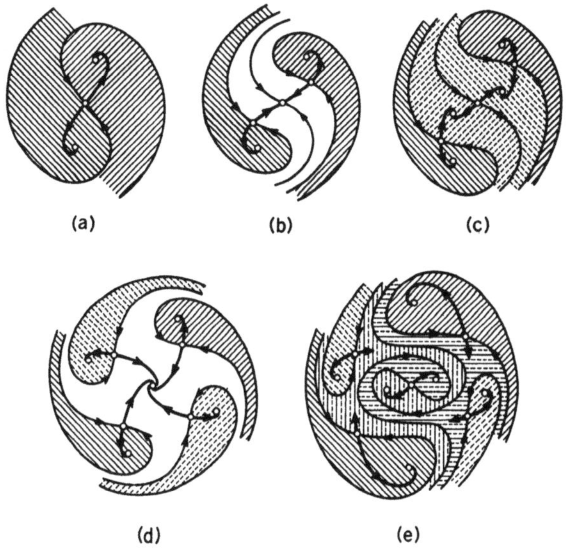

Fig. 10. Patterns of initial conditions for the 1/2-harmonic response.

external force. In this section we introduce the so-called quasi-periodic or almost periodic oscillation where the amplitude and phase of the oscillation vary slowly but periodically in the steady state. The ratio between the period for amplitude variation and the period of the external force is in general irrational, and thus there is no periodicity in the almost periodic oscillation. The solution curve tends to a stable limit cycle in this case.

It is known that this kind of oscillation occurs in a self-oscillatory system under the impression of a periodic external force [12, Chap. 13]. We take another example as shown in Fig. 11 [12, Chap. 11]. This is a kind of magnetic amplifier and, to increase the sensitivity, the capacitor C is inserted in the primary circuit which contains the saturable inductors L_1 and L_2. This system is liable to cause a low-frequency oscillation in the secondary d-c

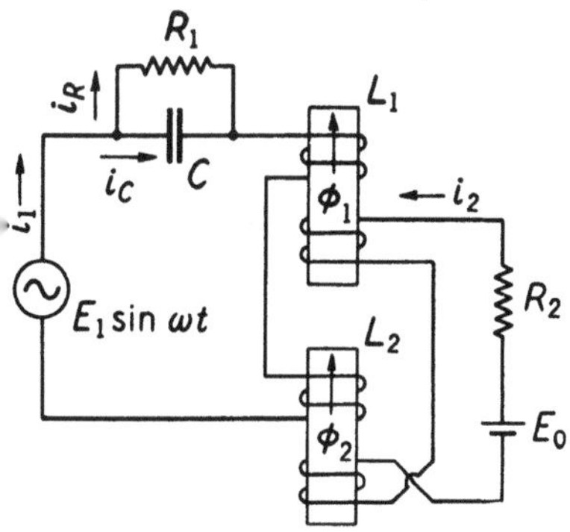

Fig. 11. Resonant circuit containing saturable-core inductors with secondary d-c windings.

windings, which gives rise to an almost periodic oscillation in the primary a-c circuit. Figure 12 explains the generation of such a phenomenon. The quantity r corresponds to the amplitude of the primary current i_1 and z to the secondary current i_2. A state of equilibrium is given by point P, but it is unstable. Also, in the figure, z decreases with time in the portion above the curve $\dot{z} = 0$, and increases below the curve $\dot{z} = 0$. Therefore, the amplitude r and the d-c current z vary along the closed curve drawn by heavy line in the figure. This results in the amplitude variation of the primary current i_1. Although not shown here, this kind of oscillation is represented by a stable limit cycle in the state space. The waveforms of these currents are shown in Fig. 13, where a, b and τ are the dimensionless quantities corresponding to i_1, i_2 and t, respectively. The successive points on the curves appear at every cycles of the external force.

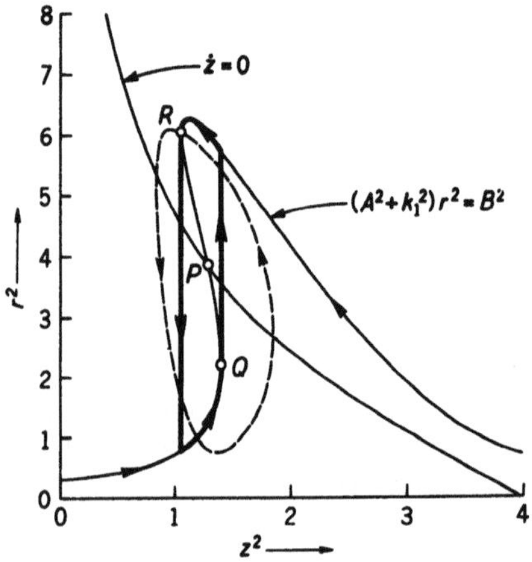

Fig. 12. Generation of an almost periodic oscillation.

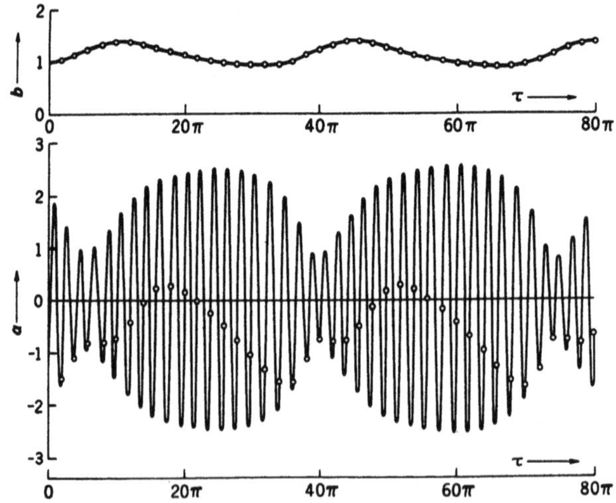

Fig. 13. Waveforms of an almost periodic oscillation.

4. Mapping procedure based on the transformation theory of differential equations [16]

The method of analysis described in Section 3 has been used extensively for the study of harmonic and subharmonic oscillations in the transient state. It is also applicable to the study of almost periodic oscillations. This method, however, has the following drawbacks. First, if the initial conditions are prescribed at values which are far different from those of the steady state, the assumption that the amplitude and phase of the oscillation vary slowly does not hold; therefore, the result obtained by this method may not be quite accurate. The second and more serious defect is that, if a number of different types of response exist, this method is practically inapplicable, since the analysis is confined only to a particular type of oscillation that is assumed beforehand. Therefore we shall, in what follows, introduce the method of analysis which is applicable under such situations.

4.1. The method of mapping

The method of mapping based on the transformation theory of differential equations has been studied by G.D. Birkhoff [13, 14], N. Levinson [15] and others. This method is useful for the study of nonlinear differential equations with periodic forcing terms. We take the differential equation of the form

$$\ddot{x} + f(x, \dot{x})\dot{x} + g(x) = e(t) \tag{4.1}$$

where $e(t)$ is periodic of period L. By setting $\dot{x} = y$, this equation becomes

$$\dot{x} = y, \quad \dot{y} = -f(x, y)y - g(x) + e(t) \tag{4.2}$$

The system (4.2) is a particular case of

$$\dot{x} = X(x, y, t), \quad \dot{y} = Y(x, y, t) \tag{4.3}$$

where $X(x,y,t)$ and $Y(x,y,t)$ are both periodic in t with period L.

Let $x(t), y(t)$ be a solution of Eqs. (4.3) which starts from a point P_0 in the xy plane. The coordinates of P_0 are then given by $x(0)$ and $y(0)$. We focus our attention toward the location of the point P_n at the instant of $t = nL$, n being 0, 1, 2, ... Let us call the transformation $P_0 \rightarrow P_1$ the mapping T and express as $P_1 = TP_0$. Accordingly, we write the iterates of the mapping by $P_2 = T(P_1) = T^2 P_0$, $P_3 = T(P_2) = T^3 P_0$, etc. We also write the inverse mapping by $P_0 = T^{-1} P_1 = T^{-2} P_2 = \ldots$. The inverse mapping proceeds in the negative direction of time t. If a solution $x(t)$, $y(t)$ has period L, then the point P_0 is a fixed point of the mapping T. If $x(t)$, $y(t)$ is a subharmonic of order $1/\nu$ ($\nu = 2, 3, 4, \ldots$), i.e. a solution of period νL, but not of period L, the points $P_0, P_1, \ldots, P_{\nu-1}$ are called periodic points. They are all fixed points of the νth iterate, T^ν, of the mapping T. Thus, in order to study the behavior of the solutions of Eqs. (4.3) as curves in the xyt space, we have only to study the successive mapping of initial points on the xy plane, or, in short, the transformation of the xy plane into itself.

4.2. Harmonic and subharmonic oscillations

To be more specific, let us again consider the harmonic oscillation and locate the fixed points which represent the periodic solutions for Eq. (1.1). By making use of a computer designed for this purpose, we obtain three fixed points, 1 to 3, as illustrated in Fig. 14. Also shown in the figure are the curves which are invariant under the mapping. Along the invariant curve drawn by fine line, successive images of the mapping move in the direction of the arrows and approach to the stable point 2 or 3. Along the other invariant curve (heavy line), successive images of the mapping approach to the point 1. This point is referred to as a directly unstable fixed point [15]. It is noted that the invariant curve drawn in heavy line divides the whole xy plane into two domains of attraction. Any initial point $[x(0), y(0)]$ prescribed in one of these domains tends, under iteration of the mapping, to the stable fixed point 2 or 3, which is located in the interior of that domain.

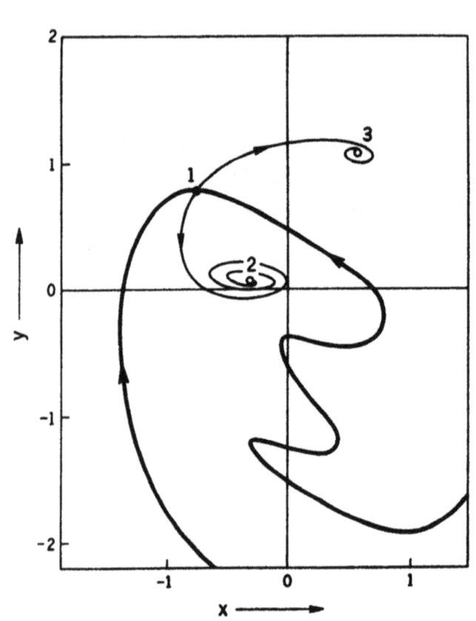

Fig. 14. Fixed points and invariant curves for the harmonic oscillation.

The above result obtained for the harmonic oscillation does not differ substantially from that of the solution curves of the preceding section (see Fig. 8). As mentioned before, the mapping procedure has the advantage that it is applicable to the analysis of a system without knowing how many and what type of response it may have.

As an example, we made an analysis to obain the fixed points and invariant curves for Eq. (1.1) with a different combination of the system parameters. The result is shown in Fig. 15. The subharmonic oscillation of order 1/3 predominantly occurs in this case. We obtain three fixed points, 1 to 3, of the mapping T and six fixed points, 4 to 9, of the mapping T^3. A pair of invariant curves which pass through the unstable fixed point 1 are also shown in the figure. One of these invariant curves (drawn in fine line) approaches all the other fixed points, 2 through 9, under iteration of the mapping T. It is worth noting that, once the directly unstable fixed point 1 is located in the xy plane, the other fixed points correlated

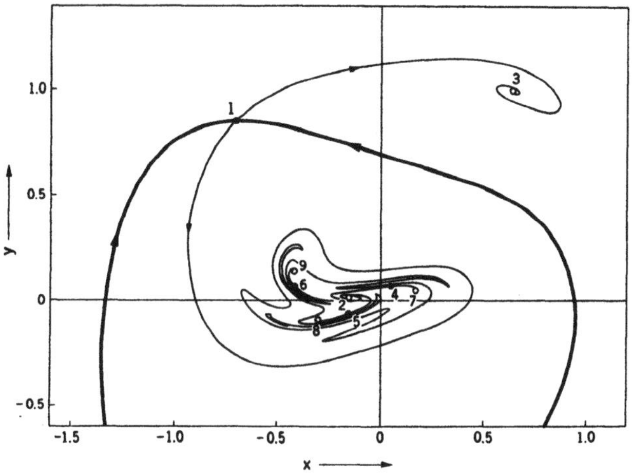

Fig. 15. Fixed points and invariant curves for the harmonic and subharmonic oscillations.

with the harmonic and subharmonic solutions are found by tracing the successive images of initial points described in the neighborhood of point 1. Thus, the mapping procedure provides an effective means for seeking periodic solutions for a given nonlinear differential equation.

The domains of attraction containing each of the stable fixed points 2, 3, 7, 8 and 9 are obtained by tracing the invariant curves of the inverse mapping T^{-1} from the unstable fixed point 1 and of the inverse mapping T^{-3} from the unstable fixed points 4, 5 and 6. The result is shown in Fig. 16. If the initial condition is given, for instance, inside the dotted area, the solution ultimately settles to the harmonic response represented by point 3. Similarly, the domains of attraction for the subharmonic response are given by shaded areas. These domains have narrowing tails as they come close to the domain of harmonic response containing point 3. These extremely narrow tails are omitted in the figure.

4.3. Doubly asymptotic points

It was shown in Fig. 14 that one of the invariant curves (heavy line) which passes through the directly unstable fixed point 1 divides the xy plane into two domains of attraction, each containing the stable fixed point 2 or 3. This configuration is schematically reproduced in Fig. 17(a). Under a certain combination of the system parameters, however, the invariant curves intersect each other, and the circumstances become very complicated as one sees in Fig. 17(b). We introduce some terminology defined by Poincaré [2]. In Fig. 17(a), there abut at point 1 four invariant branches, i.e. two α-branches (fine line) whose points converge toward point 1 on indefinite iteration of T^{-1}, and two ω-branches (heavy line) whose points converge toward point 1 on iteration of T. If we consider the totality of α- and ω-branches on the xy plane, it is readily seen that no α- (or ω-) branch can intersect another α- (or ω-) branch. However, an α-branch may intersect an ω-branch, and the points of intersection in such a case are called doubly asymptotic points.

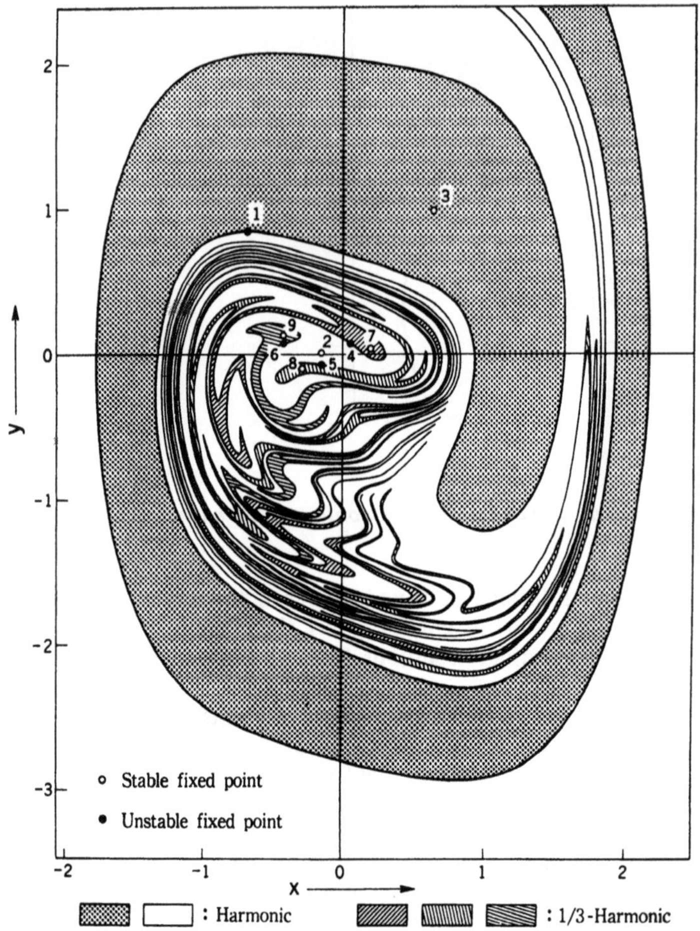

Fig. 16. Domains of attraction for harmonic and 1/3-harmonic responses.

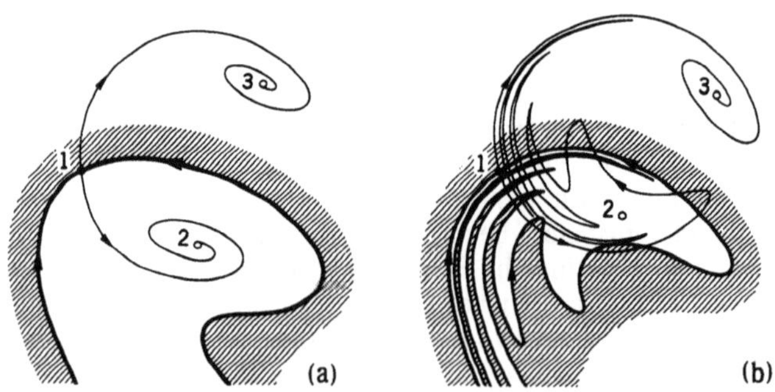

Fig. 17. Fixed points and domains of attraction. (a) Domains of attraction separated by a simple boundary. (b) Appearance of doubly asymptotic points resulting in an intricate boundary.

We consider a set of unstable periodic points obtained by successive iteration of the mapping. A doubly asymptotic point is called homoclinic if the α- and ω-branches on which it lies issue from one of these periodic points or from two points belonging to the same set (or group). For convenience, let us say that the momoclinic point of the former type is simple. In addition to this set of periodic points, let us assume another set of periodic points. Then a doubly asymptotic point is called heteroclinic if the α- and ω-branches on which it lies issue from two periodic points, each of them belonging to the different set of periodic points. In Fig. 17(b) the α-branch (fine line) and the ω-branch (heavy line) intersect repeatedly. Since they issue from the same fixed point 1, those points of intersection are simple homoclinic points. Further details of the homoclinic strucute are shown in Fig. 18.

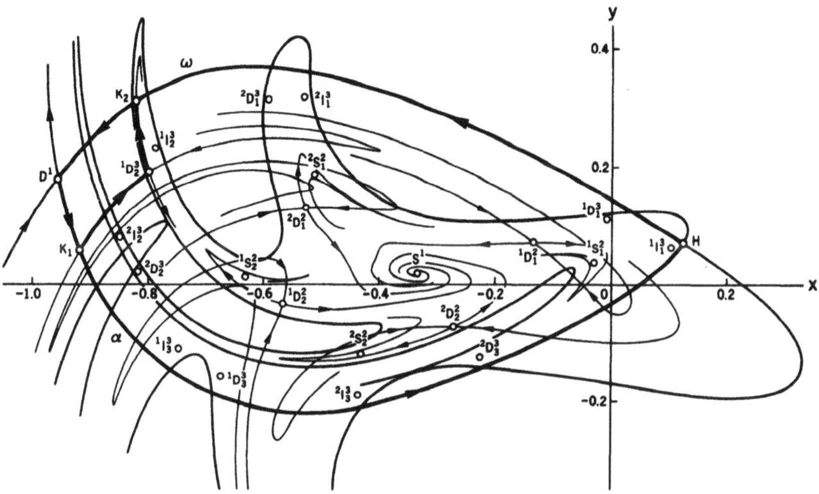

Fig. 18. Fixed points and invariant curves of the mapping for the equation: $\ddot{x} + 0.05\dot{x} + x^3 = 0.3 \cos t$.

For convenience' sake we define the notation of fixed points as follows: the symbol ${}^jD_i^n$ denotes a directly unstable n-periodic point which belongs to the jth periodic group, and

$$T({}^jD_i^n) = {}^jD_{i+1}^n, \quad T^n({}^jD_i^n) = {}^jD_i^n.$$

Similar notation applies also to an inversely unstable periodic point (I) and a stable periodic point (S) [15].

The fixed point D^1 of Fig. 18 corresponds to point 1 of Fig. 17(b). The α- and ω-branches which issue from D^1 intersect repeatedly. Besides the fixed points D^1 and S^1 of the mapping T, there are infinitely many groups of periodic points around S^1. To avoid complexity of the illustration, however, only periodic points of the orders 2 and 3 are shown in the figure, i.e.

2-periodic points: ${}^1D_i^2, {}^2D_i^2, {}^1I_i^2, {}^2I_i^2 \qquad (i = 1, 2)$

3-periodic points: ${}^1D_i^3, {}^2D_i^3, {}^1I_i^3, {}^2I_i^3. \qquad (i = 1, 2, 3)$

The invariant curves of the mapping T^2 issuing from the directly unstable 2-periodic points are also shown in Fig. 18. These curves intersect one another and also with the invariant curves of the mapping T. Thus in the end we can locate infinitely many homoclinic and heteroclinic points in the figure.

Figure 19(a) shows the waveform of a periodic oscillation in the neighborhood of the homoclinic curve $D^1 H D^1$ shown by heavy line in Fig. 18. The successive points marked on the waveform indicate the instants at which $t = 2n\pi$ ($n = 1, 2, 3, \ldots$), where 2π is the period of the external force. In this particular example the oscillation repeats the same waveform at every nine cycles of the external force; in other words, the fundamental component of the oscillation is a subharmonic of order 1/9. Another example is shown in Fig. 19(b). It is a periodic oscillation in the neighborhood of the heteroclinic curve $D^1 K_1 {}^1D_2^3 K_2 D^1$ which is also shown by heavy line in Fig. 18. The oscillation repeats the same waveform at every fourteen cyles of the external force. One sees that the oscillation is nearly harmonic when the successive images of the mapping T are in the neighborhood of D^1, and nearly subharmonic of order 1/3 when the successive images of the mapping T^3 are near ${}^1D_2^3$. The phenomenon of doubly asymptotic points was first discussed by H. Poincaré [2] in connection with the three-body problems in celestial mechanics. It is interesting to see that the similar problem occurs even in such a simple-appearing nonlinear differential equation as Duffing's equation.

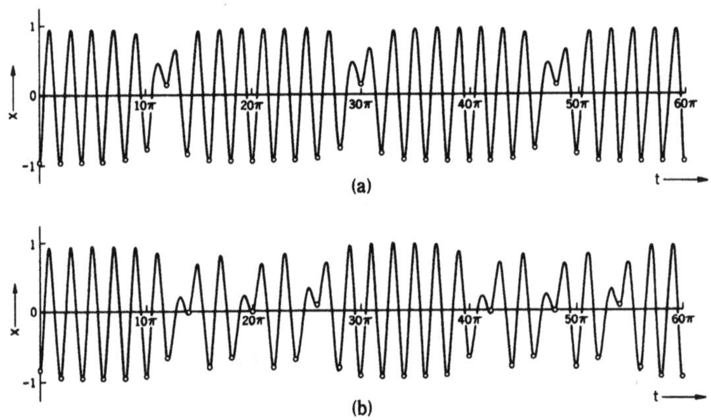

Fig. 19. Waveforms of periodic oscillations in the neighborhood of (a) homoclinic curve and (b) heteroclinic curve.

5. Concluding remarks

As mentioned earlier, this report is concerned with the research works which have been done for years in our laboratory. The scope of the study is rather limited. It deals with the forced oscillatory systems with one or two degrees of freedom. Yet, there are a number of problems unsolved to date; e.g. a special type of nonperiodic oscillation which may not be represented by a simple closed curve in the state plane, or the domains of attraction associated with the homoclinic structure of invariant curves. These problems belong, however, to a small segment of nonlinear vibrations, and there remains a wide field of researches which are now being made or opened for future investigation. Among them the following subjects are of interest to the author:

a. Analytical theory – Bifurcation, point mapping, systems with delay or heredity.
b. Stability problem – Definition, criterion (Liapunov's direct method and others).
c. Approximation methods and numerical analysis assisted by computers.
d. Higher-order systems, systems with many degrees of freedom.
e. Nonperiodic oscillations including quasi-periodic and almost periodic oscillations.

Finally, not in our field of study, but important are the subjects pertaining to random vibrations, stochastic behavior of nonlinear systems and systems described by nonlinear partial differential equations.

Acknowledgements

The author wishes to acknowledge the cooperation of his colleagues for many years. He is also grateful to the publishers for their permission in citing the figures in Refs. [12] and [16].

References

1. Load Rayleigh, *Theory of Sound*, vol. 1, Macmillan, London, 1894.
2. H. Poincaré, *Les Méthodes nouvelles de la mécanique céleste*, tom. 3, Gauthier-Villars, Paris, 1899.
3. G. Duffing, *Erzwungene Schwingungen bei veränderlicher Eigenfrequenz und ihre technische Bedeutung*, Sammlung Vieweg, Braunschweig, 1918.
4. B. van der Pol, 'On relaxation-oscillations', *Phil. Mag.*, **7-2**, 978-992 (1926).
5. A.A. Andronov, A.A. Vitt and S.E. Khaikin, *Theory of Oscillators*, Fizmatgiz, Moscow, 1959 (in Russian); English translation, Pergamon Press, Oxford, 1966.
6. N. Kryloff and N. Bogoliuboff, 'Introduction to Nonlinear Mechanics', *Ann. Math. Studies*, no. 11, Princeton, 1947.
7. N.N. Bogoliuboff and Yu. A. Mitropolsky, *Asymptotic Methods in the Theory of Nonlinear Oscillations*, Fizmatgiz, Moscow, 1963 (in Russian); English translation, Gordon & Breach, New York, 1961.
8. N. Minorsky, *Nonlinear Oscillations*, D. van Nostrand, Princeton, 1962.
9. A. Tondl, *Some Problems of Rotor Dynamics*, Publishing House of the Czechoslovak Academy of Sciences, Prague, 1965.
10. E. Philippow, *Nichtlineare Elektrotechnik*, Akad. Verlag. Geest & Portig, Leipzig, 1971.
11. C. Hayashi, *Forced Oscillators in Non-Linear Systems*, Nippon Printing & Publishing Co., Osaka, 1953.
12. C. Hayashi, *Nonlinear Oscillations in Physical Systems*, McGraw-Hill, New York, 1964.
13. G.D. Birkhoff, 'On the periodic motions of dynamical systems', *Acta Math.*, **50**, 359-379 (1927).
14. G.D. Birkhoff and P.A. Smith, 'Structure analysis of surface transformations', *J. Math.* (Liouville), **9-7**, 345-379 (1928).
15. N. Levinson, 'Transformation theory of non-linear differential equations of the second order', *Ann. Math.*, **45**, 723-737 (1944).
16. C. Hayashi, *Selected Papers on Nonlinear Oscillations*, Nippon Printing & Publishing Co. Osaka, 1975.

Developments in the Mathematical Theory of Plasticity

*H.G. Hopkins**

Summary

The mathematical theory of plasticity – one of the branches of theoretical continuum mechanics – applies to the study of stress and strain in metals in either single crystal or polycrystalline (often alloy) form. This theory is therefore quite distinct from the related physcial theory of plasticity which applies to interatomic forces and atomic movements. Here, attention is given to certain aspects of developments in the former theory, which may be taken to extend from situations of quasi-static small strain and low stress (tensile or compressive) to those of dynamic large strain and high stress (as regards its compressive isotropic component) conveniently categorized in terms of non-dimensional parameters; and the applications of the theory include engineering structures, metal-forming and impact (mechanical or explosive). The framework of the theory rests upon field equations expressing conservation and constitutive laws, and the analytical procedures are directed towards development and applications. This paper is highly selective and it gives primary attention to reviewing the types of mathematical structure that are met in specializations of the general theory. The governing equations are, mostly, intrinsically nonlinear (although normally cast in quasi-linear form), either for physical or geometrical reasons or both, and characteristics theory for systems of quasi-linear partial differential equations provides the key to their analysis. In particular, therefore, for equations that show hyperbolic features (that is, there exist real characteristics) solutions of discontinuous type exist. This is the case in many situations of interest, sometimes as a result of the adoption of various artifices and simplifications (such as similarity and quasi-steady formulations), and it is then possible to achieve a substantial measure of progress in developments and applications of the theory in particular areas. In practice, neglected physical effects act to diffuse the sharp discontinuities in field quantities, and this question is also discussed.

1. Introduction

The mathematical theory of plasticity applies to the study of stress and strain in metals in either single crystal or polycrystal form, pure or alloy. It is therefore quite distinct from the more fundamental physical theory of plasticity which applies to interatomic forces and atomic movements. However, the former macro-type theory is necessarily related to the latter micro-type theory. Mathematical plasticity theory has its origins in Saint-Venant's (1870) work; it was first significantly developed in Germany in the 1920's by Prandtl and others; it was later substantially developed, especially as regards dynamic situations of bal-

* Department of Mathematics, University of Manchester Institute of Science and Technology, Manchester, England

listics, in the U.K., U.S.A. and U.S.S.R. in the 1940's; and the early 1950's saw the publication of the books by Hill [1] and Prager and Hodge [2] and the opening up of the modern development of the subject. The theory, although of outstanding importance and interest in its own right, now represents one of the branches of the theoretical mechanics of continua which itself has connections to rational mechanics. It aims to cover a wide range of situations from those of quasi-static small strain and low stress (tensile or compressive) to those of dynamic large strain and high stress (as regards its compressive isotropic component), and these may be categorized in terms of non-dimensional parameters; and the applications of the theory may be taken to include situations as diverse as those of engineering structures, metal-forming and impact (mechanical or explosive) (see Hopkins [3] for an over-view and detailed references).

The framework of the theory rests, just as with any branch of theoretical continuum mechanics, upon field equations expressing conservation and constitutive laws, and the analytical procedures are directed towards development and application.

The content of the present paper is perforce highly selective, and primary consideration is given to reviewing the types of mathematical structure that are met in specializations of the general mathematical theory of plasticity. However, before proceeding with that particular topic it seems important to mention, albeit only briefly, some other highly important topics in which there have been significant developments. The first set of topics concerns various aspects of constitutive equations, a key issue in the entire theory. These equations involve stress and strain measures and also, generally, their material derivatives. Their precise specification is especially important for the discussion of bifurcation. A definitive account of fundamentals here has been given by Hill [4]. However, the determination of constitutive equations, aside from purely mathematical considerations of invariance as discussed in [4], that are appropriate to particular materials and physical conditions necessarily involves experimental work on test specimens and the interpretation of the recorded data within a theoretical framework is never easy and often extremely difficult especially when rate-dependent effects are involved (see [5]). The second set of topics concerns the establishment of existence and uniqueness of the equations, the formulation of variational and extremum principles in the theory, and the development of numerical analysis techniques, especially those based upon the finite element method, for solving specific problems. A very large literature indeed is available on such topics.

In proceeding, as indicated above, attention will be confined primarily to the topic of mathematical structure of the equations in the mathematical theory of plasticity. The governing equations are, mostly, nonlinear (although normally cast in quasi-linear first-order form) either for physical or geometrical reasons or both, and the characteristics theory of systems of quasi-linear first-order systems of partial differential equations provides the key to their analysis. In particular, therefore, for the important case when the equations show hyperbolic features, that is, real characteristics exist (in dynamic problems, this means that nonlinear waves exist), solutions of discontinuous type occur. In both static and dynamic problems singularities may be present. Generally, in dynamic problems, there are two possibilities: either the solution is initially discontinuous and then propagates continuously (a dispersion situation) or it is initially continuous and later becomes discontinuous (a shock situation). Of course, such occurrences are simply the result of mathematical idealizations and in practice there are present physical effects not taken into account that act to resolve the sharp discontinuities: thus, for example, work-hardening will preclude velocity discontinuities in quasi-static problems and viscosity and heat conduction will diffuse stress discontinuities in dynamic problems.

In parallel with the exact mathematical theories there are approximate engineering theories for structures (bars, beams, plates, shells) which are formulated in terms of generalized stresses (forces, moments) and strains (mean extensions, curvatures), and analogous developments have taken place often with extremely complicated practical engineering problems in mind (as, for example, in pressure-vessel design). The discussion of the structure of the equations is similar but there can be considerable complexity due to the nature of the (generalized) constitutive equations. It should be mentioned that the simplifying technique of adopting piece-wise linear yield criteria (due originally to Koiter and Prager) has proved extremely useful.

For reasons of simplicity, attention in the main body of this paper will be mostly restricted to situations involving either two space variables (quasi-static problems) or to one space variable and the time (dynamic problems). In section 2, the method of characteristics is presented in matrix form. In sections 3 and 4 some quasi-static and dynamic situations are considered in detail in order to illustrate particular points. Next, in section 5, some brief comments are made on engineering applications, including brief reference to general design aspects. Finally, in section 6, some concluding remarks are made.

2. The method of characteristics

A general system of n quasi-linear first order inhomogeneous partial differential equations in n dependent variables $u_i(x,y)$ $(i = 1, 2, ..., n)$ with x,y as two independent variables (not necessarily rectangular cartesian spatial in nature) may be written

$$L_i \equiv a_{ik}(x,y,u_j)u_{k,x} + b_{ik}(x,y,u_j)u_{k,y} + c_i(x,y,u_j) = 0, \qquad (2.1)$$

where the a_{ik}, b_{ik}, c_i are given functions of the indicated arguments, the usual conventions for subscripts in relation to summation and partial differentiation applying. Equivalent to the scalar representation (2.1) is the matrix representation

$$\mathbf{L} \equiv \mathbf{A}(x,y,\mathbf{U})\mathbf{U}_x + \mathbf{B}(x,y,\mathbf{U})\mathbf{U}_y + \mathbf{C}(x,y,\mathbf{U}) = \mathbf{O}, \qquad (2.2)$$

where $\mathbf{A},\mathbf{B}$ $(n \times n)$ and $\mathbf{C}(n \times 1)$ are given matrix functions of $x,y,\mathbf{U}$ and $\mathbf{U}(n \times 1)$ is the column vector of dependent variables.

There are two approaches to the discussion of the mathematical structure of (2.2) in terms of the method of characteristics for systems of quasi-linear first-order partial differential equations: the first may be referred to as the interior derivative approach and the second as the Hadamard (or Cauchy) approach. They are not essentially distinct from each other and will be considered in turn. It should be noted at the outset that, in general, in (2.2) $\mathbf{U}$ is not differentiated in a single direction at any point of the (x,y)-plane.

In the interior derivative approach, the objective is to reduce (2.2) to an equivalent system of scalar equations in each of which $\mathbf{U}$ (that is, all the u_i's) is differentiated in some single direction (which may, however, vary with the solution and over the plane, and also from one equation to another). The essential question is whether or not such a transformation exists and, if it does, then whether or not it is real. Thus, the initial problem is then set as the determination from $\mathbf{L} = \mathbf{O}$ of scalar ordinary differential equations in each of which $\mathbf{U}$ is differentiated in some single direction.

Without restriction it will be assumed that $\mathbf{B}$ (or $\mathbf{A}$) is non-singular, i.e. $det\mathbf{B} \neq 0$, and then

$$\mathbf{M} \equiv \mathbf{B}^{-1}(\mathbf{L}-\mathbf{C}) = \mathbf{B}^{-1}\mathbf{A}\,\mathbf{U}_x + \mathbf{U}_y = \mathbf{D}\,\mathbf{U}_x + \mathbf{U}_y, \tag{2.3}$$

where $\mathbf{D} = \mathbf{B}^{-1}\mathbf{A}$. For the moment, it is convenient to work with the differential expression $\mathbf{M}$ rather than the differential equation $\mathbf{L} = \mathbf{O}$. Now let $\mathbf{F}(n \times 1)$ be an arbitrary column vector; and let T denote the transpose of a matrix. Then,

$$\mathbf{N} = \mathbf{F}^T\mathbf{M} = \mathbf{M}^T\mathbf{F} = \mathbf{N}^T \tag{2.4}$$

is a scalar, a linear combination of the components of $\mathbf{M}$. Now, (2.4) may be written

$$\mathbf{N} = \mathbf{F}^T(\mathbf{D}\,\mathbf{U}_x + \mathbf{U}_y) = (\mathbf{U}_x^T\,\mathbf{D}^T + \mathbf{U}_y^T\,\mathbf{I})\,\mathbf{F}, \tag{2.5}$$

where $\mathbf{I}$ $(n,\times n)$ is the identity matrix. In (2.5), impose the condition on $\mathbf{F}$ that $\mathbf{U}$ is differentiated in a single direction $x_s{:}y_s$ along a curve $x = x(s)$, $y = y(s)$ (where s is some convenient parameter, such as arc length) drawn in the (x,y)-plane, that is

$$\mathbf{N} = \mathbf{U}_s^T\,\mathbf{G}, \tag{2.6}$$

where $\mathbf{G}(n \times 1)$ is an as-yet undetermined column vector, or

$$\mathbf{N} = (\mathbf{U}_x^T\,x_s + \mathbf{U}_y^T\,y_s)\,\mathbf{G}. \tag{2.7}$$

The requirement that (2.5) and (2.7) are identical gives

$$\mathbf{D}^T\mathbf{F} = x_s\mathbf{G}, \quad \mathbf{I}\,\mathbf{F} = y_s\mathbf{G}, \tag{2.8}$$

so with elimination of $\mathbf{G}$,

$$(\mathbf{D}^T - (x_s/y_s)\,\mathbf{I})\,\mathbf{F} = \mathbf{O} \quad \text{(without restriction, } y_s \neq 0) \tag{2.9}$$

or, equivalently,

$$\mathbf{F}^T\,(\mathbf{D} - (x_s/y_s)\,\mathbf{I}) = \mathbf{O}. \tag{2.10}$$

Thus, for non-trivial (non-zero) $\mathbf{F}$, the direction of differentiation x_s/y_s must be an eigenvalue of $\mathbf{D}$ (and $\mathbf{D}^T$) and $\mathbf{F}$ is the corresponding right-hand eigenvector of $\mathbf{D}^T$. Let now λ_i $(i = 1, 2, \ldots, n)$ be the eigenvalues of $\mathbf{D}$, so that they are the roots of

$$\det(\mathbf{D} - \lambda\mathbf{I}) = 0 \quad \text{or} \quad \det(\mathbf{A} - \lambda\mathbf{B}) = 0. \tag{2.11}$$

It remains to determine the form for $\mathbf{N}$ and hence $\mathbf{F}^T\mathbf{B}^{-1}\mathbf{L}$ along the characteristic directions $x_s{:}y_s = \lambda_i{:}1$. Equation (2.9) is written as

$$(\mathbf{D}^T - \lambda_i\mathbf{I})\,\mathbf{F}_i = \mathbf{O} \quad (i \text{ not summed}), \tag{2.12}$$

where λ_i and F_i are associated eigenvalues and eigenvectors. The former may be assembled into a diagonal matrix $\Lambda = \text{diag}(\lambda_i)$ and the latter into an eigenvector matrix $\mathbf{E} = [\mathbf{F}_i]$, and then (2.12) is written compactly as

$$\mathbf{D}^T\mathbf{E} = \mathbf{E}\Lambda. \tag{2.13}$$

Assuming that $\mathbf{E}$ is non-singular, then $\mathbf{D}^T = \mathbf{E}\,\Lambda\,\mathbf{E}^{-1}$ which is the expression of a well-known result in matrix algebra. Finally, it follows from the above that

$$\mathbf{O} = \mathbf{F}_i^T\,\mathbf{B}^{-1}\,\mathbf{L}\,\mathrm{d}y_i = \mathbf{F}_i^T(\mathrm{d}\mathbf{U}_i + \mathbf{B}^{-1}\,\mathbf{C}\,\mathrm{d}y_i)$$

and then, with definition of the differential operator

$$\mathrm{d}_i = x_s\partial_x + y_s\partial_y = y_s(\lambda_i\partial_x + \partial_y), \tag{2.14}$$

it follows that

$$\mathbf{F}_i^T(\mathrm{d}_i\mathbf{U} + \mathbf{B}^{-1}\,\mathbf{C}y_{i,s}) = 0 \ \text{ on } \ x_s/y_s = \lambda_i \quad (i \text{ not summed}) \tag{2.15}$$

which is the characteristic (Riemann) relation applying along the characteristic curve satisfying the differential equation $x_s/y_s = \lambda_i$, in general not fixed independently of the solution $\mathbf{U}$ but floating with it. The especially important case is that when all n characteristics are real (although not necessarily distinct) and the n real characteristic relations are distinct. In this case, which may be called fully hyperbolic, the original partial differential equations have been reduced to an equivalent system of real ordinary differential equations. Of course, even then in most cases, the characteristics have to be found in the course of solution of a given problem, and features of nonlinearity are preserved and normally lead to singularities or discontinuities, as earlier mentioned. Thus, in particular, the partial differential equations may fail locally on curves across which shock-type discontinuities in field quantities occur.

In the Hadamard (or Cauchy) approach consideration is given directly to the solution of (2.2) when data for $\mathbf{U}$ are prescribed on a curve Γ, say $x = x(s), y = y(s)$ with s equal to arc length (for convenience). In this case, there applies the known value of $\mathbf{U}_x x_s + \mathbf{U}_y y_s$ on Γ. However, rewrite (2.2) in the form

$$(\mathbf{A}x_s + \mathbf{B}y_s)\mathbf{U}_s + (-\mathbf{A}y_s + \mathbf{B}x_s)\mathbf{U}_n + \mathbf{C} = \mathbf{O}, \tag{2.16}$$

n being a coordinate normal to Γ. Equation (2.16) applies necessarily on either side of Γ and so

$$(-\mathbf{A}y_s + \mathbf{B}x_s)\,[\mathbf{U}_n] = 0 \tag{2.17}$$

where the square brackets denote the jump across Γ. If the rank of the matrix in (2.17) is n then $\mathbf{U}_n$ is continuous across Γ, and oppositely if and only if

$$\det(-\mathbf{A}y_s + \mathbf{B}x_s) = 0 \tag{2.18}$$

may $\mathbf{U}_n$ be discontinuous across Γ. In this case, Γ is called characteristic, its direction x_s/y_s being a relative eigenvalue of $\mathbf{A}$ and $\mathbf{B}$. Further, there now exists an eigenvector, say $\mathbf{F}$, and hence from (2.16) there follows

$$\mathbf{F}^T(\mathbf{A}x_s + \mathbf{B}y_s)\mathbf{U}_s + \mathbf{F}^T\mathbf{C} = \mathbf{O}, \tag{2.19}$$

the other term vanishing. Equation (2.19) means that the data for **U** on the characteristic Γ must satisfy a differential characteristic relation.

These results are consistent with but amplify those obtained via the interior derivative approach: importantly, along non-characteristic Γ, data may be arbitrarily prescribed and normal discontinuities cannot occur; and along characteristic Γ, data may not be arbitrarily prescribed and normal discontinuities may occur. The treatment of higher-order discontinuities is quite similar but leads to no essentially distinct results.

The method of characteristics may be developed for more than two independent variables. The Hadamard approach will be considered briefly for the case of three independent variables, taking the system of equations to be defined by

$$\mathbf{A}\mathbf{U}_x + \mathbf{B}\mathbf{U}_y + \mathbf{C}\mathbf{U}_z + \mathbf{D} = \mathbf{O}, \tag{2.20}$$

in the obvious notation. Suppose the solution for **U** to be known on the surface Γ, say $f(x,y,z) = 0$, then it is straightforward to prove that discontinuities in the normal derivative of **U**, say $\mathbf{U}_f$, are possible if and only if the rank of $\mathbf{A}f_x + \mathbf{B}f_y + \mathbf{C}f_z$ is less than n, that is

$$\det(\mathbf{A}f_x + \mathbf{B}f_y + \mathbf{C}f_z) = 0. \tag{2.21}$$

This is an n-th order polynomial equation in two independent ratios formed from f_x, f_y, f_z, and thus (2.21) only defines one-parameter families of surfaces and hence the characteristic surfaces are not uniquely known. There is, moreover, the complication that (2.21) unlike (2.18) cannot always be simply factored in terms of linear and quadratic terms with real coefficients. The treatment of systems involving four dependent variables is similar, as also is the question of higher-order discontinuities. Certainly, in mathematical plasticity theory, little attention has been given to characteristic (hyper-)surfaces as opposed to characteristic curves, for the implicit stated reasons.

The envelope of a family of characteristics is called a limit line across which it is not possible to continue the solution, and thus it represents normally a natural boundary. In plasticity theory such limit lines are rare but one example of their occurrence is with the Prandtl problem of a slab compressed between rough rigid plates (see [2]). Collins (unpublished work) has shown that the solution of any quasi-linear system has a square-root singularity on a limit line.

3. Quasi-static situations

For quasi-static situations and for rigid/perfectly-plastic solids, the mathematical structure of the field equations varies with the geometry of the flow and with the type of yield criterion and (associated or non-associated) flow-rule. Thus, in plane strain, the stress and velocity equations are hyperbolic for all possible yield criteria, and this is also true in torsion of prismatic bars; but, for example, in plane stress and axially-symmetric stress/strain, the field equations are not uniformly hyperbolic and changes in their structure (say, to or from hyperbolic type) occur with change in the location of the representative stress-point on the yield surface, and this applies also in bending of thin plates (there the governing equations involve moments and curvature-rates, apart from reactions). The special cases of plane strain and torsion are altogether exceptional in the uniformity and relative simplicity of the structure of the field equations, and in this regard they stand in marked contrast to most other cases. This is, in fact, the fundamental reason why, in general, most progress in the solution of problems of rigid/perfectly-plastic flow has been made for situations of plane strain and torsin (although the situation of axially-symmetric stress/strain, when the Haar-Kármán criterion is additionally adopted, is similar to that of plane strain).

In this section attention will now be confined to the situation of plane strain in order to illustrate the application of the method of characteristics given in section 2 and to indicate the differences in structure according to whether perfect plasticity or work-hardening plasticity is considered. For rigid/perfectly-plastic plane strain the governing equations are

$$\left.\begin{aligned}&\sigma_{x,x}+\tau_{xy,y}=0,\ \tau_{xy,x}+\sigma_{y,y}=0,\\&\tfrac{1}{4}(\sigma_x-\sigma_y)^2+\tau_{xy}^2=k^2,\\&u_{,x}+v_{,y}=0,\ 2\tau_{xy}/(\sigma_x-\sigma_y)=(u_{,y}+v_{,x})/(u_{,x}-v_{,y}),\end{aligned}\right\}\qquad(3.1)$$

being respectively equations of quasi-static equilibrium, yield, and flow (plastic incompressibility and the Saint-Venant condition of co-axiality of the principal stress and plastic strain-rate directions). Here and elsewhere in this paper where standard notation is being used, then it is not explicitly defined. Equations (3.1) are most easily handled when slightly re-formulated by means of the introduction of new stress variables p and ϕ defined by

$$\sigma_x=-p-k\sin 2\phi,\ \sigma_y=-p+k\sin 2\phi,\ \tau_{xy}=k\cos 2\phi\qquad(3.2)$$

so that the yield condition (3.1_3) is identically satisfied, and then the remaining equations (3.1) are written as

$$\left.\begin{aligned}&-p_{,x}-2k\cos 2\phi\ \phi_{,x}-2k\sin 2\phi\ \phi_{,y}=0,\\&-2k\sin 2\phi\ \phi_{,x}-p_{,y}+2k\cos 2\phi\ \phi_{,y}=0,\\&u_{,x}+v_{,y}=0,\ -\cot 2\phi=(u_{,y}+v_{,x})/(u_{,x}-v_{,y}).\end{aligned}\right\}\qquad(3.3)$$

Clearly, the equations (3.3) are statically determinate and have the matrix representation

$$\left.\begin{aligned}&\mathbf{AU}_x+\mathbf{BU}_y=\mathbf{0},\ \mathbf{CV}_x+\mathbf{DV}_y=\mathbf{0},\\&\text{where }\mathbf{A}=\begin{bmatrix}-1&-\cos 2\phi\\0&-\sin 2\phi\end{bmatrix},\ \mathbf{B}=\begin{bmatrix}0&-\sin 2\phi\\-1&\cos 2\phi\end{bmatrix},\ \mathbf{U}=\begin{bmatrix}p\\2k\phi\end{bmatrix}\\&\mathbf{C}=\begin{bmatrix}1&0\\\cos 2\phi&\sin 2\phi\end{bmatrix},\ \mathbf{D}=\begin{bmatrix}0&1\\\sin 2\phi&-\cos 2\phi\end{bmatrix},\ \mathbf{V}=\begin{bmatrix}u\\v\end{bmatrix}\end{aligned}\right\}\qquad(3.4)$$

Of course, the two matrix equations ($3.4_{1,2}$) may be written formally as one equation, but in this case perhaps all components should be non-dimensionalized. With the governing equations in matrix form, it is straightforward to apply the general results of section 2. Here only the final results are given and they are

$$d_s(p \pm 2k\phi) = 0 \text{ on } x_s/y_s = \cot\phi, -\tan\phi, \tag{3.5}$$

which are the well-known Hencky relations, and

$$\cos\phi d_s u + \sin\phi d_s v = 0, \ \sin\phi d_s u - \cos\phi d_s v = 0 \text{ on } x_s/y_s = \cot\phi, -\tan\phi \tag{3.6}$$

which may be re-cast as

$$d_s\bar{u} - \bar{v} d_s\phi = 0, \ d_s\bar{v} + \bar{u} d_s\phi = 0 \quad \text{on} \quad x_s/y_s = \cot\phi, -\tan\phi, \tag{3.7}$$

these being the well-known Geiringer relations. Thus, the equations are fully hyperbolic, there being two real distinct families of characteristics with four real distinct characteristic relations. The relations (3.5) to (3.7) form the basis for the solution of many problems of interest; for the simpler ones analytical solutions are possible and for the more complex ones numerical methods must be used. Generally, owing to the boundary conditions, the stress and velocity equations must be solved together and somewhat heuristic methods leading to a matrix inversion problem are involved. An elegant algebraic operational method to achieve this is due to Collins [6]. Singularities in solutions due to discontinuities in boundary conditions and discontinuities (for example, in tangential velocity) between different parts of the field in solutions are of common occurrence.

It is a striking fact that very different results indeed obtain when the perfect-plasticity above is replaced by work-hardening plasticity. The equations (3.1) now apply with a non-constant shear yield stress;

$$k = F(\int \mathrm{d}W^p), \tag{3.8}$$

where the integral is the plastic work performed on an element of material.

As before, p and ϕ may be introduced as in (3.2). Straightforward albeit somewhat lengthy analysis (see Pearce [7]) shows that the new equations may be reduced to the form

$$p_{,\alpha} + 2k\phi_{,\alpha} - k_{,\beta} = 0, \ -p_{,\beta} + 2k\phi_{,\beta} + k_{,\alpha} = 0, \tag{3.9$_{1,2}$}$$

$$hq^{-1} q_{,\gamma} = \sin 2\psi k_{,\gamma}, \tag{3.9$_3$}$$

where $q^2 = u^2 + v^2 = \bar{u}^2 + \bar{v}^2$ and $h = 2/3.H', H'$ being the slope of the equivalent stress vs plastic strain curve, and

$$\bar{u}_{,\alpha} - \bar{v}\phi_{,\alpha} = 0, \ \bar{v}_{,\beta} + \bar{u}\phi_{,\beta} = 0, \tag{3.9$_{4,5}$}$$

where subscripts α, β refer to space differentiation along the directions $x_s/y_s = \cot\phi$, $-\tan\phi$ and the subscript γ refers to space differentiation along the direction of particle flow (this direction being inclined at an angle ψ to the α-direction). If k = const. in (3.9) then it is clear that the earlier results for perfect plasticity are recovered. However, the Hencky relations are destroyed whereas the Geiringer relations are preserved in the presence of work-hardening. Also, it is apparent that the (α,β,γ)-directions are characteristic directions with corresponding three characteristic relations as the two Geiringer relations and the single work-hardening relation

$$hq^{-1}\, d_\gamma q - \sin 2\psi\, d_\gamma k = 0. \tag{3.10}$$

The remaining equations $(3.9_{1,2})$ however cannot be uncoupled ($k \neq$ const.) to form a hyperbolic system at all. In more detail, analysis shows that the system (3.9) has merely three distinct families of characteristics, and it does not permit simplification beyond a system comprising three ordinary differential equations along these characteristics and two coupled partial differential equations. The system (3.9) must therefore be treated as it stands. The results (3.9) are essentially first due to Hill (1949, unpublished work) (see [7]). Although very little progress has yet been reported with work-hardening type problems, the way forward now seems clear. Effectively, the presence of work-hardening acts to diffuse velocity discontinuities that are present in work-hardening solutions. This fact suggests that perturbation techniques should be used. Although regular perturbation techniques have been used (see [7]), it is singular pertubation techniques only that can furnish the correct basis for analysis. Collins [8] has provided an introduction to the analysis of work-hardening problems based on singular perturbation techniques involving matched asymptotic expansions. A similar kind of situation must be expected to apply for problems of axially-symmetric plastic flow and so on, and a general study of this entire question is possible and would be valuable.

4. Dynamic situations

For dynamic situations the mathematical structure of the governing equations tends to show extremely strong hyperbolic features and in general there is much complexity according to the constitutive equations and also to the geometry. However, the case when just one space variable is involved has attracted much attention and is of fundamental interest.

Apart from the early work by Donnell (1930) (see [3]) plastic wave theory stems from work undertaken in the early 1940's independently by von Kármán, Taylor and Rakhmatulin (see [3]). The problem posed and solved by Kármán is of especial interest: namely, a semi-infinite bar of elastic-plastic, strain-rate independent material subjected to a step-velocity longitudinal impact at one end. This is now discussed and for simplicity strains will be assumed small and convective effects ignored. Then the governing equations (to within certain approximations) are

$$\sigma_x = \rho v_t,\ \epsilon_t = v_x,\ \sigma = f(\epsilon) \quad \text{(with } \sigma, \epsilon > 0,\ v < 0 \text{ and necking excluded)} \tag{4.1}$$

being respectively the equation of motion, strain-velocity compatibility condition and the stress-strain relation. If σ is eliminated between equations (4.1) then the resulting system is

$$\epsilon_t - v_x = 0,\ \ \epsilon_x - c^{-2} v_t = 0,\ \ \text{where } c(\epsilon) = (f'(\epsilon)/\rho)^{1/2}, \tag{4.2}$$

so that $c(\epsilon)$ is a velocity dependent upon strain. The matrix system is very simple:

$$\mathbf{L} = \mathbf{A}\mathbf{U}_x + \mathbf{B}\mathbf{U}_t = \mathbf{O},$$

$$\text{where } \mathbf{A} = \begin{bmatrix} 0 & -1 \\ 1 & 0 \end{bmatrix}, \quad \mathbf{B} = \begin{bmatrix} 1 & 0 \\ 0 & -c^{-2} \end{bmatrix}, \quad \mathbf{U} = \begin{bmatrix} \epsilon \\ \nu \end{bmatrix} \qquad (4.3)$$

(strictly **U** should be written in non-dimensional form). Application of the general theory in section 2 shows that

$$c d_s \epsilon \mp d_s \nu = 0 \quad \text{on} \quad x_s/t_s = \pm c. \qquad (4.4)$$

Thus the problem is fully hyperbolic, with two real distinct families of characteristics. The Kármán problem is now specified by the data $\epsilon = \nu = 0$ for $x \geq 0$ at $t = -0$ and $\nu = -V_0 H(t)$ for $x = 0$ and $t \geq 0$ where H is the Heaviside unit-step function. Use of the characteristics relations shows straightforwardly that all positive characteristics are straight and signal constant values of strain, velocity and also stress. The solution involves a singularity at $x = t = 0$, and a centred fan of characteristics issues from that point. As the characteristics successively sweep through a material point, its mechanical state changes from the zero initial one to the final one given by

$$V_0 = \int_0^{\epsilon_f} c(\epsilon)\, \mathrm{d}\epsilon, \quad \nu_f = -V_0, \sigma_f = f(\epsilon_f), \qquad (4.5)$$

where the subscript f denotes final value. It is assumed specifically now that the material has a finite yield stress and decreasing hardening-rate with increasing strain and that it does not fracture. In this case, a given material particle experiences first a discontinuous elastic wave front, second a continuous plastic wave system, and finally assumes and maintains a uniform mechanical state. In retrospect it is now obvious that the Kármán problem is one of similarity (in fact, this was the basis of Kármán's own solution to his problem), the solution for strain being solely a function of x/t, as indeed simple dimensional analysis makes clear. It should be noted too that the work-hardening condition requires that $c'(\epsilon) < 0$ so that the characteristics diverge, not converge. This particular question is taken up later.

The availability of the above results prompted further studies, an especially important one being the extension to encompass strain-rate dependent mechanical behaviour which was undertaken independently by Malvern and Sokolovsky (see [3]). The above theory is now altered by the adoption of a different type of constitutive relation, say

$$E\epsilon_t = \sigma_t + g(\sigma, \epsilon), \qquad (4.6)$$

where E is Young's moddulus and g is a given function. Equation (4.6) evidently implies that for very high rates there is elastic behaviour and for very low rates there is plastic behaviour governed by $g = 0$; at intermediate rates there is a varying rate dependency. In this case, the system of governing equations may be written as

$$\mathbf{L} = \mathbf{A}\mathbf{U}_x + \mathbf{B}\mathbf{U}_t + \mathbf{C} = \mathbf{O},$$

where $\mathbf{A} = \begin{bmatrix} 0 & -1 & 0 \\ 0 & 0 & -1 \\ 0 & 0 & 0 \end{bmatrix}$, $\mathbf{B} = \begin{bmatrix} 1 & 0 & 0 \\ 0 & \rho & 0 \\ E & 0 & -1 \end{bmatrix}$, $\mathbf{C} = \begin{bmatrix} 0 \\ 0 \\ -g \end{bmatrix}$, $\mathbf{U} = \begin{bmatrix} \epsilon \\ v \\ \sigma \end{bmatrix}$.

(4.7)

It is straightforward to show that the final characteristics results are

$$\left.\begin{aligned} &Ed_s\epsilon - d_s\sigma - g = 0 \quad \text{on} \quad x_s/t_s = 0, \\ &d_s\sigma - \rho c d_s v + g = 0 \quad \text{on} \quad x_s/t_s = c, \\ &d_s\sigma + \rho c d_s v + g = 0 \quad \text{on} \quad x_s/t_s = -c, \\ &\text{where} \quad c = (E/\rho)^{1/2}. \end{aligned}\right\} \tag{4.8}$$

Thus the system remains fully hyperbolic but now has a fixed (as opposed to floating) set of characteristics. However, this latter simplicity is offset by the fact that the step-velocity impact problem is no longer one of similarity and it has to be solved explicitly at successive times. Nonetheless it is true that the solution tends at long times to the earlier one (a result which may be predicted on dimensional grounds).

Analogous problems of combined loading for circular cylindrical shells have been studied by Clifton and his associates (1966 and later) for strain-rate independent mechanical behaviour and also by Hopkins (1970, unpublished work) for strain-rate dependent mechanical behaviour; and although these involve a larger number of dependent variables, with consequent complexity, the equations are fully hyperbolic and extremely interesting results are obtained. Under the tensile stress conditions assumed above, with increasing severity of impact the situation is in practice terminated by fracture. Under compressive stress conditions, ultimately severe barrelling and possibly instability, neither of which is taken account of above, would occur. The propagation of elastic-plastic waves under compressive conditions is best studied for a semi-infinite region with its plane surface uniformly stressed. A detailed investigation of this type of problem which takes account of the nonlinear compressibility of plastic flow at very high pressures has been given by Morland [9], although numerous other studies have been made. In this case, the analysis essentially parallels that for strain-rate independent material given above, but with a most important difference. This arises because ultimately at sufficiently high values of compressive stress the stress-strain relation shows an upwards-turning curvature and this causes the characteristics to run together. The intersection of characteristics involves the local failure of the differential conservation equations and they must be replaced by finite ones. The result is that a finite leading shock wave front is built up and it propagates into undisturbed material. The pattern of characteristics now becomes quite complicated with the interaction of nonlinear loading and unloading waves. At sufficiently high compressive loading, assumed for simplicity uniform in time, it becomes possible for a plastic shock wave to supervene completely and then a uniform steady shock propagates forward into the material. This situation which may be achieved experimentally by mechanical impact or explosives techniques has much interest since it provides the means of determining constitutive equations valid at extremely high pressures. Work in this area has been extended to discuss questions of the structure of shocks in solids which involves attention to viscosity and heat conduction, the former effect normally being the most important. Of course, it must be noted that ultimately at extremely high pressures the concepts of continuum mechanics themselves will fail and then individual atomic movements must be considered. Hopkins [10] has given an account of work on shock-structure in solids.

5. Engineering applications

Mention has already been made of the parallel developments in approximate engineering theories of plasticity for structure, and an enormous literature now exists on this subject. This cannot be reviewed here, but some brief comments are made as follows. It needs to be noted that successful engineering structural design normally rests upon taking jointly into account a number (not small) of extremely diverse factors. In the present context, work on stability, shakedown (under repeated loads), geometry effects (in relation to load-carrying capacity) (see, for example, Gürkök and Hopkins [11]) and optimum design (see, for example, Prager [12]) are of great importance and interest.

6. Concluding remarks

This paper has reviewed developments in the mathematical theory of plasticity primarily from the particular viewpoint of the structure of the conservation and constitutive equations. In tracing these developments, an indication has been given of the increasing breadth and scope of the subject and of the trends taking place within it. Here, as elsewhere in scientific subjects amenable to theoretical study, it is the commonality of mathematical procedures that is important in promoting a unified and coherent theory, and the degree to which this is being achieved provides a firm test of the real progress being made.

References

1. Hill, R., *The Mathematical Theory of Plasticity*, Clarendon Press, Oxford, 1950.
2. Prager, W. and Hodge, Jr., P.G. *Theory of Perfectly Plastic Solids*, Wiley, New York, 1950.
3. Hopkins, H.G., 'Dynamic Nonelastic Deformations of Metals', *Applied Mechanics Surveys* (edited by H.N. Abramson et al.), p. 847, Spartan, Washington D.C., 1966.
4. Hill, R., 'Aspects of Invariance in Solid Mechanics', *Adv. appl. Mech.* (edited by C.-S. Yih) **18**, 1, 1978.
5. *Mechanical Properties at High Rates of Strain* (edited by J. Harding), The Institute of Physics, London and Bristol, 1974, 1979, in press.
6. Collins, I.F. 'Integral Equation Formulation of Slipline Field Problems', *Proc. ASME Symposium on Numerical Methods in Metal Forming*, 1978, In press.
7. Pearce, J.H.B., *Theory of Plane Strain for a Rigid/Work-hardening Material*, M. Sc. thesis, University of Manchester, 1970.
8. Collins, I.F., 'The Application of Singular Perturbation Techniques to the Analysis of Forming Processes for Strain-hardening Materials', *Proc. IUTAM Symposium on Metal*

Forming Plasticity (edited by H. Lippmann), 1978, In press.
9. Morland, L.W., 'The Propagation of Plane Irrotational Waves through an Elastoplastic Medium', *Phil. Trans. Roy. Soc.* **A251**, 351, 1959.
10. Hopkins, H.G., 'Shock-Structure in Solids', *Shock Tubes* (edited by I.I. Glass), p. 11, University of Toronto Press, 1970.
11. Gürkök, Ayfer and Hopkins, H.G., 'The Effect of Geometry Changes on the Load Carrying Capacity of Beams under Transverse Load', *SIAM J. appl. Math.*, **25**, 500, 1973.
12. Prager, W., 'Optimal Layout of Trusses with Finite Numbers of Joints', *J. Mech. Phys. Solids*, **26**, 241, 1978.

Models for High Temperature Fracture

*F.A. Leckie**

Summary

A study is made of the possible mechanisms which influence the growth of cracks in components operating at temperatures high enough for time-dependent effects to be important. It is suggested that in some situations the life of the component is dictated by considerations of Continuum Damage Mechanics while in others it is the growth of the crack which dominates. Expressions for the rate of crack growth in terms of the J integral are proposed.

1. Introduction

The techniques of Fracture Mechanics are used to estimate the conditions under which flaws in load-bearing components are large enough for the system to become unstable. In high strength steels, fracture occurs before the onset of extensive plastic yielding and estimates of critical crack length are obtained by applying the techniques of Linear Elastic Fracture Mechanics (LEFM). For the less strong metals, extensive plastic deformation takes place before the onset of fracture and it is then necessary to use the concepts of Post Yield Fracture Mechanics.

By comparison with the state of Linear and Post Yield Fracture Mechanics, the fracture mechanics associated with high temperature is still in a primative condition. The problem is to estimate the length of time a component with a crack can sustain load. The time element is introduced because time-dependent material damage takes place in the form of cracks and voids which nucleate and grow on the grain boundaries. To date, it would appear that the problem has been tackled by assuming that the crack grows until it reaches a critical length when classical fracture occurs. Within this approach, two different parameters have been assumed to define the crack growth. Siverns and Price [1] suggested that it is the stress intensity factor K of LEFM which is the controlling parameter. Other investigators, Webster [2], Landes and Begley [3], and Harper and Ellison [4], suggest that the controlling influence is the value of the so-called J integral introduced by Rice [5].

Another possible procedure is to attempt to extend the concepts developed in Continuum Damage Mechanics (CDM), a subject developed to study the growth of time-dependent damage which occurs in load-bearing components. In these studies, attempts have been made to measure the growth of creep damage which occurs on the grain boundaries and to include the effect of this damage on creep deformation rates. The resulting mechanics can be used to calculate the distribution and growth of damage and the stress redistribution which occurs. The predictions of this approach have been compared with the results of experiments on a variety of simple components and agreement with the predictions of theory is found to

* College of Engineering, University of Illinois at Urbana-Champaign, Urbana, Il 61801

be good. Initially, components were studied which have stress concentrations of a level normally occurring in practice (Leckie and Hayhurst [6]). The theory is also found to be satisfactory for predicting the behavior of heavily grooved bars (Hayhurst, Leckie, and Morrison [7]) which have much higher values of stress concentration factor. The theory has not been applied to components with cracks but the possibility of successfully using the theory of CDM in these extreme conditions cannot be excluded because it is observed that damage in the vicinity of cracks also occurs on the grain boundaries rather than as the transgranular damage characteristic of classical fracture. An attempt to explore this possibility will be made in this paper. To solve the problem exactly would demand a very detailed numerical study and instead of following this approach, an attempt will be made to solve the problem in an approximate way so that some insight may be gained into identifying those factors which play a dominate role.

2. Constitutive Equations

It is assumed that the total strain rate is the sum of the elastic, creep, and plastic strain rates so that

$$\dot{\epsilon}_{ij} = \dot{\epsilon}^e_{ij} + \dot{\epsilon}^c_{ij} + \dot{\epsilon}^p_{ij}. \tag{2.1}$$

The separation of irreversible strains into time-dependent creep strains and time-independent plastic strains has been justified by Onat [8] on the grounds that for straining rates in excess of 0.3/hr, the stress/strain diagram is almost independent of the strain rate. The rapid increase in strain rate for small increments of stress could be represented by some suitable function of stress, but it proves convenient analytically to make the separation into time-dependent and time-independent strain components.

The relationship between elastic strain rate and stress is

$$\dot{\epsilon}^e_{ij} = C_{ijkl}\,\dot{\sigma}_{kl} \tag{2.2}$$

where C_{ijkl} is a constant tensor. The plastic strain rate is given by the relationship

$$\dot{\epsilon}^p_{ij} = \lambda(\partial f/\partial \sigma_{ij}) \tag{2.3a}$$

where the yield condition is

$$f(\sigma_{ij}) = 0 \tag{2.3b}$$

and λ is an indeterminate scalar quantity. The creep strain rate is given by

$$\dot{\epsilon}^c_{ij}/\dot{\epsilon}_0 = (3/2)\,[(\overline{\sigma}/\psi\,\sigma_0)^{n-1}]\,(\sigma'_{ij}/\sigma_0\,\psi). \tag{2.4a}$$

In this expression, $\overline{\sigma}$ is the effective stress and σ'_{ij} the stress deviator. Also,

$$\psi = 1 - \mathrm{w}, \tag{2.4b}$$

where ω is a measure of damage on the grain boundary. In the uniaxial stress state, the equation reduces to the form

$$\dot{\epsilon}^c/\dot{\epsilon}_0 = (\sigma/\psi\,\sigma_0)^n \tag{2.4c}$$

in which $\dot{\epsilon}_0$, σ_0, and n are constants. It can be seen that as the damage ω increases, the value of ψ decreases so that in a constant stress test, the creep rate increases with time and the equation models the tertiary portion of the creep curve.

The evolution equation for ψ is given by

$$d\psi/dt = -[(\lambda - 1)/(\lambda n t_0)]\,\Delta^\eta(\sigma_{ij}/\sigma_0)\,(1/\psi_\eta), \tag{2.5a}$$

where $\eta = [\lambda n/(\lambda - 1)] - 1$ and $\lambda = [\epsilon_0/(\dot{\epsilon}_0\, t_0)]$ illustrated in Fig. 1 is known as the creep ductility.

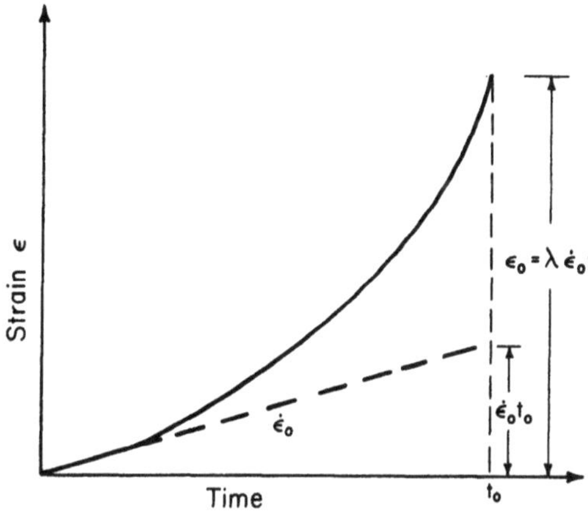

Fig. 1. Tertiary Creep Strains.

At $t = 0$, when the material is undamaged, $\psi = 1$ and at rupture, $\psi = 0$. The constant t_0 is the rupture time of an uniaxial test conducted at constant stress σ_0, and ν is the slope of the log stress/log rupture time graph. The function $\Delta(\sigma_{ij}/\sigma_0)$ is homogeneous of degree one and is referred to as the Isochronous Rupture Surface. This surface in stress space defines the constant multiaxial stresses which cause the material to fail in rupture time t_0.

The form of the function Δ varies for different materials. For example, experiment shows that for some aluminium alloys, the function

$$\Delta(\sigma_{ij}/\sigma_0) = \overline{\sigma}/\sigma_0, \tag{2.5b}$$

where $\overline{\sigma}$ is the effective stress. Copper behaves in a completely different way and the isochronous surface is given by

$$\Delta(\sigma_{ij}/\sigma_0) = 0.83\,(\sigma_1/\sigma_0) + 0.17\,(\overline{\sigma}/\sigma_0), \tag{2.5c}$$

where σ_1 is the maximum tensile stress.

The constitutive equations have been selected to fit the results of multiaxial stress experiments and no attempt is made to relate the actual physical damage to the parameter. In fact, it has been shown that a direct relation does exist but this point will not be pursued at this time.

An important prediction of the constitutive equations is that the strain to failure ϵ_Δ can be greatly dependent on the form of the applied stress and on the isochronous surface Δ. The equations are simpler to integrate if it is assumed that $n = \nu$. With this assumption, it is not difficult to obtain the result

$$\epsilon_\Delta/\epsilon_0 = [(\overline{\sigma}/\sigma_0)/\Delta]^n, \tag{2.6a}$$

where ϵ_Δ is the strain at rupture when the multiaxial stress is applied and ϵ_0 is the uniaxial rupture strain.

When $\Delta = \overline{\sigma}/\sigma_0$, as it is for aluminum, then the rupture strain is independent of stress state. For copper, however, the strain at failure is

$$\epsilon_\Delta/\epsilon_0 = [1/(0.83\,(\sigma_1/\overline{\sigma}) + 0.17)]^{5.9}. \tag{2.6b}$$

Leckie [9] compared the predictions of this formula with the experimental results of a variety of multiaxial stress states and found the comparison to be satisfactory. The important consequence of the formula (Eq. (2.6b)) is that for conditions of high constraint when $\sigma_1/\overline{\sigma}$ is high, the strain at failure can be very small indeed.

3. Discussion of Local and Continuous Damage

Experiments on plane stress plates penetrated by slits were performed by Leckie and Hayhurst [6] and demonstrated that failure was the result of damage which was continuously spread throughout the specimen. The life was independent of the acuity of the crack and suggests that the events occurring in the vicinity of the crack were unimportant. The life of the component was dictated by the laws of CDM which predicts that for plane stress specimens, the life of the plates depend only on the average stress across the minimum cross-section. Micrographic examination of the specimen further demonstrates the continuous nature of the damage.

It is not necessarily certain that the same behavior will occur in plane strain specimens since more constraint is expected and this constraint may stimulate mechanisms which cause crack growth. At the same time, continuum damage will occur throughout the loaded specimen. It would appear sensible, therefore, to divide the study of the problem into two separate parts.

In the first of these, it is assumed that the crack does not propagate and that damage occurs on a continuum scale. The growth of damage and the failure life will be dictated by the laws of CDM applied to a cracked specimen in which the crack does not propagate.

The second approach is based on the assumption that the life of the component is dictated by crack growth and that in the process of crack growth, little damage occurs on the continuum level.

In practice, however, it would be expected that both mechanics combine to influence the failure life. However, by separating the mechanism in the manner suggested, it might be possible to simplify the problem sufficiently to obtain a fuller understanding of the processes influencing failure life. This procedure will be applied to plane strain specimens with symmetric internal and external cracks (Fig. 2).

When studying the behavior of the cracked plates, it is sometimes convenient to introduce the concept of weakening and strengthening in which comparison is made with the stress σ_0 which causes time to failure t_0 in a uniaxial specimen. If the average stress at the minimum cross section requires to cause failure in time t_0 is σ_n, then strengthening is said to occur when $\sigma_n/\sigma_0 > 1$ and weakening when $\sigma_n/\sigma_0 < 1$.

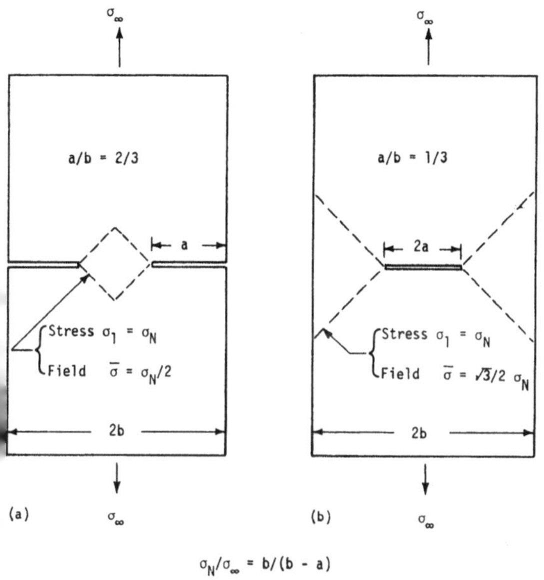

Fig. 2. Plastic Stress Fields for Plane Strain Specimens with (a) External Crack and (b) Internal Crack

4. Continuum Damage Mechanics (CDM)

Studies in CDM [6] show that the damage which occurs in the initially highly stressed portions of loaded components results in weakened material which sheds the high stress to region of lower stress. It is found that considerable stress redistribution takes place and that, as a result, good estimates of rupture life are obtained by assuming a stress field which is given by suitably scaled limit load calculations in conjunction with the appropriate isochronous surface of the material. These calculations are now performed for the specimens illustrated in Fig. 2 when made from alumimum and from copper. It is assumed that the mechanisms causing crack growth are not stimulated.

4.1. Plane Strain Specimens with Internal Crack

The plastic stress field for the internally cracked specimen is given in Fig. 2. The maximum stress $\sigma_1 = \sigma_n$ and the effective stress $\bar{\sigma} = [(3^{1/2})/2]\sigma_n$ where σ_n is the average stress at the minimum cross sectional area.

To determine the value of σ_n for the aluminum specimen to cause failure in time t_0 this stress field is substituted into the condition (Eq. (2.5b)) to give $(\bar{\sigma}/\sigma_0) = [(3^{1/2})/2](\sigma_n/\sigma_0) = 1$ from which

$$\sigma_n = 1.15\ \sigma_0. \tag{4.1}$$

The similar calculation for copper uses the rupture condition of Eq. (2.5c) so that $0.83\ (\sigma_1/\sigma_0) + 0.17\ (\bar{\sigma}/\sigma_0) = 0.977\ (\sigma_n/\sigma_0) = 1$ from which

$$\sigma_n = 1.02\ \sigma_0. \tag{4.2}$$

4.2. Plane Strain Specimen with External Cracks

The plastic stress field for the external notch in plane strain is given in Fig. 2. The maximum stress $\sigma_1 = \sigma_n$ while the effective stress $\bar{\sigma} = \sigma_n/2$.

Using these values of stress in the rupture condition (Eq. (2.5b)) for aluminum gives the results $\bar{\sigma}/\sigma_0 = \sigma_n/2\sigma_0 = 1$ from which

$$\sigma_n = 2\sigma_0. \tag{4.3}$$

The similar calculation for copper using the rupture condition (Eq. (2.5c)) gives the result $0.83\ (\sigma_n/\sigma_0) + 0.17\ (\bar{\sigma}/\sigma_0) = 0.915\ (\sigma_n/\sigma_0) = 1$ from which

$$\sigma_n = 1.09\ \sigma_0. \tag{4.4}$$

These calculations indicate that for aluminum specimens, dramatic strengthening might be expected for externally cracked members. A significant strengthening is also evident for internally cracked specimens. The strengthening factors for copper specimens are not so large and on the average are slightly greater than unity.

5. Crack Growth Models

In these studies, it is assumed that the dominant mode of failure is crack growth and that no damage takes place except in the vicinity of the crack. If it is assumed, therefore, that $\psi = 1$ in Eq. (2.4a) for the creep rate. The determination of the stress and strain fields will reduce to a standard time-dependent boundary value problem for elastic, creep, and plastic deformation.

On initial loading, the strains will be elastic and plastic but, as time advances, creep strains grow. For such problems, it has been shown by Leckie and Ponter [10] that a constant steady state stress field is established as time approaches infinity. The condition defining the stress rate σ_{ij}^{sp} is

$$\int_v \dot{D}(\sigma_{ij}^{sp})\ \mathrm{d}v = \text{minimum} \tag{5.1a}$$

subject to the condition

$$f(\sigma_{ij}^{sp}) \leq 0, \tag{5.1b}$$

where $\dot{D}$ is the creep energy dissipation rate per unit volume. If the volume of material which is yielding in the vicinity of the crack is small enough to be neglected by comparison with the non-yielding region, then the minimization process will be satisfied by the classical steady state stress field σ_{ij}^s which exists in the absence of plastic effects. This assumption, which is equivalent to the small scale yielding assumption in classical fracture, gives the result that the far distant stress and strain rate fields distance from the crack of length a are given by the formulae

$$\sigma_{ij}^s/\sigma_0 = K_\sigma (a/r)^{1/(n+1)} f_{ij}(\theta) \tag{5.2a}$$

$$\dot{\epsilon}_{ij}^s/\dot{\epsilon}_0 = K_\epsilon (a/r)^{n/(n+1)} g_{ij}(\theta) \tag{5.2b}$$

where

$$\dot{J}/(a\, \sigma_0\, \dot{\epsilon}_0\, I_n) = K_\sigma K_\epsilon = K_\sigma^{n+1} = K_\epsilon^{(n+1)/n} \tag{5.2c}$$

is the path independent integral defined by Rice [5] (Fig. 3) and I_n is a constant dependent on the value of n (Hutchinson [11]).

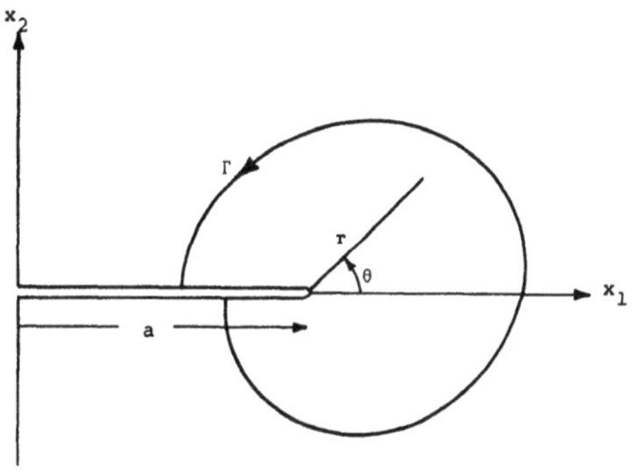

$$\dot{\bar{\epsilon}}/\dot{\epsilon}_0 = (\bar{\sigma}/\sigma_0)^n;\ \sigma_{ij}^s = \partial \dot{w}/\partial \dot{\epsilon}_{ij}^s,\ \dot{w} = \sigma_0\, \dot{\epsilon}_0 (\bar{\sigma}/\sigma_0)^{n+1}$$

$$\dot{J} = \int_\Gamma [\dot{w}\, dx_2 - \sigma_{ij}\, n_j\, (\partial \dot{u}_i / \partial x_1)\, ds]$$

$$\sigma_{ij}^s/\sigma_0 = K_\sigma (a/r)^{1/(n+1)} f_{ij}(\theta),\ \dot{\epsilon}_{ij}^s/\dot{\epsilon}_0 = K_\epsilon (a/r)^{n/(n+1)} g_{ij}(\theta)$$

Fig. 3. Stress and Strain Rate Singular Fields in Vicinity of Crack.

The plastic stress and strain fields in the immediate vicinity of a crack have been extensively studied by Rice and co-workers. The analyses indicate that there are two highly strained regions. One region consists of a concentrated shear band of length 1 at an angle of approximately 70 degrees to the crack. The value of the angle varies with the specimen geometry but the variation from 70 degrees is not great. From the solution obtained by Rice [12], the relationship between the length of the shear band 1 and the crack opening displacement is given by

$$(l/\delta) = 0.39\ (E/\sigma_y). \tag{5.3}$$

The other high strains occur ahead of the crack have been calculated by McMeeking [13] using finite element procedures. The stress and strain fields from the McMeeking analysis are shown in Fig. 4 from which it will be noted that it is the value of the crack opening displacement which dictates the extent of the plastic field ahead of the crack.

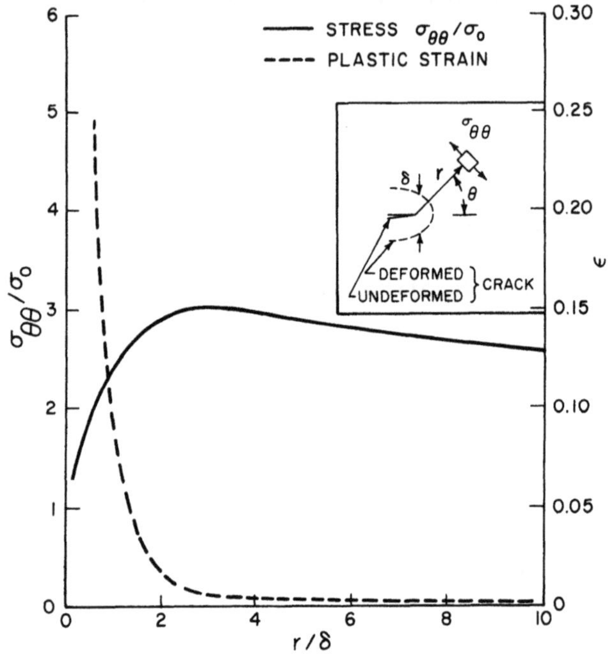

Fig. 4. Stress and Strain Fields in Front of Crack.

In order to match the plastic fields which have been described with the far distant creep field, it is assumed that the crack opening displacements from both the elastic/plastic and creep fields are equal. This is no more than a crude assumption based on the observation that kinematic constraint is likely in the regions local to the crack. For high values of creep index n, it has been suggested by Rice and Rosengren [14] that the relationship between J and crack opening displacement rate $\dot{\delta}$ is approximately

$$\dot{\delta} = \dot{J}/\sigma_0. \tag{5.4}$$

5.1. Crack Growth in Aluminum

Since it is the effective shear stress which is responsible for creep damage in aluminum, it is to be expected that the crack will grow at an angle of approximately 70 degrees to the direction of the original crack. Furthermore, since the effective stress on the slip bank is equal to the yield stress, it is to be expected that the damage will occur very rapidly. If it is assumed that creep damage is instantaneously formed, then from Eq. (5.3), $dl/dt = 0.39 (E/\sigma_y)(d\delta/dt)$ since the band is being continuously loaded. For a high strength aluminum alloy, $E/\sigma_y = 700.0$ so that, together with Eq. (5.4), the estimate of crack growth rate is

$$dl/dt = 240\,(\dot{J}/\sigma_0) \tag{5.5}$$

and grows along a line which is approximately 70 degrees to the line of the original crack.

5.2. Crack Growth in Copper

The stress and strain values in front of the crack are shown in Fig. 4. It can be seen that the maximum value of the stress ratio $\sigma_1/\overline{\sigma} = 3.0$ and occurs at a point which is about three times the crack opening from the crack point. The strain distribution has an infinite value at the crack front which drops off rapidly at distances a few times the size of the crack opening δ. Although the strain at the crack front is infinite, it is known that damage occurs on the grain boundaries and it might be expected therefore that the criterion for failure in the vicinity of the region of the crack front is fulfilled when the strain about one grain distance in front of the crack is ϵ_0 where ϵ_0 is the rupture failure strain in uniaxial tension. Using the rupture strain condition (Eq. (2.6b)) for multiaxial states at a distance from the crack front three times the crack opening, the strain to cause failure is $\epsilon_0/320$ of the strain to cause failure at the crack front. Using the McMeeking [13] results of strain distribution and for a grain size of .75 mm, it can be shown that the strain failure condition at a distance three times the crack opening is the more severe. It is possible to entertain the theory, therefore, that complete damage first occurs at a distance ahead of the crack three times the crack opening displacement. As the crack continues to grow, the damage will advance at a rate three times the crack opening until the average strain on the first grain will be approximately ϵ_0.

Hence, this model predicts a crack growth rate $dl/dt = 3(d\delta/dt)$ and, using Eq. (5.4), the expression then becomes

$$dl/dt = 3(\dot{J}/\sigma_0). \tag{5.6}$$

While this equation expresses the crack growth in a continuous form, it should be noted that the growth according to this model proceeds in discrete steps. In addition, an incubation period will elapse before crack growth becomes evident.

6. Life Predictions of Crack Advance Models

The creep crack growth models of the previous chapter predict a growth rate for aluminum

$$dl/dt = 240.0\,(\dot{J}/\sigma_0) \tag{6.1}$$

with the crack opening at an angle of approximately 70 degrees to the original crack orientation.

For copper, the crack growth is in a forward direction but at the much slower rate

$$dl/dt = 3.0\,(\dot{J}/\sigma_0). \tag{6.2}$$

With these values for crack growth rate and the current value of J, it is then possible to estimate the failure life assuming that it is the crack growth which is the dominant mechanism. Values of $\dot{J}$ for specimens with symmetric internal and external cracks have been calculated by Ranaweera and Leckie [15] so that it is possible to complete the crack growth calculation.

6.1. Life of Aluminum Specimens

Integrating Eq. (6.1), the failure time can be determined by assuming that failure occurs when the crack has advanced across the full cross section. Because $\dot{J}$ increases with crack length, the crack growth rate increases so rapidly for long cracks that the failure life is only very slightly dependent on the failure condition used. For the external cracked specimen

$$t_R = 0.0408/\dot{\epsilon}_0\,[(3^{1/2}/2)(\sigma_n/\sigma_0)]^{n+1}. \tag{6.3}$$

To find the strengthening factor, the failure time, t_R, is selected to be t_0. It is also known from definition that $\epsilon_0 = \lambda\,\dot{\epsilon}_0\,t_0$. For an aluminum with $\epsilon_0 = 0.07$, $\lambda = 2.5$, and $n = 10$, Eq. (6.3) gives a strengthening factor

$$\sigma_n/\sigma_0 = 1.21. \tag{6.4}$$

Repeating the same calculation for the internally cracked specimen gives the strengthening factor

$$\sigma_n/\sigma_0 = 0.90. \tag{6.5}$$

6.2. Life of Copper Specimens

The same procedure can be followed for the copper specimens. For copper, $\epsilon_0 = 0.07$, $\lambda = 4$, and $n = 5.9$. For the externally cracked geometry, the strengthening factor is

$$\sigma_n/\sigma_0 = 1.78 \tag{6.6}$$

while for the internally cracked specimen

$$\sigma_n/\sigma_0 = 1.34. \tag{6.7}$$

7. Conclusions

Two mechanisms of failure have been studied for load-bearing components operating at temperatures sufficiently high for time-dependent effects to be important. One mechanism of failure is described by CDM and the other by Crack Growth.

Strengthening factors have been determined applying both theories to components which are penetrated internally and externally with symmetric cracks. The results are given in Table 1. From this table, it can be deduced that in aluminum specimens, it is the crack growth mechanisms which dominates over the continuum damage. In copper, the opposite effect is apparent with the CDM being the more dominant. A more comprehensive theory will include the coupling effects of the two mechanisms but this is a much more difficult problem and the above investigation gives some pointers for a controlled experimental investigation.

Table 1

Material	*Al*		*Cu*	
Model	*CDM*	*C.G.*	*CDM*	*C.G.*
Internal Crack	1.15	0.90	1.02	1.34
External Crack	2.0	1.21	1.09	1.78

Bibliography

1. Siverns, M.J., and A.T. Price, *Nature*, 1970, pp. 228, 760.
2. Webster, G.A., *Conf. Mech. and Physics of Fracture*, 1975, Paper No. 18, Inst. of Physics, London.
3. Landes, J.D., and J. Begley, 1974 Westinghouse Scientific Paper, 74-1E7-FESGT-P1.
4. Harper, M.P., and E.A. Ellison, 1977 Strain Analysis, 12(3), p. 167.
5. Rice, J.R., 1967, *J. Applied Mech.*, Vol. 34, p. 237.
6. Leckie, F.A., and D.R. Hayhurst, 1974, *Proc., Roy. Soc. of London*, A340.
7. Hayhurst, D.R., F.A. Leckie, and C.J. Morrison, *Proc., Roy. Soc. of London*, 1978, A360, p. 243.
8. Onat, E.T., 1976, Yale University Report, 'Representation of Inelastic Behavior,' ORNL-SUB-3863-2.
9. Leckie, F.A., 1978, *Trans. Roy. Soc. of London*, Vol. 288, p. 27.
10. Leckie, F.A., and A.R.S. Ponter, 1970, *J. Applied Mech.*, Vol. 37, 1p. 426.
11. Hutchinson, J.W., 1968, *J. Mech. Phys. Solids*, 13 Cond., p. 337.
12. Rice, J.R., 1974, *J. Mech. Phys. Solids*, Vol. 22, p. 17.
13. McMeeking, R.M., 1977, *J. Mech. Phys. Solid*, Vol. 25, p. 357.
14. Rice, J.R., and G.F. Rosengren, 1968, *J. Mech. Phys. Solids*, Vol. 16, p. 1.
15. Ranaweera, M.D., and F.A. Leckie, 1979, to appear in *Comp. Meth., Applied Mech. and Engr.*

Some Mathematical Problems Connected with Solid Mechanics

*J.L. Lions**

Summary

Some general mathematical considerations, directed towards applications, are given. The methods are abstract, based on functional analysis. Two methods are discussed, that of the Variational Inequalities and that of the Asymptotic Expansions for periodic structures. They are applied in the classical obstacle problem, in plasticity theory and at perforated materials.

Introduction

In this paper, which is of an expository nature, we present a *few examples* of applications of functional analysis to problems arising in Mechanics and more precisely in *Solid Mechanics*(1).

We present here *two methods* which have proven to be useful, namely the Variational Inequalities(2) and the *Asymptotic expansions for periodic structures*. In section 1 we formulate a very simple looking problem, the classical *obstacle problem*: a weak solution of this problem can be obtained by using the V.I. technique; we give a brief indication on the *penalty method* which can be used to prove a regularity result.

In section 2 we present some remarks on problems arising from *elasto-plasticity*. They lead to V.I. (where there are still some open questions) and in the case of perfect plasticity to the need of introducing *new functional spaces* (cf. Section 2.3).

In section 3 we explain how a properly chosen ansatz can lead to the solution of a number of questions arising in materials with a periodic structure (composite materials, perforated materials); one of the first works along these lines for perforated materials, by analytic computations, is due to Koiter [1]. As we said, this is *a very small part* of the interesting mathematical problems which arise in connection with solid mechanics. To the references already given above we wish to add here the book by P.H. Rabinowitz, ed. [1] (and the paper M. Crandall and P.H. Rabinowitz [1]), on bifuraction theory and by Ph. Ciarlet [1] on elasticity theory and on theory of plates and shells, a subject where Prof. Koiter has contributed so much.

1. The obstacle problem

Let Ω be an open set of $I\!R^n$ ($n = 2$ or $n = 3$ in the applications) with boundary Γ; set in a formal fashion, we want to find a function $u = u(x)$ defined in Ω such that:

* College de France and Laboria

(1) The notes preceed the bibliography.

$$\left.\begin{array}{l} -\Delta u - f \geq 0, \quad u - \psi \geq 0, \\ \\ [-\Delta u - f]\,[u - \psi] = 0 \text{ in } \Omega \end{array}\right\} \quad (1.1)$$

subject to:

$$u = 0 \text{ on } \Gamma, \quad (1.2)$$

where in (1.1) f and ψ are given functions[3]. Formally, if $n = 2$, (1.1) (1.2) describes the equilibrium of a membrane, subject to forces f, fixed on Γ, and *subject to be above the obstacle represented by* ψ. In general, there will be a region Ω_0 in Ω (the *contact region*) where $u = \psi$; in the other part (the *equilibrium region*) we shall have $u > \psi$ and $-\Delta u = f$. The interface S between the two regions is a *free surface*. In a number of applications[4] one meets the following generalization of (1.1) (1.2); let A be a second order elliptic opterator given by:

$$A = -\frac{\partial}{\partial x_i}\left(a_{ij}(x)\,\frac{\partial}{\partial x_j}\right),^{(5)} \quad (1.3)$$

where:

$$\left.\begin{array}{l} a_{ij} \in L^{\infty}(\Omega)^{(6)}, \\ \\ a_{ij}(x)\,\xi_i\xi_j \geq \alpha\,\xi_i\xi_j, \forall \xi_i \in I\!R, \alpha > 0, \;\; \text{a.e. in } x. \end{array}\right\} \quad (1.4)$$

There one looks for u solution of:

$$\left.\begin{array}{l} Au - f \geq 0, \quad u - \psi \geq 0, \\ \\ [Au - f]\,[u - \psi] = 0 \;\text{ in }\; \Omega \end{array}\right\} \quad (1.5)$$

with the Boundary condition (1.2). It is again a *free boundary problem*.

We now proceed with the solution of (1.5) (1.2), by *using the technique of* V.I. Let us recall first a general result about *V.I.*[7]. Let V be a real Hilbert space; its norm is indicated by $\| \; \|$; let $a(u,v)$ be a continuous bilinear form on V, which is coercive, i.e.:

$$a(v,v) \geq \alpha\,\|v\|^2, \;\alpha > 0, \;\forall v \in V; \quad (1.6)$$

let K be a non empty closed convex subset of V; let $v \to (f,v)$ be a continuous linear form on V; then, *there exists a unique element u such that*:

$$\left.\begin{array}{l} u \in K, \\ \\ a(u, v-u) \geq (f, v-u) \;\; \forall\, v \in K; \end{array}\right\} \quad (1.7)$$

(1.7) is what is called a V.I.

Remark 1.1 In the *symmetric case* ($a(u,v) = a(v,u) \;\forall\, u,v$), (1.7) is a characterization of the unique element u in K which minimizes over K the functional:

$$J(v) = \frac{1}{2} a(v,v) - (f,v). \tag{1.8}$$

Remark 1.2 If $K = V$, (1.7) reduces to:

$$a(u,v) = (f,v) \; \forall \, v \in V \tag{1.9}$$

and the existence and uniqueness of u in V satisfying (1.9) is the Lax-Milgram's lemma.

We now indicate how problem (1.5) (1.2) *fits into the general framework of* V.I. We introduce[8]:

$$H^1(\Omega) = \{v \mid v, \frac{\partial v}{\partial x_1}, \dots, \frac{\partial v}{\partial x_n} \in L^2(\Omega)\}, \tag{1.10}$$

provided with the Hilbertian norm:

$$\|v\| = (\int_\Omega [v^2 + |\Delta v|^2] \, dx)^{1/2}, \tag{1.11}$$

$$\Delta v = \left\{ \frac{\partial v}{\partial x_1}, \dots, \frac{\partial v}{\partial x_n} \right\},$$

and we define:

$$H_0^1(\Omega) = \{v \mid v \in H^1(\Omega), \quad v = 0 \text{ on } \Gamma\}^{(9)}. \tag{1.12}$$

For $u, v \in H_0^1(\Omega) = V$, we define:

$$a(u,v) = \int_\Omega a_{ij}(x) \frac{\partial u}{\partial x_j} \frac{\partial v}{\partial x_i} \, dx. \tag{1.13}$$

We assume that:

$$f \in L^2(\Omega), \tag{1.14}$$

$$\psi \in H^1(\Omega), \; \psi \leq 0 \text{ on } \Gamma. \tag{1.15}$$

We set:

$$(f,v) = \int_\Omega f v \, dx \tag{1.16}$$

and we define:

$$K = \{v \mid v \in H_0^1(\Omega) = V, \; v \geq \psi \text{ a.e. in } \Omega\}. \tag{1.17}$$

By virtue of (1.15), $0 \in K$ so that $K \neq \phi$ and K is a *closed convex subset of* V. By virtue of (1.4), condition (1.6) is satisfied; and therefore *one can apply* (1.7). It is now a simple matter to check the following:

$$\left.\begin{array}{l}\textit{if } u \textit{ is a smooth solution of } (1.5)\,(1.2) \textit{ then it satisfies } (1.7)\\ \textit{for the choices } V = H_0^1(\Omega), \textit{ and } a(u,v), (f,v) \textit{ and } K \textit{ given by}\\ (1.13), (1.16), (1.17)\end{array}\right\} \tag{1.18}$$

Reciprocally, if u is solution of (1.7) and $u \in H^2(\Omega)$[10] then it satisfies (1.5) (1.2). Let us now summarize where we stand:

1. we know the existence and uniqueness of u satisfying (1.7) for the choices (1.13), (1.16), (1.17); *this function can be thought of as a weak solution* of[11] (1.5), (1.2);
2. if moreover one can prove that the solution of (1.7) belongs to $H^2(\Omega)$[12], then u is solution of (1.5), (1.2). Let us give briefly an indication on this property[13].

We use here *the penalty method*[14]: for every $\epsilon > 0$, we consider the (non linear) equation:

$$\left.\begin{array}{l} Au_\epsilon - \dfrac{1}{\epsilon}(u_\epsilon - \psi)^- = f^{(15)},\\ u_\epsilon = 0 \text{ on } \Gamma. \end{array}\right\} \tag{1.19}$$

Since the operation:

$$v \to -\frac{1}{\epsilon}(v-\psi)^- = \beta(v)$$

is monotone (i.e.

$$(\beta(v) - \beta(\hat{v}),\ v - \hat{v}) \geq 0),$$

the equation (1.19) admits a unique solution. One can prove (cf. the paper quoted above) that:

$$u_\epsilon \to u \text{ in } H_0^1(\Omega) \text{ as } \epsilon \to 0^{(16)}. \tag{1.20}$$

Let us show how one can obtain from (1.9) a *regularity theorem*. We take the scalar product of (1.19) with $(u_\epsilon - \psi)^-$; on $\Gamma (u_\epsilon - \psi)^- = (-\psi)^- = 0$ since $\psi \leq 0$, so that (1.19) gives (after being written $A(u_\epsilon - \psi) - \frac{1}{\epsilon}(u_\epsilon - \psi)^- = (f - A\psi)$):

$$a(u_\epsilon - \psi, (u_\epsilon - \psi)^-) - \frac{1}{\epsilon}\,|(u_\epsilon - \psi)^-|^2 = (f - A\psi, (u_\epsilon - \psi)^-)^{(17)}. \tag{1.21}$$

But:

$$a(v, v^-) = -\,a(v^-, v^-) \tag{1.22}$$

so that (1.21) gives:

$$a((u_\epsilon - \psi)^-, (u_\epsilon - \psi)^-) + \frac{1}{\epsilon}\,|(u_\epsilon - \psi)^-|^2 = -(f - A\psi, (u_\epsilon - \psi)^-) \leq |f - A\psi|\,|(u_\epsilon - \psi)^-| \tag{1.23}$$

hence it follows that, if $A\psi \in L^2(\Omega)$,

$$|\frac{1}{\epsilon}(u_\epsilon - \psi)^-| \leq C. \tag{1.24}$$

Therefore (1.19) and (1.24) show that, as $\epsilon \to 0$,

$$Au_\epsilon \text{ remains in a bounded set of } L^2(\Omega). \tag{1.25}$$

Consequently, if $A\psi \in L^2(\Omega)$, *then*:

$$Au \in L^2(\Omega).$$

Therefore, *if the coefficients of A are smooth enough, together with* Γ, *it follows from* (1.26), (1.2) *and classical results about the regularity of the Dirichlet problem, that*:

$$u \in H^2(\Omega). \tag{1.27}$$

We end this section with a few remarks:

Remark 1.3 In the (weak) formulation (1.7) by *V.I.* techniques, *the free boundary 'disappears' from the formulation*. Therefore the next question is: how to recover informations on the free boundary, and, in particular, on *the regularity of the free boundary*. For the obstacle problem, this problem has been studied first in H. Lewy and G. Stampacchia [1]. General methods for the study of the regularity of free boundaries are given in D. Kinderlehrer and L. Nirenberg [1].

Remark 1.4 Problems of type (1.5), (1.2) (and also with other boundary conditions) arise in (in fact are equivalent to) *optimal stopping time problems*. Cf. A. Bensoussan and J.L. Lions [1] [2] [3].

Remark 1.5 For the numerical approximation of problems considered in this section and the following one, we refer to R. Glowinski, J.L. Lions, R. Tremolières [1] and to the Bibliography therein.

2. Problems connected with plasticity theory

2.1. Another V.I.

Problems of elastic-plastic torsion lead (in dimension 2) to the following *V.I.* (Cf. H. Lanchon [1], Ting [1], G. Duvaut and J.L. Lions [1]). We define:

$$K = \{v \mid v \in H_0^1(\Omega, \mid \nabla v(x) \mid \leq 1 \text{ a.e.}\} \tag{2.1}$$

K is a non empty closed convex subset of $H_0^1(\Omega)$)

$$a(u,v) = \int_\Omega \nabla u . \nabla v \, dx \tag{2.2}$$

and we consider *the solution* of the *V.I.*

$$a(u, v-u) \geq (f, v-u) \quad \forall v \in K, \quad u \in K. \tag{2.3}$$

Like in all problems of *V.I.* there is a 'hidden' free surface in the formulation (2.3): *there are two regions in* Ω (in general); *in the plasticity region*, one has:

$$|\nabla u(x)| = 1 \tag{2.4}$$

and the free boundary consists in the boundary of the plastic region. The *regularity of the solution* is here more difficult than for the obstacle problem. We refer to H. Brézis and G Stampacchia [1], and to L.A. Caffarelli and A. Friedman [1], [2], L.A. Caffarelli and N Rivière [1].

2.2 An elastoplastic plane stress problem

In the paper L.C. Evans and B.F. Knerr [1] the study of a thin elastoplastic plate is reduced to the study of the following *fourth order V.I.* One defines in $\Omega \subset \mathbb{R}^2$

$$K_\lambda = \{v \mid v \in H^2(\Omega), v = h, \frac{\partial u}{\partial \nu} = g \text{ on } \Gamma^{(18)}, \mathscr{F}(u) \in \lambda \text{ a.e. in } \Omega\} \tag{2.5}$$

where:

$$\mathscr{F}(u) = \left(\frac{\partial^2 u}{\partial x_1^2}\right)^2 + \left(\frac{\partial^2 u}{\partial x_2^2}\right)^2 + \left(\frac{\partial^2 u}{\partial x_1^2} - \frac{\partial^2 u}{\partial x_2^2}\right)^2 + 6\left(\frac{\partial^2 u}{\partial x_1 \partial x_2}\right)^2. \tag{2.6}$$

We *assume* that

$$K_\lambda \neq \phi; \tag{2.7}$$

K_λ is *closed* and *convex* in $H^2(\Omega)$. We define:

$$a(u, v) = (\Delta u, \Delta v) \tag{2.8}$$

and we consider the V.I.

$$a(u, v-u) \geq 0 \quad \forall v \in K_\lambda, \quad u \in K_\lambda, \tag{2.9}$$

which again admits a *unique solution*. The *regularity* of the solution is a difficult question. By using non trivial penalty arguments, it is shown in Evans-Knerr, loc. cit., that if h and g are C^2 on Γ, then:

$$u \text{ is } \textit{locally} \text{ in } H^3(\Omega). \tag{2.10}$$

Remark 2.1 Many problems are still open for the regularity of the solution and for the properties of the free boundaries, for *V.I.* for second order *systems* of elliptic operators. Let us indicate along these lines the problem of the regularity of the free boundary *in Bingham flows* (Cf. G. Duvaut and J.L. Lions [1]).

2.3. **Perfect plasticity and space BD(Ω)**

Let us take now $\Omega \subset \mathbb{R}^3$. In the case of *linear elasticity* the classical elliptic system is coercive on the space of vector functions $V \in (L^2(\Omega))^3$ such that:

$$\epsilon_{ij}(v) \in L^2(\Omega) \quad \forall i,j \tag{2.11}$$

where $\epsilon_{ij}(v)$ is the deformation tensor:

$$\epsilon_{ij}(v) = \frac{1}{2}\left(\frac{\partial v_i}{\partial x_j} + \frac{\partial v_j}{\partial x_i}\right). \tag{2.12}$$

Roughly speaking, in the *elasto-plastic* situations, one can work with *V.I.* over *convex subsets on the space defined by* (2.11). But this is *not* the case anymore for *perfectly plastic* problems where the '*right* space' to work in is much more tricky to handle; essentially, the functional:

$$\int_\Omega \epsilon_{ij}(v)\,\epsilon_{ij}(v)\,\mathrm{d}x$$

is *replaced by*

$$\int_\Omega (\epsilon_{ij}(v)\,\epsilon_{ij}(v))^{1/2}\,\mathrm{d}x. \tag{2.13}$$

This is an analogous situation to the one met with *minimal surface* theory where the 'Dirichlet's integral' is replaced by:

$$\int_\Omega (1 + (\nabla v)^2)^{1/2}\,\mathrm{d}x. \tag{2.14}$$

It has been shown by M. Miranda [1] that a fundamental space to use in connection with (2.14) is

$$BV(\Omega) = \{v \mid v \in \mathscr{D}'(\Omega)^{(19)}, \frac{\partial v}{\partial x_i} \in \mathscr{M}(\Omega)^{(20)}\}. \tag{2.15}$$

It is then an obvious thing to introduce the space(21)

$$BD(\Omega) = \{v \mid v \in (\mathscr{D}'(\Omega))^3, \ \epsilon_{ij}(v) \ \in \mathscr{M}(\Omega)\}. \tag{2.16}$$

The non trivial point is to show that one *can define traces on* Γ *for elements of* $BD(\Omega)$ (Suquet [1]), the best result along these lines being that (G. Strang and R. Temam [1]):

elements of $BD(\Omega)$ *have traces on* Γ *which belong to* $(L^1(\Gamma))^3$. (2.17)

This is one of the basic tools for the study of problems of perfect plasticity and of the V.I. connected with them.

2.4. Various remarks

Remark 2.2 In *non linear elasticity*, the theory of monotone, or pseudo-monotone, operators has been used; cf. J.L. Lions [1] and the Bibliography therein. If one considers the case where the problem can be put in the framework of Calculus of Variations, situations which can be solved where one cannot apply the theory of monotone operators have been studied by J.M. Ball [1].

Remark 2.3 We refer to G. Duvaut and J.L. Lions [1] for other applications of V.I. to Mechanics (unilateral problems, problems with friction (cf. also J.J. Kalker [1]), etc.).

Remark 2.4 Another technique, related to V.I. and which can be useful in some problems of Mechanics, is the technique of *Quasi Variational Inequalities* (Q.V.I.), introduced in A. Bensoussan and J.L. Lions [4] [2]. It has been, in particular, used by C. Baiocchi [1] (Cf. also C. Baiocchi and A. Capelo [1], Vol. 2) for infiltration problems.

3. Perforated materials

Let us consider a domain Ω_ϵ which consists of an open set Ω minus a set of 'small' holes (or obstacles), of 'size' ϵ, and which are arranged in a periodic manner (periodically perforated materials). More precisely we consider in $\mathbb{R}^n$ the unit cube Y with a hole θ such that $\overline{\theta} \subset Y$; let S be the boundary of θ. We consider the set ϵY with the hole $\epsilon\theta$ and we cover $\mathbb{R}^n$ with all the translated sets by length $k\epsilon$, $k \in \mathscr{Z}$, in all directions; and we translate the holes accordingly; let $\tau(\epsilon\overline{\theta})$ be the set of all these (closed) holes. Then we define:

$$\Omega_\epsilon = \Omega \setminus (\Omega \cap \tau(\epsilon\overline{\theta})) \tag{3.1}$$

and we consider the following problem: find u_ϵ in Ω_ϵ such that:

$$-\Delta u_\epsilon = f \text{ in } \Omega_\epsilon \tag{3.2}$$

(where f is given on Ω), subject to Boundary conditions:

$$u_\epsilon = 0 \text{ on } \Gamma_\epsilon \text{ (i.e. } \Gamma \setminus \text{intersection of } \Gamma \text{ with holes),} \tag{3.3}$$

and [22]

$$\frac{\partial u_\epsilon}{\partial \nu} = 0 \text{ on } S_\epsilon \text{ (i.e. union of boundaries of the holes inside } \Omega). \tag{3.4}$$

For every $\epsilon > 0$ this problem admits a unique solution. We are looking for *an asymptotic expansion of u_ϵ in terms of ϵ.*

The ansatz **(Cf. A. Bensoussan, J.L. Lions and G. Papanicolaou [1], J.L. Lions [2] [3])**

We 'try' u_ϵ as given by the ansatz:

$$u_\epsilon = u_0 + \epsilon u_1 + \ldots \tag{3.5}$$

where:

$$\left.\begin{array}{l} u_j = u_j\,(x,y), \\ u_j\,(x,y) \text{ is defined for } x \in \Omega, y \in \mathscr{Y} = Y \backslash \theta \\ u_j\,(x,y) \text{ is periodic in } y \text{ (with period 1 in all } y\text{'s variables);} \end{array}\right\} \tag{3.6}$$

and where in (3.5) *we replace* y *by* x/ϵ. In order to compute $\Delta u_j(x,y)$ it amounts to the same thing to apply the operator

$$\epsilon^{-2}\,\Delta_y + 2\epsilon^{-1}\,\Delta_{xy} + \epsilon^0\,\Delta_x \tag{3.7}$$

where:

$$\Delta_{xy} = \frac{\partial^2}{\partial x_i\,\partial y_i}, \tag{3.8}$$

and to replace y by x/ϵ in the result. Condition (3.4) becomes:

$$\left(\epsilon^{-1}\,\frac{\partial}{\partial\nu(y)} + \nu_j(y)\,\frac{\partial}{\partial x_j}\right) u_\epsilon = 0 \tag{3.9}$$

i.e.

$$\frac{\partial u_0}{\partial\nu(y)} = 0, \quad x \in \Omega, y \in S \tag{3.10}$$

$$\frac{\partial u_j}{\partial\nu(y)} + \nu_i(y)\,\frac{\partial u_{j-1}}{\partial x_i} = 0 \text{ for } x \in \Omega,\ y \in S,\ j \geq 1. \tag{3.11}$$

Using (3.5), (3.7) in (3.2) gives:

$$-\Delta_y u_0 = 0, \tag{3.12}$$

$$-\Delta_y u_1 - 2\,\Delta_{xy} u_0 = 0, \tag{3.13}$$

$$-\Delta_y u_2 - 2\,\Delta_{xy} u_1 - \Delta_x u_0 = f, \text{ etc.} \tag{3.14}$$

But (3.12), (3.10) and the periodicity of u_0 with respect to y imply that:

$$u_0 = u_0\,(x) \text{ does not depend on } y. \tag{3.15}$$

Then (3.13) and (3.11) with $j = 1$ give:

$$\left|\begin{array}{l} -\Delta_y u_1 = 0, \\ \dfrac{\partial u_1}{\partial\nu(y)} = -\,\nu_j(y)\,\dfrac{\partial u_0}{\partial X_j}\,(x). \end{array}\right. \tag{3.16}$$

If we *define* χ^j by [23]:

$$\left|\begin{array}{l} -\Delta_y \chi^j = 0 \text{ in } \mathscr{Y}, \\ \dfrac{\partial \chi^j}{\partial \nu(y)} = \nu_j(y) \text{ on } S, \\ \chi^j \text{ is periodic in } y, \int_{\mathscr{Y}} \chi^j(y)\,\mathrm{d}y = 0 \end{array}\right. \tag{3.17}$$

then:

$$u_1(x,y) = -\chi^j(y) \frac{\partial u_0}{\partial X_j}(x) + \hat{u}_1(x); \tag{3.18}$$

(3.14) and (3.11) with $j = 2$ give:

$$\left|\begin{array}{l} -\Delta_y u_2 = f + \Delta_x u_0 - 2 \dfrac{\partial \chi_j}{\partial y_i} \dfrac{\partial^2 u_0}{\partial X_i \partial X_j}, \text{ in } \mathscr{Y} \\ \dfrac{\partial u_2}{\partial \nu(y)} = \nu_i(y) \chi^j(y) \dfrac{\partial^2 u_0}{\partial X_i \partial X_j} - \nu_i(y) \dfrac{\partial \hat{u}_1}{\partial X_i}, \text{ on } S, \\ u_2 \text{ periodic in } y. \end{array}\right. \tag{3.19}$$

Problem (3.19) admits a solution u_2 if

$$-\int_S \frac{\partial u_2}{\partial \nu(y)} \mathrm{d}S = \int_{\mathscr{Y}} \left[f + \Delta_x u_0 - 2 \frac{\partial \chi^j}{\partial y_i} \frac{\partial^2 u_0}{\partial X_i \partial X_j} \right] \mathrm{d}y$$

i.e.

$$\mathscr{A} u = f \text{ in } \Omega, \tag{3.20}$$

where:

$$\mathscr{A} = -\frac{1}{|\mathscr{Y}|} \int_{\mathscr{Y}} \delta_i^j - \frac{\partial \chi^j}{\partial y_i}(y) \quad \mathrm{d}y, \tag{3.21}$$

$|\mathscr{Y}|$ = volume of $\mathscr{Y}$.

One shows that $\mathscr{A}$ is *elliptic. It is called the homogenized operator*. One adds to (3.20) the boundary condition:

$$u = 0 \text{ on } \Gamma. \tag{3.22}$$

One shows then that u_ϵ *converges in some sense to* u as $\epsilon \to 0$. In order to make this statement precise, one defines (cf. D. Cioranescu [1]) an *extension operator* P_ϵ from $H^1(\Omega_\epsilon)$

into $H^1(\Omega)$, and one has:

$$P_\epsilon u_\epsilon \to u \quad \textit{in } H^1(\Omega) \textit{ weakly}. \tag{3.23}$$

Remark 3.1 The proof of (3.23) uses also ideas of *L.* Tartar [1].

Remark 3.2 One can proceed with the above computation and compute $u_1, u_2, \ldots$ There are difficulties with the boundary condition on Γ.

Remark 3.3 An expansion valid to any order in ϵ can be computed (cf. J.L. Lions [2]) when one deals with *Dirichlet's* boundary conditions on S_ϵ, and assuming that $f \equiv 0$ near Γ (otherwise one meets probably some kinds of *boundary layers*, but this is an open question).

Remark 3.4 The idea of the ansatz *of the type* (3.5) has been systematically used for *composite materials* in A. Bensoussan, J.L. Lions and G. Papanicolaou [1].

Remark 3.5 When dealing with visco-elastic materials (with short or long range memory) in the *evolution case*, one meets non local homogenized operators *which are not partial differential operators*; they can be expressed in terms of *Partial Differential Operators* if we use simultaneously u_0 and u_1; u_1 plays then the role of a hidden variable(24); cf. J.L. Lions [3] [4].

Remark 3.6 Problems of elasto-plasticity for composite materials lead to interesting questions. The 'homogenization' of the V.I. analogous to (2.1) for a composite material has been solved by L. Carbone [1], using, among other things, the general approach of E. de Giorgi [1].

Remark 3.7 For a study of the *behaviour of the spectrum* in the homogenization process, we refer to S. Kesavan [1] and to M. Vanninathan [1].

Notes

1. It goes without saying that no attempt is made, by any means, to give an exhaustive presentation of the subject. For many other aspects we refer to G Fichera [1], to the book G.M. de la Penha and L.A. Medeiros, ed. [1].
2. In short V.I.
3. Hypothesis on f and on ψ are made precise later on. We assume that $\psi \leq 0$ on Γ, for compatibility reasons.
4. In particular in the *theory of infiltration through porous media*; cf. C. Baiocchi [1], C. Baiocchi and Capelo [1], and the bibliography therein.
5. We use summation convention.
6. Space of bounded and measurable functions in Ω. We emphasize the fact that the a_{ij}'s *are not necessarily symmetric*.
7. Cf. J.L. Lions and G. Stampacchia [1].
8. These are the classical *Sobolev spaces*.
9. Taken in appropriate sense. Cf. for instance J.L. Lions and E. Magenes [1].
10. Sobolev space of functions in $H^1(\Omega)$ such that $\dfrac{\partial^2 u}{\partial x_i \partial x_j} \in L^2(\Omega)\ \forall i,j$.
11. This is *exactly* the analogous of the weak solution of the classical Dirichlet or Neumann problems.
12. This is a *regularity property*.
13. The main ideas for the study of regularity properties come from H. Brezis and G.

Stampacchia [1], H. Brézis [1].

14. Cf. J.L. Lions [1]. Other penalty arguments are given in A. Bensoussan and J.L. Lions [1] [2], R. Glowinski and J.L. Lions [1] (where numerical examples are presented).
15. We set: $\phi^+ = \sup(\phi, 0), \phi^- = \sup(-\phi, 0)$. The term $-\frac{1}{\epsilon}(u_\epsilon - \psi)^-$ is the *penalty term*.
16. In fact, if $A\psi \in L^2(\Omega)$, one can prove (cf. A. Bensoussan, J.L. Lions [1]) that $\|u_\epsilon - u\| \leq C\epsilon^{1/2}$.
17. We have set $|\phi| = (\phi, \phi)^{1/2}$.
18. With the notations of J.L. Lions and E. Magenes [1], we assume that h (resp. g) is given in $H^{3/2}(\Gamma)$ (resp. in $H^{1/2}(\Gamma)$).
19. $\mathscr{D}'(\Omega)$ = distributions on Ω.
20. $\mathscr{M}(\Omega)$ = bounded Radon measures on Ω; BV = Bounded Variation.
21. BD = Bounded Deformation.
22. $\frac{\partial}{\partial \nu}$ = normal derivative.
23. A solution exists because $\int_S \nu_i(y)\, dS = 0$.
24. As indicated to the A. by E. Sanchez Palencia.

Bibliography

C. Baiocchi, [1] 'Free boundary problems in the theory of fluid flow through porous media', *Proc. Int. Congress Math.*, Vancouver, 1974, Vol. II, p. 237-243.

C. Baiocchi and A. Capelo, [1] *Disequazioni variazionali e quasi variazionali. Applicazioni a problemi di frontiera libera.* Bologne, 1978; Vol. 1 Problemi variazionali. Vol. 2 Problemi quasi variazionali.

J.M. Ball, [1] 'Convexity conditions and existence theorems in non linear elasticity', *A.R.M.A.* **63** (1977), 337-403.

A. Bensoussan and J.L. Lions, [1] *Applications des inéquations variationnelles en contrôle stochastique* – Dunod 1978; [2] *Contrôle impulsionnel.* Dunod, 1980; [3] 'Problèmes de temps d'arrêt optimal et inéquations variationnelles paraboliques', *Applicable Analysis* **3** (1973), 267-294; [4] 'Nouvelles méthodes en contrôle impulsionnel', *J.A.M.O.*, **1** (1975), 289-312.

A. Bensoussan, J.L. Lions and G. Papanicolaou, [1] *Asymptotic Analysis for periodic structures*, North Holland, 1978.

H. Brezis [1], 'Problèmes unilatéraux', *J.M.P.A.*, **51**, (1972), 1-168.

H. Brezis and G. Stampacchia, [1] 'Sur la régularité de la solution d'inéquations elliptiques', *Bull. S.M.F.*, **96** (1968), 153-180.

L.A. Caffarelli and A. Friedman, [1] 'The free boundary for elastic plastic torsion problem', to appear; [2] 'Reinforcement problems in elasto-plasticity', to appear.

L.A. Caffarelli and N.M. Riviere [1] 'On the Lipschitz character of the stress tensor of an elastic plastic bar', to appear.

L. Carbone, [1] 'Sur un problème d'homogénéisation avec des contraintes sur le gradient', *J.M.P.A.*, 1979.

Ph. Ciarlet, [1] *Topics in Mathematical Elasticity*, North Holland, 1980; [2] 'A justification of the Von Karman equations', *A.R.M.A.* 1979.

D. Cioranescu, [1] Thesis Paris, 1977.

M. Crandall and P.H. Rabinonitz, [1] 'Bifurcation from simple eigenvalues', *J.F. Analysis*, 8 (1971), p. 321-430.

G. Duvaut and J.L. Lions, [1] *Les inéquations en Mécanique et en Physique*, Dunod 1972 (English translation; Springer, 1976).

L.C. Evans and B.F. Knerr, [1] 'An elasto plastic stress problem', *J.A.M.O.*, 1979.

G. Fichera, [1] 'Existence theorems in elasticity. Boundary value problems of elasticity with unilateral constraints', in *Encyclopedie of Physics* (S. Flugge, chief Ed.), Vol. VI a/2, Mechanics of Solids II (C. Trauesdell, Ed.), Springer 1972, p. 372-424.

E. de Giorgi, [1] 'Convergence problems for functionals and operators', in *Metodi recenti di Analisi non lineare e applicazioni*, Rome, May 1978, Lecture notes in Math., E. Magenes and U. Mosco, ed.

R. Glowinski and J.L. Lions, [1] To appear in *Calcolo*, 1979.

R. Glowinski, J.L. Lions and R. Tremolières [1] *Analyse numérique des inéquations variationnelles*, Dunod, 1976, vol. 1, Théorie Générale, Vol. 2: Applications aux phénomènes stationnaires et d'évolution. English translation, North Holland, 1980.

J.J. Kalker, [1] 'A minimum principle for the law of dry friction with application to elastic cylinders in rolling contact', *J. Appl. Mech.* **38**, E (1971), 875-880 and 881-887.

S. Kesavan [1] Thesis, Paris, 1980.

D. Kinderlehrer and L. Nirenberg, [1] 'Regularity in free boundary problems'. *Ann. S.N.S. Pisa*, **4**, (1977), p. 373-391.

W.T. Koiter, [1] 'Some general theorems on doubly periodic and quasi periodic functions', *Proc. K. Neder. Akad. Wet.*, Amsterdam, 1959, 62, p. 120-128; [2] 'General theorems for elastic plastic solids', *Progress in Solid Mechanics*, North Holland, 1960, p. 165-221; [3] 'Stress distribution in an infinite elastic sheet with a doubly-periodic set of equal holes', *Boundary Problems Diff. Eq.*, Madison, Univ. Wisconsin Press, (1960), p. 191-213.

M. Lanchon, [1] 'Torsion élastoplastique d'une barre cylindrique de section simplement ou multiplement connexe'. *J. de Mécanique* **13** (1974), 267-320.

G.M. de La Penha and L.A. Medeiros, ed., [1] *Contemporary Developments in Continuum Mechanics and Partial Differential Equations*, North Holland Math. Studies, 30, 1978 (Rio de Janeiro Symp. 1977).

M. Leroy and G. Stampacchia, [1] 'On the regularity of the solution of a variational inequality', *Comm. P.A.M.*, **XXII** (1969), 153-188.

J.L. Lions, [1] *Quelques méthodes de résolution des problèmes aux limites non linéaires*, Dunod, Gauthier – Villars, Paris, 1969; [2] 'Asymptotic expansions in perforated media with a periodic structure', *Rocky Mountains Journal*, 1979; [3] 'Remarks on some asymptotic problems in composite materials and in perforated materials', *I.U.T.A.M. Symposium*, Northwestem University, September 1978; [4] Homogénéisation non locale', in *'Metodi Recenti di Analisi non lineare e applicazioni'*, Rome, May, 1978, Lecture Notes in Math., E. Magenes and U. Mosco, ed.; [5] 'Introductory remarks to asymptotic analysis of periodic structures'. Lecture Porabka – Kozubnik, September 1977.

J.L. Lions and E. Magenes, [1] *Problèmes aux limites non homogènes et applications*. Paris, Dunod, Vol. 1, 1968, (English translation, Springer, 1970).

J.L. Lions and G. Stampacchia, [1] 'Variational inequalities', *C.P.A.M.*, **XX** (1967) 493-519.

M. Miranda, [1] Comportamento dell successioni convergenti di frontiere minimali, *Rend. Sem. Univ. Padova*, (1967), 238-257.

P.H. Rabinowitz ed., [1] *Applications of bifurcation theory*. Acad. Press, 1977.

G. Strang and R. Temam, [1] 'Functions of bounded deformation', to appear.

P. Suquet, [1] 'Existence et régularité des solutions des équations de la plasticité parfaite', *C.R.A. Sc. Paris*, **286**, D (1978), 1201-1204.

L. Tartar, [1] 'Quelques remarques sur l'homogénéisation', in *Functional Analysis and Numerical Analysis*, (1976) Japan-France Seminar, Tokyo and Kyoto, M. Fujita, Ed. (Japan Society for the Promotion of Science, 1978), p. 469-481.

T.W. Ting, [1] 'Elastic plastic torsion of square bar', *Trans. A.M.S.* **123** (1966), 369-401.

M. Vanninathan [1] Thesis, Paris, 1979.

Variational Methods in Optimization of Structures

F. Niordson and N. Olhoff *

Summary

Variational calculus plays an essential role in the optimalization of structures. It not only delivers an elegant formulation of the problem but also gives a good insight into the restrictions imposed by the formulation.

A number of aspects is discussed, a.o. the existence of solutions, the situation and character of singularities, especially in optimalizing beam-structures and structures which may buckle, also were it is necessary to consider two interacting buckling modes.

Finally for a beam of fiber reinforced material the distribution and direction of the fibers is determined such that at a given load optimal stiffness is reached.

1. Introduction

From a practical point of view any structure or structural element is sufficiently well described by a *finite* set of real numbers. Furthermore each real number is with sufficient accuracy selected from a *finite* set of numbers. Hence a problem of optimal design may through a process of discretization be reduced to a *problem of multiple choice*, namely of finding the best solution among a finite set of solutions. However, since the number of possible solutions may be very large, even so, sophisticated mathematical methods of selection are usually needed to identify the best solution. Such methods, which mostly fall into the group of algorithms referred to as 'mathematical programming' are well developed and well suited for solving such problems.

It must however be realized that by discretizing an intrinsically continuous structure, the problem is basically changed and certain fundamental aspects of the original problem of optimal design are distorted and others are excluded from the analysis. First of all the question of existence of a solution. From the engineering point of view this may seem to be of little consequence and in fact may sometimes seem to be of advantage. Thus if the method employed yields better and better structures the engineer will probably be happier than if he only had the correct answer to the problem, which may be that there is no optimum. Nevertheless as we shall see, there is important knowledge to be gained from a correct non-discrete formulation of a basically continuous structure.

Such a formulation will lead to a variational problem and this paper is devoted to some important knowledge gained from the employment of the calculus of variations to optimization of structures.

Important questions regarding the existence of solutions as related to the actual formula-

* Department of Solid Mechanics, The Technical University of Denmark, Lyngby, Denmark

tion of the problem will be discussed. This includes questions regarding continuity, singularities, jump-conditions, and bimodality. In addition an approach to optimization of reinforced structures using a smear-out process will be introduced.

Concerning other questions of structural optimization, we direct the readers attention to general reviews and expository articles [1-4], with their effective keys to the exhaustive literature of the field. As early special papers behind the developments surveyed in the present paper, Refs. [5-8] should be mentioned.

2. Existence of solutions

In contrast to the mathematician the engineer is rarely concerned with the existence of solutions probably because his mathematical models are well explored and seldom yield totally surprizing results. In other words his preconceived idea of a solution and its general properties is hardly ever wrong. His problem is therefore in most cases reduced to quantifying a quantity, he knows a priori exists.

In optimal design however the situation is different. To illustrate this, let us recall the well known problem of finding a minimal surface connecting two coaxial circular boundaries (Fig. 1).

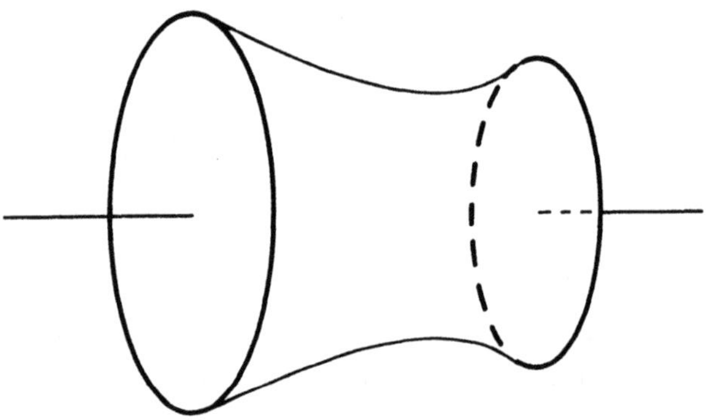

Fig. 1

Our preconceived idea of a solution to this problem suddenly fails, when the axial distance between the circular regions exceeds a critical value. It may be wize to keep this in mind when one deals with problems of optimal design.

A similar fallacy is encountered in the optimal design of vibrating plates [9] [10]. If no thickness constraints are considered, the variational analysis of the problem leads to a structure which is smooth but certainly not a global optimum. In Ref. [10] it was demonstrated that there is no upper bound to the first natural frequency of any simply supported plate of given volume. By conceiving the stiffness properly bell-shaped the original formulation of the problem using sufficiently smooth functions is quite adequate and the situation is therefore as follows:

With no restriction on the thickness of a supported plate there is no upper bound for the natural frequencies whether the thickness of the plates is allowed to vary discontinuously or not. It is suprizing and to some extent misleading to find that the variational formulation yields smooth shapes that are nothing but local optimal solutions stationary to small variations in the thickness function.

But we gain the important knowledge that in a proper formulation of the problem of optimal design of plates of given volume and domain, the thickness of the plate should be restricted. This rather natural restriction completely changes the problem. Work in this direction by Mr. Keng-Tung Cheng under the guidance of the authors, although at a very preliminary stage seems to lead to properly stated problems with solutions giving designs which seems logical, see Fig. 2. This figure shows a quarter of a clamped, solid square plate which is subjected to a uniform, static load distribution, and optimized for maximum stiffness (minimum compliance). The volume, plate domain and material are given, and both minimum and maximum constraints are prescribed for the plate thickness. In order to study parametrically the automatic development of integral plate stiffeners, a so-called *optimal structural remodeling* formulation (see [11]) is actually used.

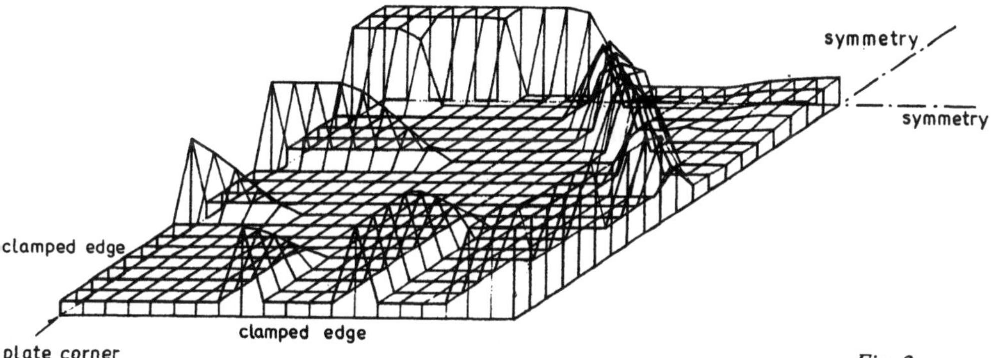

Fig. 2

3. Singularities

For continuous structures, variational methods are necessary for determining the greatest possible merit and the associated design. For a given purpose this optimal design constitutes the upper bound solution for all admissible designs. Such upper bound solutions are not only of considerable interest for the scientist, but also for the engineer, because they provide a basis for measuring the efficiency of any other design, and indicate the path to be taken towards a better economy of material.

Since any formulation in optimal design with restrictions on the design variables generally lead to solutions with lower merits, it is often necessary to consider the particular problem in geometrically unconstrained form, in order to obtain the appropriate upper bound solution. Although a formulation with restrictions on the design variables is more complex than a corresponding formulation of the unconstrained problem, the equations for the upper bound solution emerging from the latter formulation, are generally more difficult to handle numerically. This is not a trivial issue, because with exception of some optimization problems with linear relationship between the structural stiffness and the design variable, closed form solution are generally not obtainable, and it is therefore necessary to solve the governing equations numerically in a discretized form.

The numerical difficulties in connexion with geometrically unconstrained formulations for optimal design of single-purpose structures, are due to the occurrence of points [5-8, 12-15] (or curves in the case of two-dimensional structures [9, 10, 16]) where the stiffness vanishes in the optimal solution, and where the deflection function and/or its derivatives exhibit singular or discontinuous behaviour. Thus, in order to determine the solution and to obtain accurate results, these singularities must be treated properly. This requires knowledge of the particular type of singularity present, and in the following we shall pay some attention to this problem.

Singularities generally occur at points (or curves) of zero bending moment in the optimal solution, and as we approach the singular point the stiffness and the bending moment will tend towards zero in one of two essentially different ways, either in a way typical of a 'hinge' [7, 16], or else in a manner typical of a free end [12, 16].

In fact, the occurrence of singular points in geometrically unconstrained, one-dimensional optimization problems, is intimately connected with the general rule that solutions to such problems are statically determinate (provided that only a single deflections function is considered in the formulation). Thus, when an *a priori* statically redundant one-dimensional structure is optimized without geometric constraints, the statical determinacy of the optimal solution is due to the development of inner hinges or separations.

In optimal design of elastic Bernoulli-Euler beams, it is usually assumed that the relationship between the cross-sectional area $A(x)$, e.e. the design variable, and the second area moment of inertia $I(x)$ is of the form

$$I(x) = cA^p(x), \tag{3.1}$$

where x is the coordinate along the axis of the beam, and where the factor c and the exponent p are given, positive constants. Then, for $p = 3$ (3.1) models beams of rectangular cross-sections of fixed width and variable height, for $p = 2$, it models beams with variable solid cross-sections of geometrically similar shape, and for $p = 1$ beams with solid sections of fixed height and variable width. The case $p = 1$ also includes constant-width sandwich cross-sections of lightweight cores of uniform height and zero stiffness covered by two identical, thin sheets of variable thickness.

The two types of singular behaviour, (I and II), that occur in geometrically unconstrained, optimal Bernoulli-Euler beams are both associated with zero bending moment and stiffness, but in one type (I) the shear force is finite, while in the other type (II) the shear force also vanishes at the singularity.

The singular behaviour can be determined analytically by expanding the solution in a power series (see e.g. [6, 7, 12-15]) in the vicinity of the singular point $x = x_i$, which may either be an inner point or an end point of the beam. Let us in the sequel conceive $x = x_i$ as an arbitrary inner singular point (singular boundary points are comprised as special cases).

The leading terms of the power series expansions at a Type I singularity for the deflection $y(x)$ and the design variable (cross-sectional area) $A(x)$ are given by [17]

$$y(x) = \begin{cases} a_0 + a_1^+ (x - x_i) + b_1^+ (x - x_i)^r + \ldots & x \geq x_i \\ a_0 + a_1^- (x_i - x) + b_1^- (x_i - x)^r + \ldots & x \leq x_i \end{cases} \tag{3.2}$$

(which reflects continuity of y), and

$$A(x)=\begin{cases} g_1^+ (x-x_i)^s + \ldots & x \geq x_i \\ \\ g_1^- (x_i-x)^s + \ldots & x \leq x_i \end{cases} \tag{3.3}$$

respectively, where a_0, a^+_1, a_1^-, b^+_1, b_1^-, g^+_1 (>0) and g_1^- (>0) are coefficients. Depending on the value of p, the powers r and s in (3.2) and (3.3) attain the following values,

$$\begin{aligned} p &= 1 : r = 2 \quad, s = 1 \\ p &= 2 : r = 5/3, s = 2/3 \\ p &= 3 : r = 3/2, s = 1/2. \end{aligned} \tag{3.4}$$

For $p = 1$, higher powers than $r = 2$ are generally not present in (3.2). For $p = 2$ and $p = 3$, r is the leading non-integer power of the series, which implies that the second and higher order derivatives of y are unbounded in the neighbourhood of the singularity.

The finite shear force $q(x) = -cE(A^p y'')'$ at a singularity of type I ensures non-zero values of the coefficients b_1^+ and b_1^- in (3.2) as well as of the coefficients g_1^+ and g_1^- in (3.3). Except in buckling problems and problems in which an external transverse force acts at the point $x = x_i$, the shear force $q(x)$ will be continuous at $x = x_i$, and therefore we have $b^+_1 = -b_1^-$ in (3.2) and $g^+_1 = g_1^-$ in (3.3). Note that the latter relation implies that the slopes $A'(x_i^+)$ and $A'(x_i^-)$ are equal numerically but of opposite signs at the point $x = x_i$ of vanishing A. The slopes are finite for $p = 1$, but infinite for $p = 2$ and $p = 3$.

The behaviour of y given by Eqs. (3.2)-(3.4) for $p = 1, 2$, and 3 at any inner point $x = x_i$ of zero cross-sectional area $A(x)$ and bending moment $m(x) = cEA^p y''$ is characterized by a continuous deflection function y, a possible finite jump in the slope y' ($a_1^+ \neq a_1^-$ in general), and a continuous non-zero shear force $q(x) = -cE(A^p y'')'$. The behaviour of the beam is therefore characterized kinematically and physically at this point by a *hinge*. The functional behaviour in the neighbourhood of a type I singular point $x = x_i$ is depicted in Fig. 3 for the case $p = 2$, assuming $q(x)$ to be continuous. Singularities of Type I, i.e. hinges will appear in a variety of geometrically unconstrained optimization problems, as e.g. optimal design for maximum stiffness (minimum compliance), maximum strength, maximum Euler buckling load, maximum frequency of transverse vibrations, etc. Precisely the same type of singular behaviour is found in geometrically unconstrained plate problems (for $p = 1$ or $p = 3$), for example along simply supported edges, [9, 16], or along interior curves that may develop in the optimal solution [10]. In fact, expanding the deflection function y and the design variable A (which for $p = 3$ identifies the thickness of solid plates) along any normal to the simply supported edged or interior hinge curve, Eqs. (3.2)-(3.4) are re-established.

Let us now discuss singularities of the second type in which the behaviour is associated with zero shear force $q(x)$ in addition to zero bending moment $m(x)$. At such an interior singularity the deflection y as well as the slope y' may exhibit a discontinuity and the point therefore corresponds to an *inner separation* of the beam.

Singularities of this type are not found in the same broad class of problems as is the case with singularities of the first type. In optimal design for maximum buckling load for example, inner singularities of type II are not admitted by the field equations, which by the way seems obvious from physical grounds. However, a singularity of this second type is found at a free end of a column [6], when buckling instability due to its own weight is considered.

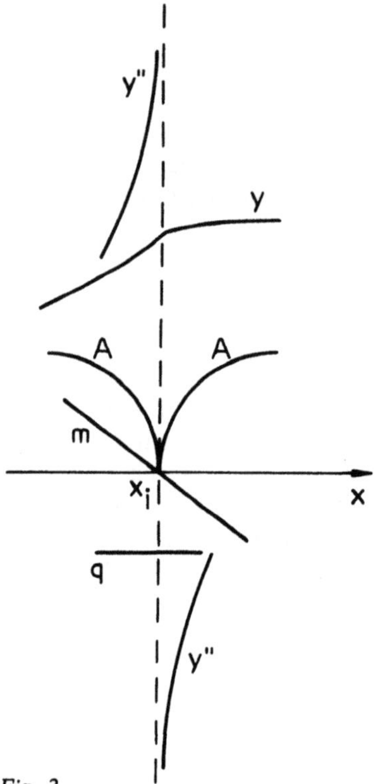

Fig. 3

In optimal design for maximum stiffness of transversely loaded beams, the field equations do admit inner singularities of type II, but as one may expect, optimal beams of this kind are more likely to have singularities of the first type, i.e. hinges.

However, a priori statically indeterminate beams optimized for maximum fundamental frequency of transverse vibrations, and also statically determinate beams optimized for maximum values of higher order transverse vibration frequencies, or for maximum distances between consecutive higher order frequencies, will often exhibit singularities of the second type for $p = 2$ and 3 [15]. (For $p = 1$ such problems are usually degenerate [12, 18, 19], and will not be discussed further here). When we investigate the equations for optimal transversely vibrating beams by means of power series expansions, we find that the leading terms of the deflection $y(x)$ and the design variable, that is, the cross-sectional area $A(x)$, are found to be [15, 17],

$$y(x) = \begin{cases} k^+ (x - x_i)^n + \ldots x > x_i \\ \\ k^-(x_i - x)^n + \ldots x < x_i \end{cases} \tag{3.5}$$

and

$$A(x) = g\,(x - x_i)^s + \ldots x \geq x_i, \;\; x \leq x_i, \tag{3.6}$$

respectively, in the vicinity of an arbitrary inner point $x = x_i$ of a singularity of the second type. In Eqs. (3.5) and (3.6), k^+, k^- and g are non-vanishing coefficients (g is positive) and the powers n and s of the leading terms attain the following values,

$$p = 2 : n = -2, s = 4$$
$$p = 3 : n = -1, s = 2. \qquad (3.7)$$

This type of singular behaviour is seen to be quite pronounced in optimal, transversely vibrating beams, in that the deflection function $y(x)$ and all its derivatives tend to infinity at $x = x_i$, at which point the design variable $A(x)$ as well as the bending moment $m(x) = cEA^p y''$ and the shear force $q(x) = -cE(A^p y'')'$ vanish. Fig. 4 illustrates the behaviour of these functions in the vicinity of the point $x = x_i$ for $p = 2$.

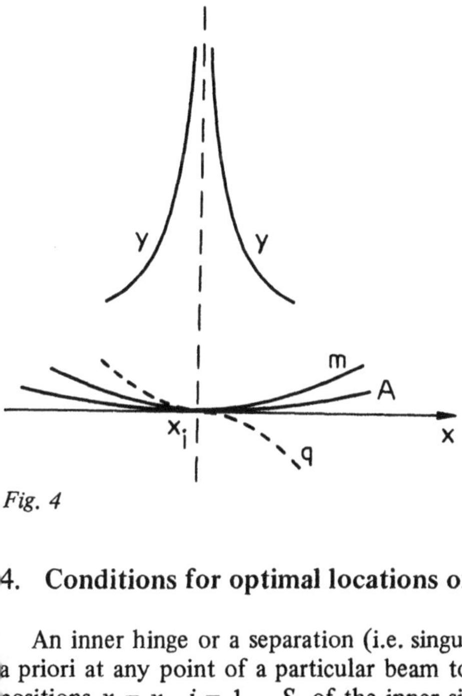

Fig. 4

4. Conditions for optimal locations of inner singularities

An inner hinge or a separation (i.e. singularity of first or second type) may be prescribed a priori at any point of a particular beam to be optimized. However, in most problems, the positions $x = x_i$, $i = 1, ...S$, of the inner singularities are not specified or known, and the determination of their locations is then an integral part of the optimization problem. In such problems, the positions $x = x_i$ of the singularities, and in some cases even their types, are treated as design variables in addition to $A(x)$. While the design variable $A(x)$ and the deflection function $y(x)$ are governed by a coupled, non-linear set of differential equations (the field equations) for optimality, there are additional conditions governing optimal positions $x = x_i$ of the inner singularities.

Conditions for singular points were first found in [20, 21], where it was established that a necessary condition for an inner hinge (singularity of type I) to be optimally placed in an elastic beam for maximum stiffness and for maximum frequency of transverse vibration, is that the slope y' of the deflection is continuous at the singular point, i.e.

$$y'(x_i^+) - y'(x_i^-) = 0. \tag{4.1}$$

Recalling the expansion for $y(x)$ given by Eq. (3.2) for a type I singularity, condition (4.1) implies that $a_1^+ = -a_1^-$ in (3.2), whenever the point $x = x_i$ is optimally located.

Fig. 5 and Fig. 6 illustrate the zero slope jump condition (4.1) for optimally located hinges. The doubly clamped beam in Fig. 5 is acted on by a uniformly distributed static load, and is optimized for maximum stiffness (minimum compliance), while the clamped/simply supported beam in Fig. 6 is optimized for maximum fundamental frequency of transverse vibrations. Both beams have geometrically similar cross-sections ($p = 2$), and the figures display to a suitable scale the shapes, i.e. the linear dimensions $\pm\sqrt{A}$ of the optimal beams. The deflection functions are denoted by y.

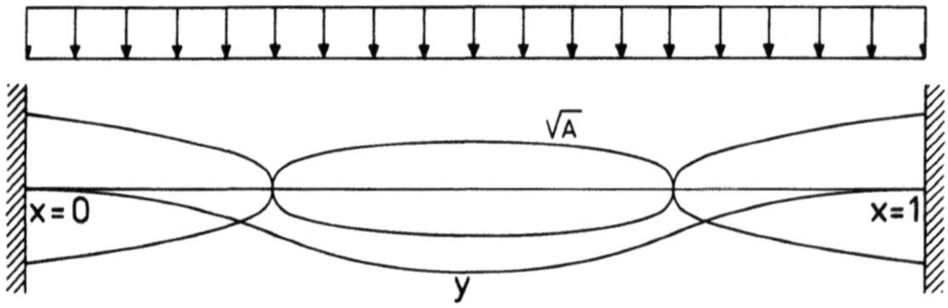

Fig. 5

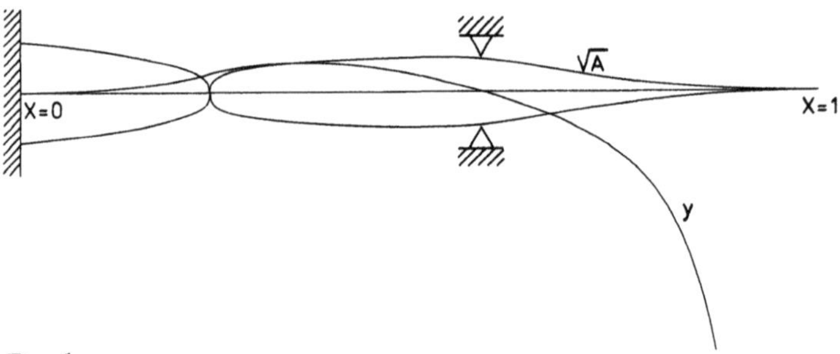

Fig. 6

In the case of optimal design of columns for maximum Euler buckling load, the condition for optimal location $x = x_i$ of an inner singularity of first type (which is the only possible type of inner singularity in this kind of problem) is more complex. Thus, it was recently found in [17] that the complete condition is

$$\begin{aligned} &\textit{either} \quad y'(x_i^+) - y'(x_i^-) = 0 \\ &\textit{or} \qquad y'(x_i^+) + y'(x_i^-) + 2\,\frac{y(x_i) - y(x_{i+1})}{x_{i+1} - x_i} = 0, \end{aligned} \tag{4.2}$$

for optimal position $x = x_i$ of an inner singularity (hinge). Here, x_{i+1} ($> x_i$) denotes the position of an adjacent inner hinge or hinged end point of the column, and the latter condition presumes that the column has no support between the points $x = x_i$ and $x = x_{i+1}$.

Fig. 7 shows the optimal design for maximum Euler buckling load of a doubly clamped column ($p = 2$) with inner hinges. In this example, the optimal position of the left hand hinge is governed by the second equation in (4.2), while the position of the right hand hinge is governed by the first equation. If, for example, the first equation in (4.2) were used for both inner hinges, their positions would change, and a slightly different design be obtained. This design would maximize the *second* buckling eigenvalue (with a symmetric mode), see Ref. [22], but it would have a much lower fundamental Euler buckling eigenvalue than the design shown in Fig. 7, and hence not be optimal in the sense of maximizing the buckling load.

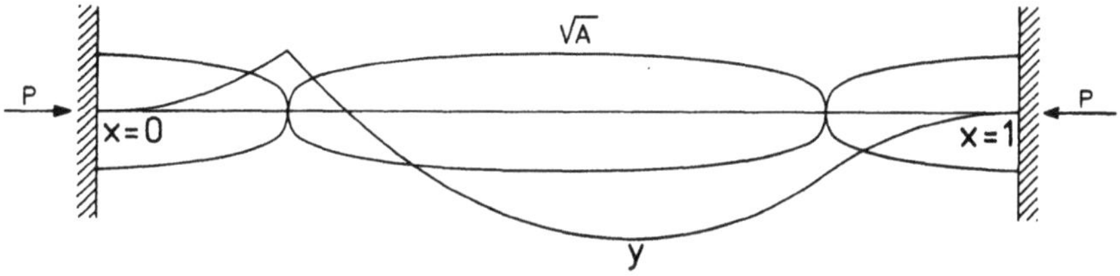

Fig. 7

Conditions for optimally placed inner singularities of the second type (inner separations) are considered next. Results were first developed in [15, 17] for such singularities.

For optimal, transversely vibrating beams (with $p = 2$ or 3), the condition for optimal location of an inner type II singularity is, cf. Eq. (3.5),

$$k^{+} \pm k^{-} = 0, \tag{4.3}$$

that is, the coefficients of the leading singular terms in the expansions of the vibration mode $y(x)$ are either equal or equal with opposite signs at an optimally placed inner singularity of second type. Fig. 8 illustrates the optimal design of a beam ($p = 2$) optimized for maximum fundamental frequency of transverse vibrations. The beam, which is clamped at one end and equipped with four inner simple supports, has two optimally placed inner singularities of the second type (separations).

At a inner singularity of the second type of transversely, statically loaded elastic beams optimized for maximum stiffness (minimum compliance), the deflection $y(x)$ is finite (whereas derivatives are singular), and the condition for the optimal location of the point $x = x_i$ is simply [17]

$$y(x_i^{+}) - y(x_i^{-}) = 0. \tag{4.4}$$

i.e. continuity of y, if the transverse load distribution is non-vanishing and continuous at the singularity.

It is worth emphasizing that the slope y' will in general be discontinuous at an optimally placed singularity of the second type. This means that condition (4.1) for a singularity of the first type is incompatible with conditions (4.3) or (4.4) for a singularity of the second type, which in turn implies that not only the *positions* but also the *types* of the singularities are design variables in problems were both types of singular behaviour are admitted by the field equations. In order to determine the optimal solution, it may therefore be necessary to solve

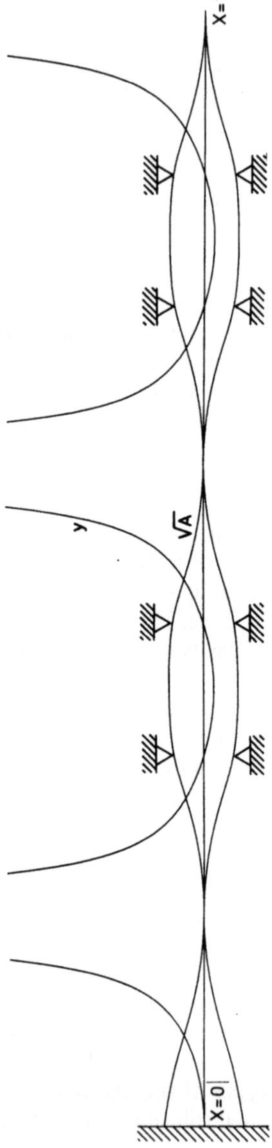

Fig. 8

the problem assuming different sets of singularities, and selecting the optimal solution from among the candidate solutions.

5. Bimodal optimal design

In this section, we shall discuss a problem in which a bimodal rather than a single mode formulation must be considered in order to arrive at the correct optimal design.

The problem is to maximize the Euler buckling load P of a doubly clamped, solid, elastic column ($p = 2$) of given volume V and length L. This problem was first considered in Ref. [23], but as pointed out in [22] an erroneous solution was arrived at. The design shown in Fig. 7 constitutes the correct solution within the premises of a single mode formulation of the problem, and it replaces the design from [23], which also was obtained on the basis of a single mode formulation.

However, the design shown in Fig. 7 is only optimal within the class of doubly clamped columns with *two inner hinges*. It is quite obvious that the column would obtain a greater Euler buckling load for the same volume, length and material, if it were made to buckle in a symmetric fundamental buckling mode with a continuous slope throughout. This could easily be achieved by restributing the given material slightly so that the hinges became locked. The problem is, however, that the field equations of the geometrically unconstrained column do predict zero cross-section (singular behaviour of first type) at points of vanishing bending moment [22], and two such points are necessarily present in a clamped-clamped column whenever a single mode formulation is used.

This clearly indicates that the single mode formulation is inadequate for the problem under consideration, which motivated a reformulation of the problem in [22], leading to a new differential equation for optimality, that does not necessarily lead to vanishing cross-section at points of zero bending moment.

In Ref. [22], the problem was reformulated by assuming the fundamental buckling load of the optimal design to be a *double* eigenvalue. The development is based on a functional defined as the potential energy of the column in its buckled state in two simultaneous modal shapes, including terms representing volume and minimum area constraints, respectively. The latter constraint was included in the formulation because it made it convenient to perform a parametric study of the problem in the entire range from uniform (fully constrained) design to geometrically unconstrained optimal design. By variational analysis of the functional, a *bimodal* set of governing equations was derived for the optimal design problem.

Solving the bimodal set of equations subject to a sufficiently small geometric minimum constraint, this constraint turned out to become inactive in the optimal design. Hence, the geometrically unconstrained solution to the problem was found to have *finite* variable cross-section throughout, with a minimum magnitude of 0.226 V/L.

This bimodal, geometrically unconstrained solution, is symmetrical within double precision computational accuracy, and is shown in Fig. 9 to suitable scale by the linear dimensions $\pm\sqrt{A}$. Fig. 9 also shows modes y_1 and y_2 for the double optimal buckling eigenvalue. Purely symmetrical and antisymmetrical linear combinations of these modes can be constructed. The optimal, double buckling load is found to be 32.62% higher than the buckling load of a uniform, doubly clamped column of the same volume, length and material.

The result shown in Fig. 9 (and obtained in [22]) is a noteworthy counterexample to the generally accepted claim that the optimal solution to a geometrically unconstrained, one-dimensional, single-purpose, structural optimization problem is statically determinate.

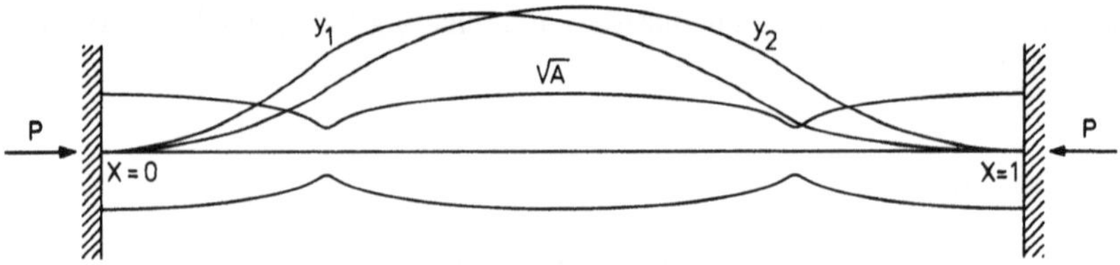

Fig. 9

Recent papers [17, 24-26] and current research [27-30] are devoted to optimal design with bi- or even multimodal optimal eigenvalues for structures of some complexity or statical indenterminacy.

6. Discrete structures

The variational approach to problems of structural optimization yields results, some of which cannot be obtained by discretizing the structure. But in addition the variational formulation is sometimes very useful in generating equations and algorithms for solving the problem. In fact, even some basically discrete structures can advantageously be treated by variational methods after a smear-out process.

Plates with a large number of identical or similar stiffeners like those used in aerospace structures or ships hulls and fiber-reinforced materials including reinforced concrete may often be treated as continuous structures with good accuracy.

A review of the literature on optimal fiber-reinforced plates and grillage-like continua is given by Rozvany in Ref. [31], pp. 171-176. Work by Mroz [32], Prager [33] Rozvany [34] and many others have lead to a series of interesting and important results. In a recent work at our Department by S. Højlund Rasmussen [35] the problem of fibre-reinforced elastic structures has been treated in more generality using a continuous description. Since the work has not been published we shall give a short account of some results obtained.

We shall assume the presence of a finite number N of fields of fibres, each field being smeared out and characterized by its intensity $B_{(s)}$ and direction $b^i_{(s)}$ $(s = 1, ..., N)$ where $B_{(s)}$ and $b^i_{(s)}$ are arbitrary smooth functions of the space variables. The fibres are embedded in an elastic, homogeneous and isotropic matrix, characterized by its shear modulus G_M and its Poisson's ratio ν.

The problem we are interested in is to distribute a given amount of fibres in a geometrically defined matrix such that the rigidity of the structure for given external loads is optimal.

In order to avoid the complexity of the exact (smeared-out) equations of a fibre-reinforced material and to concentrate on the central problem we shall use a simple model for the reinforced material, assuming its total elastic energy W to be

$$W = W_M + \sum_{s=1}^{N} \tfrac{1}{2} B_{(s)} E_{F(s)} \epsilon^2{}_{(s)} \tag{6.1}$$

where W_M is the elastic energy density of the matrix, and the sum the contribution to the elastic energy from the fibres. Here $E_{F(s)}$ is a material constant ($\sim$ Young's modulus) for the fibres in the field s and $\epsilon_{(s)}$ is the normal strain in the direction of the fibres of field s.

This simple model permits us to derive the equations of elasticity in comparatively simple terms. The expression (6.1) will in general be a good approximation if the volume content of the fibres is small everywhere.

The stress-strain relations derived from (6.1) are

$$\sigma^{ij} = 2G_M \left(\epsilon^{ij} + \frac{\nu}{1-2\nu}\,\epsilon^k_k g^{ij}\right) + \\ + \sum_{s=1}^{N} B_{(s)} E_{F(s)}\,\epsilon_{kl}\, b^i_{(s)} b^j_{(s)} b^k_{(s)} b^l_{(s)} \tag{6.2}$$

with the inversion

$$\epsilon^{ij} = \frac{1}{2G_M}\left\{\sigma^{ij} - \tilde{\nu}\sigma^k_k g^{ij}\right\} - \frac{1}{2G_M}\left\{\sigma^k_k \left(\alpha g^{ij} + \sum_{s=1}^{N} \alpha_s b^i_{(s)} b^j_{(s)}\right) + \\ + \sigma_{kl} \sum_{s=1}^{N} b^k_{(s)} b^l_{(s)} \left(\beta_s g^{ij} + \sum_{t=1}^{N} \beta_{st} b^i_{(t)} b^j_{(t)}\right)\right\} \tag{6.3}$$

where α, α_s, β_s and β_{st}, are the solutions to the linear equations

$$\begin{Bmatrix} \frac{1}{\tilde{\nu}} & 1 & 1 & \cdots & 1 \\ 1 & 1+\frac{2G_M}{(BE_F)_{(1)}} & \cos^2(2,1) & \cdots & \cos^2(N,1) \\ 1 & \cos^2(1,2) & 1+\frac{2G_M}{(BE_F)_{(2)}} & \cdots & \cos^2(N,2) \\ \vdots & \vdots & \vdots & \vdots & \vdots \\ 1 & \cos^2(1,N) & \cos^2(2,N) & \cdots & 1+\frac{2G_M}{(BE_F)_{(N)}} \end{Bmatrix} \times \\ \begin{Bmatrix} \alpha+\tilde{\nu} & \alpha_1 & \alpha_2 & \cdots & \alpha_N \\ \beta_1 & \beta_{11} & \beta_{12} & \cdots & \beta_{1N} \\ \beta_2 & \beta_{21} & \beta_{22} & \cdots & \beta_{2N} \\ \vdots & \vdots & \vdots & & \vdots \\ \beta_N & \beta_{N1} & \beta_{N2} & \cdots & \beta_{NN} \end{Bmatrix} = \begin{Bmatrix} 1 & 0 & 0 & \cdots & 0 \\ 0 & 1 & 0 & \cdots & 0 \\ 0 & 0 & 1 & \cdots & 0 \\ \vdots & \vdots & \vdots & & \vdots \\ 0 & 0 & 0 & \cdots & 1 \end{Bmatrix} \tag{6.4}$$

where

$$\cos^2(s,t) = b^i_{(t)}\, b_{(s)i} b^j_{(t)}\, b_{(s)j} \tag{6.5}$$

The material constant $\tilde{\nu}$ is

$$\tilde{\nu} = \begin{cases} \dfrac{\nu}{1+\nu} & \text{for the threedimensional case} \\[2ex] \dfrac{\nu}{1+\nu} & \text{for plane stress} \\ \nu & \text{for plane strain} \end{cases}$$

When formulating the optimality conditions we must take into account a number of side-conditions. If the amount of fibres in each field is prescribed, we have

$$\int_\Omega B_{(s)}\, \mathrm{d}V = B^0_{(s)}, \quad s = 1, ..., N \tag{6.6}$$

or, if the total amount is given

$$\sum_{s=1}^{N} \int_\Omega B_{(s)}\, \mathrm{d}V = B^0 \tag{6.7}$$

Furthermore, since the concentration of fibres cannot be negative,

$$B_{(s)} = \geq 0, \qquad s = 1, ..., N$$

or (6.8)

$$B_{(s)} = h^2_{(s)}, \qquad s = 1, ..., N$$

and since $b^i_{(s)}$ is a unit vector

$$b^i_{(s)}\, b_{(s)i} = 1, \quad s = 1, ..., N \tag{6.9}$$

Finally, there may be cases where no fibres are permitted to end inside the structure. In this case the divergences

$$D_i B^i_{(s)} = 0, \qquad s = 1, ..., N \tag{6.10}$$

Taking all these conditions into account we may construct the following functional

$$\begin{aligned} W = &\int_\Omega G_M \left\{ \epsilon_{ij}\, \epsilon^{ij} + \frac{\nu}{1-2\nu}\, \epsilon^k_k\, \epsilon^l_l \right\} \mathrm{d}V \\ &+ \frac{1}{2} \sum_{s=1}^{N} \int_\Omega (BE_F)_{(s)}\, \epsilon_{ij}\, \epsilon_{kl}\, b^i_{(s)}\, b^j_{(s)}\, b^k_{(s)}\, b^l_{(s)}\, \mathrm{d}V \\ &- \sum_{s=1}^{N} \xi_{(s)} \left\{ \int_\Omega B_{(s)}\, \mathrm{d}V - B^0_{(s)} \right\} - \xi \left\{ \sum_{s=1}^{N} \int_\Omega B_{(s)}\, \mathrm{d}V - B_0 \right\} \end{aligned} \tag{6.11}$$

$$- \sum_{s=1}^{N} \int_{\Omega} \eta_{(s)} \left\{ h^2_{(s)} - B_{(s)} \right\} \mathrm{d}V$$

$$+ \sum_{s=1}^{N} \int_{\Omega} \lambda_{(s)} \left\{ 1 - b^i_{(s)} b_{(s)i} \right\} \mathrm{d}V + \sum_{s=1}^{N} \int_{\Omega} \mu_{(s)} D_i B^i_{(s)} \mathrm{d}V$$

From (6.11) it can be shown that a necessary condition for the reinforcement to be optimal is that the directions of the fibres coincide with the directions of principal strain. Thus there is no need for more than three fields in the general three-dimensional case, or two fields in the plane cases. The optimal fields are of course then orthogonal.

As an example, Fig. 10 shows the optimal reinforcement of a cantilever beam that is loaded with a distributed shear load at the right-hand end.

The dashed lines indicate the division lines between the zones with two, one, and zero fields of fibre reinforcements.

This example will suffice to illustrate the results obtained by variational analysis of smeared-out discrete structures.

7. Concluding remarks

In optimization of structures, the calculus of variations plays a fundamental role. Besides being a natural and convenient way to formulate the problem mathematically, the variational approach leads to a deeper understanding of the limitations imposed by the formulation. In this paper we have discussed some of the problems encountered by this approach and some obstacles that must be removed before the solution can be fully understood and utilized. But we have also indicated an extended field for the application of the variational method to intrinsically discrete structures.

References

1. Wasiutyński, Z., Brandt, A., 'The Present State of Knowledge in the Field of Optimum Design of Structures', *Appl. Mech. Rev.*, **16**, pp. 344-350, 1963.
2. Sheu, C.Y., Prager, W., 'Recent Developments in Optimal Structural Design', *Appl. Mech. Rev.*, **21**, pp. 985-992, 1968.
3. Niordson, F.I., Pedersen, P., 'A Review of Optimal Structural Design', *Proc. 13th Int. Congr. Th. Appl. Mech.* (Ed. E. Becker and G.K. Mikhailov), Moscow, pp. 264-278, Springer-Verlag 1973.
4. Rozvany, G.I.N., Mróz, Z., 'Analytical Methods in Structural Optimization', *Appl. Mech. Rev.*, **30**, pp. 1461-1470, 1977.
5. Keller, J.B., 'The Shape of the Strongest Column', *Arch. Rational Mech. Anal.*, **5**, pp. 275-285, 1960.
6. Keller, J.B., Niordson, F.I., 'The Tallest Column', *J. Math. Mech.*, **16**, pp. 433-446, 1966.
7. Niordson, F.I., 'On the Optimal Design of a Vibrating Beam', *Quart. Appl. Mech.*, **23**, pp. 47-53, 1965.
8. Prager, W., Taylor, J.E., 'Problems of Optimal Structural Design', *J. Appl. Mech.*, **35**, pp. 102-106, 1968.

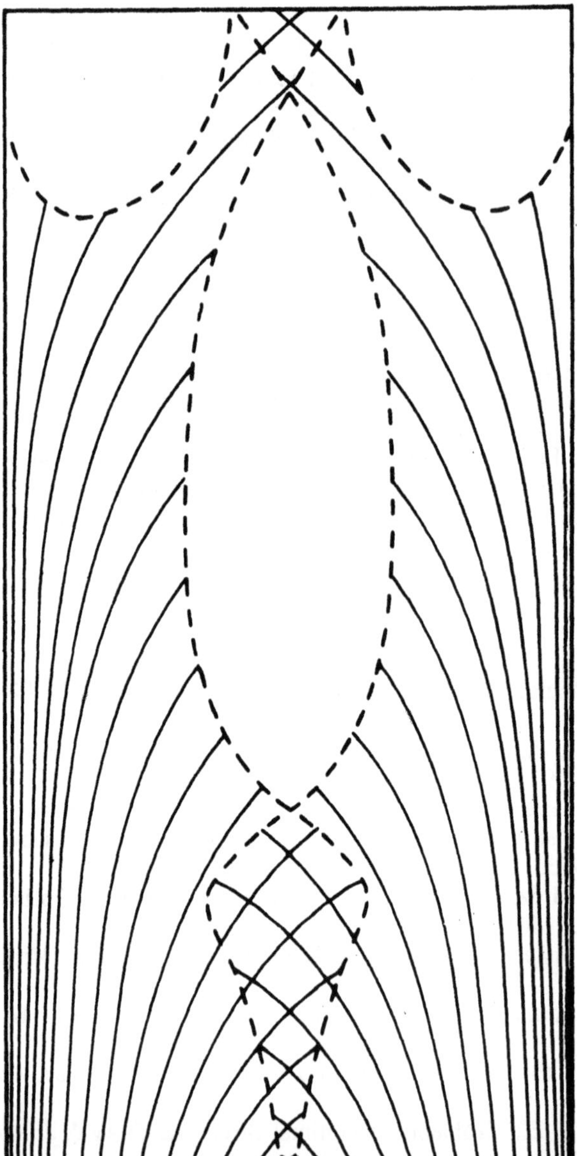

Fig. 10

9. Olhoff, N., 'Optimal Design of Vibrating Rectangular Plates', *Int. J. Solids Struct.*, **10**, pp. 93-109, 1974.
10. Olhoff, N., 'On Singularities, Local Optima and Formation of Stiffeners in Optimal Design of Plates', *Proc. IUTAM Symposium on Optimization in Structural Design* (Ed. A. Sawczuk and Z. Mróz), Warsaw, pp. 82-103, Springer-Verlag 1975.
11. Olhoff, N., Taylor, J.E., 'On Optimal Structural Remodeling', *J. Optimization Theory and Appl.*, **27**, pp. 571-582, 1979.
12. Karihaloo, B.L., Niordson, F.I., 'Optimum Design of Vibrating Cantilevers', *J. Optimization Theory and Appl.*, **11**, pp. 638-654, 1973.
13. Karihaloo, B.L., Niordson, F.I., 'Optimum Design of Vibrating Beams under Axial Compression', *Arch. of Mech*, **24**, pp. 1029-1037, 1972.
14. Karihaloo, B.L., Niordson, F.I., 'Optimum Design of a Circular Shaft in Forward Precession', *Proc. IUTAM Symp. on Optimization in Structural Design* (Ed. A. Sawczuk and Z. Mróz), Warsaw, pp. 142-151, Springer-Verlag 1975.
15. Olhoff, N., 'Optimization of Vibrating Beams with Respect to Higher Order Natural Frequencies', *J. Struct. Mech.*, **4**, pp. 87-122, 1976.
16. Olhoff, N., 'Optimal Design of Vibrating Circular Plates', *Int. J. Solids. Struct.*, **6**, pp. 139-156, 1970.
17. Olhoff, N., Niordson, F.I., 'Some Problems Concerning Singularities of Optimal Beams and Columns', *ZAMM*, **59**, pp. T.16-26, 1979.
18. Brach, R.M., 'On the Extremal Fundamental Frequencies of Vibrating Beams', *Int. J. Solids Struct.*, **4**, pp. 667-674, 1968.
19. Brach, R.M., 'On Optimal Design of Vibrating Structures', *J. Optimization Theory and Appl.*, **11**, pp. 662-667, 1973.
20. Masur, E.F., 'Optimality in the Presence of Discreteness and Discontinuity', *Proc. IUTAM Symposium on Optimization in Structural Design* (Ed. A. Sawczuk and Z. Mróz), Warsaw, pp. 441-453, Springer-Verlag 1975.
21. Masur, E.F., Written discussion to Ref. [10]. *Loc. cit.*
22. Olhoff, N., Rasmussen, S.H., 'On Single and Bimodal Optimum Buckling Loads of Clamped Columns', *Int. J. Solids Struct.*, **13**, pp. 605-614, 1977.
23. Tadjbakhsh, I., Keller, J.B., 'Strongest Columns and Isoperimetric Inequalities for Eigenvalues', *J. Appl. Mech.*, **9**, pp. 159-164, 1962.
24. Masur, E.F., Mróz, Z., 'On Non-stationary Optimality Conditions in Structural Design', *Int. J. Solids Struct.* (to appear).
25. Masur, E.F., Mróz, Z., 'Singular Solutions in Structural Optimization Problems', *Proc. IUTAM Symposium on Variational Methods in the Mechanics of Solids* (Ed. S. Nemat-Nasser), Evanston, Ill., USA, Pergamon 1979 (to appear).
26. Prager, S., Prager, W., *A Note on Optimal Design of Columns*, (to appear).
27. Taylor, J.E., Private communication.
28. Haug, E.J., Private communication.
29. Lurie, K.A., Private communication.
30. Claudon, J.L., Private communication.
31. Rozvany, G.I.N., *Optimal Design of Flexural Systems,*, Pergamon Press, Oxford, 1976.
32. Mróz, Z., 'On the Design of Non-Homogeneous Technically Orthotropic Plates', *Proc. IUTAM Symposium on Non-Homogeneity in Elasticity and Plasticity*, Warsaw, pp. 191-202, Pergamon, Oxford, 1958.
33. Prager, W., *Transmission Optimale des Charges par Flexion, Annales de l'Institut Technique du Batiment et des Travaux Publics*, Serie: *Theories et Methodes de Calcul*, No.

193, pp. 83-91, 1976.

34. Rozvany, G.I.N., 'Grillages of Maximum Strength and Maximum Stiffness', *Int. J. Mech. Sci.*, **14**, pp. 651-666, 1972.
35. Rasmussen, S.H., *Optimization of Fibre-reinforced Structures* (in Danish), Ph.D. Thesis, Dept. of Solid Mechanics, Technical University of Denmark, 1978.

On the Construction of Models of Continua Interacting with an Electromagnetic Field

*L.I. Sedov, A.G. Tsypkin**

Summary

This paper presents a general theory of constructing models for continuous media in the presence of interaction of material bodies with an electromagnetic field with allowance made for electric currents, polarization, and magnetization. The treatment proceeds from a basic variational equation. A closed system of equations is derived for continuous motions, external conditions being given; the system includes Maxwell's equations and equations of state describing polarization, magnetization, and internal mechanical stresses. (It is known [1, 2] that the basic variational equation also yields the conditions at second-order (strong) discontinuities.)

In the presence of electromagnetic fields, too, the basic variational equation for actual phenomena is shown to reduce, locally, to the first and the second laws of thermodynamics. Explanations are offered for a number of important questions arising as one passes from the equations of the first and the second laws of thermodynamics to the universal variational equation. These questions include the meaning of the partial time derivatives used and the variations of tensor components, the concept of energy of an electromagnetic field as a four-dimensional scalar, the selection of the scalar function for the Lagrangian, the choice of the nonzero functional δW^*, the expressions for uncompensated heat for actual processes and those subject to variation, etc. Typical examples of specified models of solid and fluid continua interacting with an electromagnetic field are discussed.

1. Introduction

During the past few years there appeared many papers on the construction of models of continua taking into account electric polarization, magnetization, divided movable charges, and conductivity currents. However, until the present time there has been no reasonably substantiated approach to modelling based upon thermodynamic methods with a minimum of the simplest assumptions. It is essential that these assumptions, which are always inevitable, be introduced on the basis of universal physical principles.

As far back as 1965, L.I. Sedov [3] showed that to obtain a closed macroscopic system of equations for the simplest typical reversible processes in continua basing upon the first and the second laws of thermodynamics (valid for continuous processes), one should specify, in addition to the external conditions, the internal energy of the field and the material medium in terms of the polarization-magnetization tensor, of the variables governing mechanical motion as well as the internal thermodynamic defining parameters. For the models con-

* Moscow University; Mathematical Institute of the USSR Academy of Sciences

sidered this approach provides all the equations of state, particularly those for the laws of polarization and magnetization.

A number of papers [4-7] deal with the further development of this theory using the basic variational equation and with its extension to the processes with weak and second-order discontinuities when the arguments in the Lagrangian involve higher derivatives of the functions sought with respect to time and coordinates. It might be well to point out that for the variational equations applied in the presence of an electromagnetic field there has been no thermodynamic interpretation of the Lagrangian density.

According to the principal idea, in the case of real processes the basic variational equation for a small volume element of the medium and the electromagnetic field is to be reduced, locally, to a total equation of balances for the increments of all sorts of energy generated by interactions in the fields and the material media during the processes under study. This fact can serve as a physical guideline in establishing the form of the functionals entering into the basic variational equation. The latter, however, may also involve terms vanishing for real processes, e.g. elementary influx of energy of a gyroscopic nature or special terms associated with irreversibility of the processes subject to variation, and so on.

The main objects of the present paper are:

i. To clarify the question of local reduction of the basic variational equation to the energy equation for the system 'electromagnetic field plus material medium' considered as a single whole. Up to now, there has been doubt that such a reduction is possible. The discussion below shows that even for reversible processes in electromagnetic fields the structure of the basic variational equation proves to be complicated due to the consideration of interactions of a small volume of the field and the medium with the neighbouring elementary volumes. In the case of irreversible phenomena the occurence of additional terms in the expression for variated increment of uncompensated heat also leads to complications.

ii. To present a closed system of equations including equations of state for the system 'electromagnetic field plus material medium' within the framework of the special theory of relativity.

2. Main symbols and coordinate systems

Let $x^1, x^2, x^3, x^4 = ct$ be coordinates in a chosen inertial Cartesian reference system of the observer in the four-dimensional pseudo-Euclidean space, $ds^2 = (cdt)^2 - (dx^1)^2 - (dx^2)^2 - (dx^3)^2$, where c is the velocity of light in vacuum, t is the time. Let $\xi^1, \xi^2, \xi^3, \xi^4 = c\tau$ be Lagrangian coordinates of the points of the medium in a moving concomitant coordinate system considered as frozen into the medium (a comoving coordinate sytem). We assume, by way of definition, that the quantity $d\tau$ along the world line $\xi^d =$ const. equals the increment of the intrinsic (proper) time. Denote the covariant components of the metric tensor in the reference system of the observer and in the comoving system by g_{ij} and $\hat{g}_{ij}$ ($\hat{g}_{44} = 1$), respectively; $ds^2 = \hat{g}_{ij} d\xi^i d\xi^j$.

It is assumed here and in what follows that the small Roman indices take on the values of 1, 2, 3, 4 while the small Greek ones the values of 1, 2, 3; the summation is carried out over the coincident super- and subscripts. The hat ^ above vector and tensor components indicates that these components are taken with respect to the comoving system of coordinates.

Assuming $ds \neq 0$, denote the contravariant components of the dimensionless 4-dimensional unit velocity vector of the points of the medium by $u^i = \dfrac{dx^i}{d\xi^4} = \dfrac{dx^i}{ds}$; the mass density of the medium is, by definition,

$$\rho = \rho_0\,(\xi^\mu)\,[\det \| \hat{g}_{ij} - \hat{u}_i \hat{u}_j \|]^{-1/2}. \tag{2.1}$$

A similar expression defines the density of free charge, ρ_e. It will readily be seen from the definition (2.1) that the mass density ρ and the density of charge ρ_e introduced in the comoving frame of reference obey the four-dimensional equations of continuity

$$\nabla_i(\rho u^i) = 0, \quad \nabla_i(\rho_e u^i) = 0, \tag{2.2}$$

where ∇_i is the 4-dimensional operator of covariant differentiation in any coordinate system.

Let $\ni_i$ denote basis vectors in the reference system of the observer in the 4-dimensional space and $\hat{\ni}_i$ those in the comoving coordinate system. In addition to both systems, let us introduce, locally,* for every point M of the medium the intrinsic inertial system x^i with the basis tetrad $\ni_i^*$ such that Christoffel three-index symbols Γ_{sj}^{*k} be equal to zero at the point M, the three-dimensional velocity of this point of the medium with respect to the intrinsic coordinate system being zero. The 4-dimensional velocities, $\mathbf{u}^*$, of the points of the intrinsic system will exactly be equal to the 4-dimensional velocity, $\mathbf{u}$, of the point M of the medium. At the neighbouring point M of the medium, in general, $\mathbf{u}^* \neq \mathbf{u}$.

Owing to the choice of the comoving coordinate system ($\hat{g}_{44} = 1$), the following equalities hold at the point M:

$$\mathbf{u}^* = \mathbf{u} = \ni_4^* = \hat{\ni}_4. \tag{2.3}$$

Hence it follows that the axis of time t for the intrinsic system of coordinates is tangent to the world line of the medium at the point M; consequently, $dt^* = d\tau^* = d\tau$.

Since, generally speaking, it is impossible to make $\hat{g}_{4\alpha}(\xi^\alpha, \tau)$ vanishing at every point of the medium, one easily obtains that the equality $\ni_\alpha^* = \ni_\alpha$ cannot be satisfied for all the points of the medium. However, this equality may be fulfilled at any separate point by a proper choice of Lagrangian system of coordinates. The set of intrinsic reference systems with basis tetrahedrals $\ni_i^*$ for every possible point M represents a nonholonomic family; in other words, there is no global reference system with inertial coordinate tetrahedrals $\ni_i^*$ introduced above.

Along the world lines of the points of the medium we have

$$\begin{aligned} &\frac{d\mathbf{u}^*}{d\tau} = \mathbf{0}, \quad \frac{d\ni_i^*}{d\tau} = \mathbf{0}, \quad \frac{d\ni_i^*}{dx^k} = 0, \\ &\frac{d\mathbf{u}}{d\tau} = \mathbf{a}, \quad \frac{d\hat{\ni}_i}{d\tau} = \hat{\Gamma}_{i4}^k \hat{\ni}_k = \frac{d\bar{u}}{d\xi^i} = \hat{\nabla}_i \hat{u}^k \hat{\ni}_k \neq 0, \end{aligned} \tag{2.4}$$

* In the Riemannian space the fact that this is being done locally is essential; in the Minkowski universe the intrinsic (proper) coordinate system can be introduced for every point M as a global system of reference

where **a** is the absolutely defined 4-dimensional acceleration vector for the points of the medium which is, in general, non-vanishing. If the world line is isotropic, then $g_{44} = 0$, $d\tau = 0$ and in this case one can present relationships analogous to (2.4) where $d\tau$ should be substituted by the increment, $d\lambda$, of the corresponding parameter λ defined along the isotropic world line.

To describe interaction effects of an electromagnetic field with a polarizable and magnetizable medium, introduce antisymmetric tensor of electromagnetic field with the components F_{ij} and H^{ij} (the notation coincides with that accepted in the papers [1, 3, 4]):

$$F_{ij} = \left\| \begin{matrix} 0 & B^3 & -B^2 & E_1 \\ -B^3 & 0 & B^1 & E_2 \\ B^2 & -B^1 & 0 & E_3 \\ -E_1 & -E_2 & -E_3 & 0 \end{matrix} \right\|, \quad H^{ij} = \left\| \begin{matrix} 0 & H_3 & -H_2 & -D^1 \\ -H_3 & 0 & H_1 & -D^2 \\ H_2 & -H_1 & 0 & -D^3 \\ D^1 & D^2 & D^3 & 0 \end{matrix} \right\| \tag{2.5}$$

It should be noted that if $\hat{\ni}_i = \ni_i^*$, the components of the tensors F_{ij} and H^{ij} at a corresponding point in the intrinsic and the comoving coordinate systems are identical due to the way of choice of the intrinsic coordinate system.

3. Energy equations for the electromagnetic field and the medium

The equation of energy for the electromagnetic field in any inertial (in particular, intrinsic) system of coordinates can be written as follows:

$$\frac{\partial \mathcal{E}}{\partial \tau} = -\operatorname{div} S - F, \tag{3.1}$$

where, by way of definition, $\mathcal{E} = (\mathbf{BH} + \mathbf{DE})/(8\pi)$ is the energy of the unit volume of the electromagnetic field (a three-dimensional scalar), $S = (c/4\pi)\mathbf{E} \times \mathbf{H}$ is the three-dimensional Poynting's vector, F is the influex of energy from the field to the medium governed by the processes of Joule heat emission, polarization, and magnetization of the continuum.

As may be seen, the quantity $(1/8\pi)(\mathbf{BH} + \mathbf{DE})$, like kinetic energy in Newtonian mechanics, depends on the choice of the inertial system of coordinates. For two different inertial coordinate systems moving with respect to one another with a constant three-dimensional velocity the inequality $\frac{1}{8\pi}(\mathbf{B^*H^*} + \mathbf{D^*E^*}) \neq \frac{1}{8\pi}(\mathbf{B'H'} + \mathbf{D'E'})$ holds, where asterisks and primes indicate, respectively, that the corresponding vectors are calculated in the intrinsic coordinate system and in an arbitrary inertial system. On the other hand, the following scalar relationship is always valid:

$$\frac{1}{8\pi}(\mathbf{B^*H^*} + \mathbf{D^*E^*}) = \frac{1}{16\pi}(F_{ij}H^{ij} - 4u^{*k}u^{*s}F^j_kH_{sj}). \tag{3.2}$$

The 4-dimensional invariant in the right-hand side of (3.2) represents the expression for the electromagnetic field energy in any system of inertia. Here u^{*i} identifies the components of the 4-dimensional velocity of the points of the intrinsic coordinate system in any reference system where the components of the tensors $F_k^{\ j}$ and H_{sj} are expressed in terms of the vectors **E**, **H**, **D**, **B**. Formula (3.2) gives a definition for the energy of the electromagnetic field as a four-dimensional scalar.

In the absence of a medium (i.e. in 'vacuum') there appears, at first sight, some arbitrariness in the choice of the initial ('intrinsic') inertial system of coordinates. This arbitrariness is obviated if it is remembered that the concepts of intensities of electric and magnetic fields acquire meaning only when we place 'test' bodies into vacuum, by which means the introduction of comoving and intrinsic coordinate systems becomes feasible.

In a fixed intrinsic inertial system of coordinates for the point M the expression (3.2) for the electromagnetic field energy may be considered not only at the chosen point M where $\mathbf{u}=\mathbf{u}^*$, but also at the neighbouring points M' where $\mathbf{u}(M') \neq \mathbf{u}^*(M')$. In this case it is important that in Eq. (3.2) u^{*i} stands for the components of the 4-dimensional velocity of points in the intrinsic reference system corresponding to the fixed point M. When differentiating the expression for energy, as given by Eq. (3.2), with respect to 4-dimensional coordinates (particularly, with respect to time), account should be taken of relations (2.4) which are valid in any coordinate system.

The equation of energy for an electromagnetic field has the form (3.1) in any inertial system of coordinates (not only in the intrinsic one). But in an arbitrary coordinate system the quantities $\mathcal{E}$ and F no more represent the electromagnetic field energy in the medium and the influx of energy from the electromagnetic field to the medium, respectively. Making use of the Poynting's equation, one can write energy equation (3.1) in the form [1]

$$\frac{\partial \mathcal{E}^*}{\partial \tau} = \frac{1}{4\pi}\left(\mathbf{H}^* \frac{\partial \mathbf{B}^*}{\partial \tau} + \mathbf{E}^* \frac{\partial \mathbf{D}^*}{\partial \tau}\right) + \mathbf{j}^*\mathbf{E}^* - F.$$

After simple calculations the above equation is reduced to

$$\frac{\partial}{\partial \tau}\left(\frac{\mathbf{B}^*\mathbf{H}^* - \mathbf{D}^*\mathbf{E}^*}{8\pi}\right) = \frac{1}{4\pi}\left(\mathbf{H}^* \frac{\partial \mathbf{B}^*}{\partial \tau} - \mathbf{D}^* \frac{\partial \mathbf{E}^*}{\partial \tau}\right) + \mathbf{j}^*\mathbf{E}^* - F. \tag{3.3}$$

Using 4-dimensional tensors of electromagnetic field, introduce the 4-dimensional invariants

$$\frac{\mathbf{B}^*\mathbf{H}^* - \mathbf{D}^*\mathbf{E}^*}{8\pi} = \frac{1}{16\pi} F_{ij} H^{ij} = L, \quad \mathbf{j}^*\mathbf{E}^* = c I^k F_{ki} u^i,$$
$$\frac{1}{4\pi}\left(\mathbf{H}^* \frac{\partial \mathbf{B}^*}{\partial \tau} - \mathbf{D}^* \frac{\partial \mathbf{E}^*}{\partial \tau}\right) = \frac{c}{8\pi} H^{ij} u^k \nabla_k F_{ij}. \tag{3.4}$$

Here $\mathbf{j}^*\mathbf{E}^* = cI^k F_{ki} u^i$ is Joule heat, I^k is the 4-dimensional electric current with components $I^\alpha = \frac{1}{c} j^\alpha$, $I^4 = \rho_e$. The relations (3.4) allow one to write the equation of energy for the elctromagnetic field (3.3) in the invariant form

$$dL = \frac{1}{8\pi} H^{ij} \nabla_k F_{ij}\, dx^k + I^k F_{ki}\, dx^i - F\, d\tau. \tag{3.5}$$

Here the influx of energy from the medium to the electromagnetic field, $F = F_4$ is defined in the intrinsic system of coordinates.

The equation of energy for a volume element of the continuum can be written in the intrinsic coordinate system as follows:

$$dU = F d\tau + dK, \tag{3.6}$$

where U is a 4-dimensional scalar equal to the total energy of the unit volume taken in intrinsic reference system; F is the influx of energy from the electromagnetic field to the continuum involving Joule heat as well as the influx caused by polarization and magnetization processes; dK is the additional influx of energy to the medium. Now we take dK as $dQ^{(e)} - Q_i dx^i + dW_0$, where $dQ^{(e)}$ is the influx of heat to the medium during the time interval $d\tau$, Q_i is the density of an external body force, dW_0 is an extra influx of energy across the particle boundary as well as owing to the change in structural parameters. Note that if the electromagnetic field is known, Eq. (3.5) can be used to calculate the influx of energy $F d\tau$ from the field to the medium in terms of the field characteristics. If, on the contrary, the motion of the continuum is given (or known), the influx of energy, $-F d\tau$, can be determined from (3.6) in terms of the characteristics of the motion and then substituted into Eq. (3.5).

In an arbitrary reference system the equation of energy for the system 'continuum plus electromagnetic field' can be written as

$$-d(L+U) + \frac{1}{8\pi} H^{ij} \nabla_k F_{ij}\, dx^k + I^k F_{ki}\, dx^i + dQ^{(e)} +$$

$$+ dW_0 - Q_i\, dx^i = 0. \tag{3.7}$$

Using Eqs. (3.5), (3.6) and (3.7) one must take into account the following. Let infinitesimal scalar $d\epsilon$ be expressed as a product of L, U or $L + U$ and a corresponding invariable three-dimensional elementary volume taken in inertial reference system, for example

$$d\epsilon = L\, dV_{03} \quad \text{and} \quad d\,d\epsilon = \frac{dL}{d\tau}\, dV_{03}\, d\tau$$

While operating with intrinsic and concomitant reference system in which basic unit vectors $\ni_i = \overset{*}{\ni}_i$ at point M and the corresponding 'fluid' volumes $d\hat{V}_3$ in concomitant system are different at different points, in point M one may write

$$d\epsilon = \hat{L}\, d\hat{V}_3 \quad \text{and} \quad d\,d\epsilon = \frac{d}{d\tau}\,(\hat{L}\, d\hat{V}_3)\, d\tau$$

At every point of the volume $d\hat{V}_3$ and dV_{03} under consideration at corresponding fixed moments of proper time $\overset{*}{\tau}$ and τ we assume that

$$d\overset{*}{\tau} = d\tau;\ dV_{03} = d\hat{V}_3;\ L = \hat{L} \quad \text{and} \quad \frac{d\,d\epsilon}{d\tau} = \frac{dL}{d\tau}\, dV_{03} = \frac{d(\hat{L}\, dV_3)}{d\tau} \tag{3.8}$$

The volume $d\hat{V}_3$ is variable while $dV_{03} = d\hat{V}_3$ at the given point M, so the following equalities

$$\frac{d}{d\tau}\left(\frac{d\epsilon}{dV_{03}}\right)=\frac{\partial L}{\partial \tau}=\frac{d\hat{L}}{d\tau}+\hat{L}\,\frac{1}{d\hat{V}_3}\,\frac{dd\hat{V}_3}{d\tau}$$

are valid. At the intrinsic reference system the equality

$$\frac{\delta d\hat{V}_3}{d\hat{V}_3}=\frac{\delta d\hat{V}_4}{d\hat{V}_4}=\nabla_i\,\delta x^i$$

holds when $\delta\tau$ is constant at every point of the elementary volume $d\hat{V}_3$ and $\delta x^\alpha, \delta\overset{*}{\tau}=\delta\tau$ are any displacements. So at every point M because of constancy dV_{03} and variation of $d\hat{V}_3$ in any intrinsic reference systems we have

$$L=\hat{L} \quad \text{but} \quad \delta L=\delta\hat{L}+\hat{L}\,\nabla_i\,\delta x^i \tag{3.9}$$

where

$$\hat{L}=\frac{1}{16\pi}\,H^{ij}F_{ij}=\frac{1}{16\pi}\,\hat{H}^{ij}\hat{F}_{ij}$$

is 'tied' to its own reference system in this point.

It is obvious that after multiplication of Eq. (3.9) by $dV_{03}\,d\tau=d\hat{V}_3\,d\tau$ for real or thinkable displacements one can use equalities*

$$\begin{aligned}\delta(L+U)dV_{04}&=[\delta(\hat{L}+\hat{U})+(\hat{L}+\hat{U})\nabla_i\delta x^i]d\hat{V}_4=\delta\,[(\hat{L}+\hat{U})dV_4]=\\&=\delta\,[(\hat{L}+\hat{U})\sqrt{\hat{g}}/\sqrt{\overset{*}{g}}\,]dV_4=\delta\,[(\hat{L}+\hat{U})\,|\overset{*}{x}{}^i{}_j\,|]d\hat{V}_4\end{aligned} \tag{3.10}$$

Below Eq. (3.7) is taken as a main basis for construction of basic variational Eq. (4.9) with using invariants $\hat{L}$ and $\hat{V}$ which are introduced with a help of intrinsic and comoving reference sytems. In this connection equalities (3.10) are very important.

* Equality

$$\frac{d\hat{V}_3}{dV_{03}}=\frac{d\hat{V}_4}{dV_{04}}=\sqrt{\frac{\hat{g}}{\overset{*}{g}}}=|\overset{*}{x}{}^i{}_j\,|$$

where $\hat{g}=|\,\hat{g}_{ij}\,|$ and $\overset{*}{g}=|\overset{*}{g}_{ij}\,|$ can be obtained in a such way. Without limiting community at point M one can assume that $\hat{\ni}_i=\ni_i$. So at this point we have $d\overset{*}{x}{}^i=d\xi^i;\ \hat{g}_{ij}=\overset{*}{g}_{pq}\,\overset{*}{x}{}^p{}_i\,\overset{*}{x}{}^q{}_j$ and further for invariant volumes $d\hat{V}_4$ and dV_{04} one may write

$$d\hat{V}_4=\sqrt{\hat{g}}\,d\xi^1\,d\xi^2\,d\xi^3\,d\xi^4,\quad dV_{04}=\sqrt{\overset{*}{g}}\,d\xi^1\,d\xi^2\,d\xi^3\,d\xi^4$$

hence we have

$$\frac{d\hat{V}_4}{dV_{04}}=\sqrt{\frac{\hat{g}}{\overset{*}{g}}}=|\overset{*}{x}{}^i{}_j\,|$$

and at point M $\overset{*}{x}{}^i{}_j=\delta^i{}_j$ and at neighbouring points M' – $\overset{*}{x}{}^i{}_j\neq\delta^i{}_j$.

4. Definition of variations of tensor functions and the variational equation for the system 'continuum plus electromagnetic field'

Let η^A be a certain scalar or a component of some tensor (here A is a combined symbol representing indices for tensors of different ranks). Given two reference systems, viz. that of the observer, x^i, and the comoving one, ξ^k, the tensor components η^A characterizing the continuum of the electromagnetic field may be looked upon either as functions of coordinates x^i or as functions of ξ^k.

Variations of scalar or tensor functions can be defined by different formulae. For example, one can introduce variations of invariant functions as non-invariant functions depending on the choice of coordinate systems. But variations of scalars and tensors can be defined, respectively, as invariant scalar or tensor functions of coordinates with infinitely small components.

Consider a few possible definitions of variations. Let $\hat{\eta}^A = \hat{\eta}^A(\xi^k)$ be the components of some tensor in the basis $\hat{\ni}_i$ (the former can, in particular, be a scalar). $\hat{\ni}_i$ and ξ^k being fixed, define variations $\delta\hat{\eta}^A$ as infinitely small components of the tensor in one and the same basis $\hat{\ni}_i$ by the formula

$$\delta\hat{\eta}^A = \hat{\eta}'^A(\xi^k) - \hat{\eta}^A(\xi^k), \tag{4.1}$$

where $\hat{\eta}^A$ and $\hat{\eta}'^A$ are respective components of the real and variated tensors in the basis $\hat{\ni}_i$. In an analogous way, assuming η^A to be functions of coordinates x^i define variations $\partial\eta^A$ (with x^i and ∂_i constant) by the formula

$$\partial\eta^A = \eta'^A(x^i) - \eta^A(x^i), \tag{4.2}$$

where η'^A and η^A are the components of tensors in the basis $\ni_i$. Note that the formulae (4.1) and (4.2) define the variations $\delta\hat{\eta}^A$ and $\partial\eta^A$ as arbitrary tensor or scalar functions of Lagrangian or Eulerian coordinates, respectively. This arbitrariness is explained by the fact that the tensors $\hat{\eta}'^A$ and η'^A subjected to variation may be chosen in a variety of ways. The variations $\delta\hat{\eta}^A$ and $\ni\eta^A$ can be considered not only in the bases $\hat{\ni}_i$ and $\ni_i$, respectively, but in any other bases as well.

Since the relation between Eulerian and Lagrangian coordinates of a point of a continuum provides the law of motion of the medium, $x^i = x^i(\xi^k)$, which is also subject to variation, then the variation of the law of motion will be defined by the formulae

$$\delta x^i = x'^i(\xi^k) - x^i(\xi^k).$$

At every given point $x_0^i = x^i(\xi_0^k)$ the variations δx^i, $\delta\hat{\eta}^A$, and $\partial\eta^A$ introduced above can be related by equations resulting from the definition of the tensor η : $\eta = \hat{\eta}^A\hat{\ni}_A = \eta^A\ni_A$ and from the definition

$$\delta\eta = \partial\eta, \tag{4.3}$$

where $\hat{\ni}_A$ and $\ni_A$ are polyadic products made up of the vectors of the bases $\hat{\ni}_i$ and $\ni_i$, respectively.

Based on Eq. (4.3) and the formula $\delta\hat{\ni}_i = \hat{\nabla}_i\delta x^s\hat{\ni}_s$, the following sequence of equalities can be written:

$$\delta\eta = \delta_1 \hat{\eta}^k \hat{\ni}_k = (\delta_2 \hat{\eta}^k + \hat{\eta}^i \hat{\nabla}_i \delta \hat{x}^k) \hat{\ni}_k = \partial_1 \eta^k \ni_k = (\partial_2 \eta^k + \delta x^i \nabla_i \eta^k) \partial_k. \tag{4.4}$$

Here $\delta_1, \delta_2, \partial_1, \partial_2$ denote various possible types of variations of the components of the tensor η. Equalities (4.4) also yield relations between these variations. For example, if $\hat{\ni}_i = \ni_i$, one has

$$\delta_1 \hat{\eta}^k = \partial_2 \eta^k + \delta x^i \nabla_i \eta^k, \quad \delta_2 \hat{\eta}^k = \partial_1 \eta^k - \eta^i \nabla_i \delta x^k. \tag{4.5}$$

When $\hat{\ni}_i \neq \ni_i$, the above equalities can be rewritten as follows:

$$\begin{aligned} \delta_1 \hat{\eta}^j x^k{}_j &= \delta_1 \eta^k = \partial_2 \eta^k + \delta x^i \nabla_i \eta^k, \\ \delta_2 \hat{\eta}^j x^k{}_j &= \delta_2 \eta^k = \partial_1 \eta^k - \eta^i \nabla_i \delta x^k. \end{aligned} \tag{4.6}$$

Similar formulae for covariant components η_k can readily be obtained with allowance made for the equation $\delta \hat{\ni}^i = -\nabla_s \delta \hat{x}^i \hat{\ni}^s$. General formulae for tensors of any rank with an arbitrary position of the indices can be written in analogous fashion.

Concurrent with the variations $\delta_i \hat{\eta}^A$ and $\partial_i \eta^A$ $(i = 1, 2)$ which are tensor components, one can also consider variations which are not components of tensors. To take an illustration, one can introduce variations $\delta'_1 \eta^A$ of the tensor components η^A in the following way:

$$\delta'_1 \eta^A = \partial_2 \eta^A + \delta x^s \cdot \partial \eta^A / \partial x^s, \tag{4.7}$$

where the variation $\partial_2 \eta^A$ has the same meaning as in Eq. (4.5).

Consider a linear scalar form $B_A \delta_1 \eta^A + P_i \delta x^i$, where B_A are components of some tensor of the same rank as that of the tensor $\delta \eta^A$; P_i are components of a vector. The quantities B_A and P_i can acquire specific physical or geometrical meaning pertaining to the concrete phenomenon under study. Using Eqs. (4.4) and (4.7), the above scalar linear form can be transformed into $C_A \delta'_1 \eta^A + Q_i \delta x^i$. Unlike B_A and P_i, the coefficients C_A and Q_i are no longer components of tensors and vectors. The consideration above shows that the use of the variation of the (4.4) type proves to be preferable in a geometrical or a physical sense to the variation of the (4.7) type.

In addition to variations introduced mentally, one can consider actual increments corresponding to solutions of some problems. In this case we shall agree to substitute the variations δ_1 and ∂_2 of the parameters η^A by actual local increments of the parameters η^A in compliance with the formulae

$$\delta_1 \eta^A = d\eta^A = cu^i \nabla_i \eta^A \, d\tau = \dot{\eta}^A \, d\tau, \quad \partial_2 \eta^A = 0. \tag{4.8}$$

The variations δ_1 and ∂_2 designated through δ and ∂ are used below. The equation of energy for the system 'continuum plus electromagnetic field' written in the four-dimensional invariant form considered in Section 3 does not make it possible to obtain an integral equation of energy for a finite volume of the continuum. In the general case of a moving medium in the special and the general theory of relativity it is impossible to introduce the intrinsic time and vector properties common to the whole of the body or a finite portion of it. Therefore, the equation of energy and conservation laws, generally speaking, have no physical meaning for finite volumes. However, the equation of energy (3.7) for a small volume of the continuum suggests the idea of the form of the terms to be specified in the basic variational equation.

The basic variational equation put forward by L.I. Sedov [4] is of the form

$$\delta \int_{V_4} \Lambda \, \mathrm{d} V_4 + \delta W^* + \delta W = 0. \tag{4.9}$$

Here $\mathrm{d} V_4$ is a four-dimensional element of an arbitrary volume V_4 in the pseudo-Euclidean space bounded by a three-dimensional surface Σ_3; Λ is density of the Lagrange function; δW^* is a functional to be specified (for continuous processes it is a space integral over V_4); δW is a functional which is an integral over the three-dimensional surface Σ_3. In the first term of the variational equation (4.9) the integrand is subject to variation while the domain of integration V_4 in considered theory is not varied.

Now we introduce the major assumption that the functions L and U for the system 'continuum plus electromagnetic field' are specified as functions of the following defining parameters:

$$x^i{}_j, s, F_{ij}, \nabla_k F_{ij}, K^B. \tag{4.10}$$

Here $x^i{}_j = \partial x^i / \partial \xi^j$, K^B are constants or given functions of Lagrangian coordinates ξ^k, s is entropy. The quantities K^B may involve, among other things, components of metrical tensor g_{ij} (if the metric of the space is given) and tensor or scalar constants characterizing geometrical and physical properties of the medium. Such an assumption is quite sufficient for constructing many important models of continua.

In the cases when a model constructed has to describe various physical phenomena (such as gyromagnetic effect, irreversibility of magnetization or of deformation, and the like), the set of defining parameters (4.10) should be supplemented with scalar of tensor quantities μ^A (sometimes with their derivatives, too) which are responsible for additional internal degrees of freedom of the physical model under study. The subsequent formulae and conclusions are easily generalized to these cases. Instead of the components of the tensor F_{ij}, the components of the polarization-magnetization tensor may be introduced: $P_{ij} = (1/4\pi)(F_{ij} - H_{ij})$, which in many cases leads to the same models [3, 4, 5, 7].

Let us introduce into the Lagrangian Λ, in addition to $x^i{}_j$ and F_{ij}, components of a 4-dimensional vector $\mathbf{A} = (A^i)$ (which will be found to be the vector potential of the field) and components of the tensor H^{ij} as the functions sought for.

For the Lagrangian Λ we take an expression composed of two summands; one of them is equal to the term $-(L + U) \equiv -(\hat{L} + \hat{U})\sqrt{g/\overset{*}{g}}$. entering into the lefthand side of Eq. (3.7) and taking into account (3.10) under the sign of the differential, another one represents a supplementary term Λ' making possible for Maxwell's equations to be obtained from the variational equation. Thus, we put

$$\Lambda = -(L + U) + \Lambda', \quad \text{where} \quad \Lambda' = \frac{1}{8\pi} F_{ij} H^{ij} - \frac{1}{4\pi} H^{ij} \nabla_i A_j. \tag{4.11}$$

It is shown below that for actual processes the term Λ' equals zero identically by virtue of Euler's equations.

Assuming that the processes considered permit introducing temperature T of the medium, the second law of thermodynamics for the system 'continuum plus field' can be written as $\rho T ds = dQ^{(e)} + dQ'$, where s is entropy per unit mass of the material medium, dQ'

is uncompensated heat. For simplicity, we assume that the uncompensated heat is brought about by two mechanisms only: the emission of Joule heat and dissipative processes governed by the tensor $\tau_i{}^k$

$$dQ' = I^k F_{ki} dx^i + \tau_i{}^k \nabla_k dx^i$$

Irreversible phenomena concerning magnetization and polarization of the medium are not taken into account here.

Further we accept a major assumption that for the processes subjected to variation the following equations are valid*:

$$\delta Q' = I^k F_{ki} \delta x^i - I^k \partial A_k + \tau^k{}_i \nabla_k \delta x^i \qquad (4.12)$$

Making use of Eqs. (4.12), the Second Law of thermodynamics can be presented as follows:

$$\rho T \delta s = \delta Q^{(e)} - I^k \partial A_k + \tau_i{}^k \nabla_k \delta x^i \qquad (4.13)$$

This equation can be considered as the definition of $\delta Q^{(e)}$. The functional δW^* is a space integral of the left-hand side of Eq. (3.7) with the exception of the terms $-d(L + U)$ (which enters into Λ) and dW_0 which transforms into δW, the variation δW^* being obtained upon substitution of mental increments for actual ones:

$$\delta W^* = \int\limits_{V_4} \left[\frac{1}{8\pi} H^{mn} \nabla_i F_{mn} \delta x^i + \rho T \delta s + I^k \partial A_k - \tau_i{}^k \nabla_k \delta x^i - Q_i \delta x^i \right] \mathrm{d} V_4 \qquad (4.14)$$

5. The system of equations of mechanics and electrodynamics

Let us derive a system of coupled equations of continuum mechanics and electrodynamics by means of the variational equation (4.9). We take the expression for the Lagrangian Λ in the form (4.11) and that for the functional δW^* in the form (4.14). Variation of the first term in (4.9) will be carried out with allowance made for the equality

$$\delta \int\limits_{V_4} \Lambda \,\mathrm{d} V_4 = \int\limits_{V_4} [\partial \Lambda + \delta x^i \nabla_i \Lambda] \,\mathrm{d} V_4$$

Assuming that the variations ∂H^{ij}, ∂A_k, δx^i, δs, ∂F_{ij} are continuous and linearly independent, from the variational equation (4.9) one obtains a system of coupled equations of mechanics and electrodynamics valid in an arbitrary coordinate system. This system of equations is arrived at by equating to zero the coefficients in independent variations in the space integral of Eq. (4.9).

The coefficients in the variations ∂H^{ij} and ∂A_k lead to Maxwell's equations:

$$F_{ij} = \nabla_i A_j - \nabla_j A_i, \quad \nabla_i H^{ki} = 4\pi I^k. \qquad (5.1)$$

* The occurence of the term $-I^k \partial A_k$ can be justified, for example, by the fact that it yields Maxwell's equations which summarize experimental data. In this regard attention should be also drawn to the non-conservative character of equations for the electromagnetic field in the presence of conduction currents

The coefficients in the variations δx^i (with the account of (5.3) and (5.4)) lead to the equations of

$$\nabla_k \left[\frac{\partial(L+U)}{\partial x^i_{\ j}} x^k_{\ j} \right] - \nabla_k \left[\frac{\partial(L+U)}{\partial \nabla_k F_{mn}} \nabla_i F_{mn} \right] + \nabla_k \tau_i^k = Q_i + \frac{\partial(L+U)}{\partial K^B} \nabla_i K^B \tag{5.2}$$

The coefficients in the variation δs lead to the formula for the temperature T:

$$\rho T = \frac{\partial(L+U)}{\partial s} \tag{5.3}$$

The coefficients in independent variations $\partial F_{ij}(i>j)$ in this case lead to the equations of state for H^{ij}, the arguments $F_{ij}(i<j)$ in the function $L+U$ being replaced by $-F_{ij}(i>j)$

$$\frac{1}{4\pi} H^{ij} = \frac{\partial(L+U)}{\partial F_{ij}} - \nabla_k \frac{\partial(L+U)}{\partial \nabla_k F_{ij}} . \tag{5.4}$$

If the scalar $L+U$ is independent of the argument $\nabla_k F_{ij}$, the equations of state (5.4) take a simpler form:

$$\frac{1}{4\pi} H^{ij} = \frac{\partial(L+U)}{\partial F_{ij}} \quad (i>j).$$

In addition to Eqs. (5.1) to (5.4), the variational equation yields the following expression for the functional δW:

$$\delta W = \int_{\Sigma_3} [P_i^k \, \delta x^i + N^{ik} \, \partial A_i + M^{ijk} \, \partial F_{ij}] \, \eta_k \, d\sigma_3 \tag{5.5}$$

where

$$P_i^k = \frac{\partial(L+U)}{\partial x^i_{\ j}} x^k_{\ j} + \tau_i^k$$

$$N^{ik} = -(1/4\pi) H^{ik}, \quad M^{ijk} = \partial(L+U)/\partial(\nabla_k F_{ij}). \tag{5.6}$$

It is obvious that

$$\nabla_k P_i^k = Q_i + \frac{\partial(L+U)}{\partial K^B} \nabla_i K^B + \nabla_k \left[\frac{\partial(L+U)}{\partial \nabla_k F_{mn}} \nabla_i F_{mn} \right] \tag{5.7}$$

As a consequence of Euler's equations one can obtain the following relation

$$\rho T \frac{ds}{d\tau} = \frac{dU}{d\tau} + Q_i \frac{dx^i}{d\tau} + \nabla_k S_i^k \frac{dx^i}{d\tau} - \nabla \tau_i^k \frac{dx^i}{d\tau} + I^k F_{ki} u^i + - \nabla_k \left[\frac{dx^i}{d\tau} x^k_{\ j} \frac{\partial(L+U)}{\partial x^i_{\ j}} \right] \tag{5.8}$$

Here $s_i^{\ k}$ are the components of Minkowski tensor. If $\delta K^B = 0$ then $dK^B/d\tau = 0$, generally $\delta x^i \nabla_i K^B \neq 0$.

It is easy to see that Eq. (5.8) will be the same if one assumes that for phenomena being varied $\partial K^B = 0$. Consequently in the reference system of observer for the given functions $K^B(\xi^\alpha, \tau)$ one has

$$\delta K^B = \delta x^i \nabla_i K^B$$

and for actual phenomena one gets $\delta k^B = dk^B$ while $dk^B/d\tau$ may be different from zero.

With $\partial K^B = 0$ upon varying in the equations of impulses (5.2) and (5.7), one will have in the right part $Q_i + [\partial(L + U)/\partial k^B] \nabla_i K^B$ instead of Q_i, and thus $[\partial(L + U)/\partial K^B] u^i \nabla_i K^B$ will be absent in the Eqs. (5.8) and (5.8′).

Equality (5.8) together with the Second Law of thermodynamics after some transformations results in the equation of heat influx

$$\frac{d(L+U)}{d\tau} = \frac{dQ^{(e)}}{d\tau} - Q_i \frac{dx^i}{d\tau} + I^k F_{ki} \frac{dx^i}{d\tau} + \frac{1}{8\pi} [H^{jk} \nabla_i F_{jk}] \frac{dx^i}{d\tau} + \\ + \nabla_k \left(\tau_i^{\ k} \frac{dx^i}{d\tau}\right) + \nabla_k \left[\frac{dx^i}{d\tau} x^k_{\ j} \frac{\partial (L+U)}{\partial x^i_{\ j}}\right]. \qquad (5.8')$$

It is easy to see that Eq. (5.8′) is an explicit form of Eq. (3.7). It was implied throughout the preceding equations (5.1) to (5.9) that L and U are the functions (of the arguments (4.10)) to be specified. For the variational equation (4.9) to be turned locally into energy equation, it is necessary that the function L be defined by Eq. (3.4); the latter, in view of Eq. (5.4), results in

$$\frac{1}{8\pi} H^{ij} F_{ij} = L = \frac{1}{2} \left[\frac{\partial(L+U)}{\partial F_{ij}} - \nabla_k \frac{\partial(L+U)}{\partial \nabla_k F_{ij}}\right] F_{ij} \quad (i > j) \qquad (5.10)$$

This relation provides a physical constraint on the functions L and U to be specified.

6. On various models of solid, liquid, and gaseous media

Proceeding from the general equations (5.1), (5.2) and the equations of state (5.3), (5.4), one can obtain concrete models of continua. To this end it is necessary to specify the functions L, U of the defining parameters (with allowance made for Eq. (5.10)) and to fix physical laws governing irreversible processes for the 4-dimensional vector of current I^k (relations of the Ohm's law type), for tensor components $\tau_i^{\ k}$ (relations of the type of the laws for viscous stresses). In addition to this, information concerning external forces Q_i as well as physical or geometrical parameters K^B governing macroscopic or microscopic structure of the medium should be available.

In taking this approach, the system of relations obtained includes, in particular, the model of a non-linearly elastic body involving polarization and magnetization effects, the models of ideal or viscous liquids or gases as well as of magneto- and electrodynamics, the models of ferromagnetic liquids and a great variety of other models, both already investigated and to be constructed to fit particular classes of phenomena.

Consider some properties of the equations of state for H^{ij} (5.4), the equations of state (5.6) which provide expressions for the components of energy-momentum tensor $P_i^{\,k}$. One can assume, in particular, that the function L is a quadratic form of antisymmetric components of the tensor F_{ij}:

$$L = (1/16\pi)\, c^{ijkl} F_{ij} F_{kl}. \tag{6.1}$$

Here tensor components c^{ijkl} are some specified functions of the defining parameters $x^i{}_j$, entropy s, and of K^B. It follows from (6.1) that $c^{ijkl} = c^{klij} = -c^{jikl} = c^{ijlk}$.

Moreover, we assume, in compliance with (5.10), that U is independent of F_{ij} and $\nabla_k F_{ij}$. Then the equations of state (5.4) for the electromagnetic field are

$$H^{ij} = c^{ijkl} F_{kl}, \quad \text{where} \quad c^{ijkl} = c^{ijkl}(x^p{}_q, s, K^B). \tag{6.2}$$

If c^{ijkl} depend only on g_{ij}, ρ, u^{*i}, s, and, perhaps, on some other scalar quantities, then there is spatial isotropy and hence the components c^{ijkl} will be determined as*

$$c^{ijkl} = \frac{1}{2}\left[\frac{1}{\mu}(g^{ik}g^{jl} - g^{il}g^{jk}) + \left(\epsilon - \frac{1}{\mu}\right)(g^{ik}u^{*j}u^{*l} - g^{jk}u^{*i}u^{*l} + \right.$$
$$\left. + g^{jl}u^{*i}u^{*k} - g^{il}u^{*j}u^{*k})\right]. \tag{6.3}$$

Here ϵ and μ are the dielectric constant and magnetic permeability, respectively, which may depend on ρ and s or on ρ and T and, perhaps, on some variable or constant scalar parameters in more general cases. In the intrinsic system of coordinates Eq. (6.2) reduces, by virtue of (6.3), to the widely used formulae $D = \epsilon E$, $B = \mu H$.

In the case when the material medium is neither magnetized nor polarized, $\epsilon = 1, \mu = 1$. If the medium is polarized but not magnetized, then $\mu = 1$, $\epsilon \neq 1$. Finally, if the medium is magnetized but not polarized, then $\epsilon = 1, \mu \neq 1$. Such a simple situation is inherent in the approach considered with the chosen set of defining parameters (4.10) and the supplementary assumptions accepted above.

In general case in the intrinsic system of coordinates calculating the components of tensor $P_i^{\,k}$ by formula (5.6) in accordance with formula (3.13) we must put

$$L + U = \left[\frac{1}{16\pi} \hat{g}^{ik}\hat{g}^{jl}\hat{F}_{ij}\hat{H}_{kl} + \hat{U}\right]\sqrt{\frac{\hat{g}}{\overset{*}{g}}}$$

while

$$\overset{*}{g}{}^{pq} = \hat{g}^{mn}\,\overset{*}{x}{}^p{}_m\,\overset{*}{x}{}^q{}_n \quad \text{and} \quad \partial\overset{*}{g}{}^{pq}/\partial x^i{}_j = 0.$$

As a result we shall receive a formula in which we shall not write asterisks as this formula remains the same in the arbitrary system of coordinates

* For particular kinds of anisotropy other formulae can readily be presented in place of (6.3) [1]

$$P_i^{\,k} = \frac{\partial \hat{U}}{\partial x^i{}_j} x^k{}_j + \hat{U}\delta^k{}_i + \tau_i^{\,k} + \frac{1}{2}(S_i^{\,k} + S^k{}_i) + \frac{1}{16\pi} g^{mn} g^{pq} \hat{F}_{mp} \frac{\partial \hat{H}_{nq}}{\partial x^i{}_j} x^k{}_j. \tag{6.4}$$

If polarization and magnetization of media are absent, then $\hat{H}_{nq} = \hat{F}_{nq}$ and $\partial \hat{H}_{nq}/\partial x^i{}_j = 0$ and Minkowski tensor becomes symmetrical to $S^{ik} = S^{ki}$ and turns into Maxwell tensor. In this case formula (6.4) gets simplified and takes the form

$$P_i^{\,k} = \frac{\partial \hat{U}}{\partial x^i{}_j} + \hat{U}\delta_i^{\,k} + \tau_i^{\,k} + S_i^{\,k} \tag{6.5}$$

while $-\nabla_k S_{i\cdot}^{\cdot k} = F_i$, where F are components of vector of Lorentz four-dimensional force.

If polarization and magnetization take place and equation of state (5.4) in which ϵ and η do not depend on $x^i{}_j$ and, in particular, on density ρ, but can depend on entropy s and other independent parameters, is valid, so the last member in formula (6.4) differs from zero and equals $(S_{i\cdot}^{\cdot k} - S^{k\cdot}_{\cdot i})/2$. In this case Minkowski tensor is generally asymmetrical and formula (6.4) becomes analogous to formula (6.5) and takes a form

$$P_i^{\,k} = \frac{\partial \hat{U}}{\partial x^i{}_j} x^k{}_j + \hat{U}\delta_i^{\,k} + \tau_i^{\,k} + S_{i\cdot}^{\cdot k} \tag{6.6}$$

If coefficients ϵ and μ depend on density ρ, so additional member $\widetilde{P}_i^{\,k}$ appears in the right-part of formula (6.6) and for this member the following formula

$$\widetilde{P}_i^{\,k} = \frac{1}{8\pi}\left[\frac{1}{2}\,\frac{\rho}{\mu^2}\,\frac{\partial \mu}{\partial \rho} F_{mn}F^{mn} - \rho\left(\frac{1}{\mu^2}\,\frac{\partial \mu}{\partial \rho} + \frac{\partial \epsilon}{\partial \rho}\right) F_{mn}F^n{}_s u^m u^s\right](\delta_i^{\,k} - u_i u^k)$$

is valid. In the intrinsic system of coordinates this member may be rewritten in the form

$$\widetilde{P}_\alpha^{\,\beta} = -\frac{\rho}{8\pi}\left(\frac{\partial \mu}{\partial \rho} H^2 + \frac{\partial \epsilon}{\partial \rho} E^2\right)\delta_\alpha^{\,\beta} \quad (\alpha, \beta = 1, 2, 3)$$

while $\widetilde{P}_\alpha^{\,4} = \widetilde{P}_4^{\,\beta} = \widetilde{P}_4^{\,4} = 0$

Tensor $\widetilde{P}_\alpha^{\,\beta}$ represents additional pressure.

To determine tensor components $\tau_i^{\,k}$ related to dissipation laws, additional assumptions are required which may be of a different nature, but this problem is not treated here.

On the face of it, the above method for constructing models with the aid of the basic variational equation (4.9) may appear rather sophisticated and artificial, but this sophistication is unavoidable since it is inherent in the problems at hand. Furthermore, presently the use of variational 'principles' (without δW^* and δW) has already become dominant and, perhaps, the only source for constructing examples of new models in the theory of relativity and other physical theories.

It should be emphasized that the question of establishing new physical-mechanical models presents a problem of fundamental importance which has to be investigated and solved only once for numerous separate applications later on. In a similar manner, one introduces the simplest models of an ideal fluid and an elastic body, with ensuing applications to hydro- and aerodynamics, structural mechanics and other fields.

In practice, when solving particular problems numerically, one can directly use the equation (4.9); for the models considered it contains not only the closed system of Euler's equations, but the additional initial and boundary conditions as well. Moreover, owing to the integral form of this equation, the conditions at second-order discontinuities are satisfied automatically.

References

1. Sedov L.I., *A Course in Continuum Mechanics*, vols. 1-4, Wolters-Noordhoff Publ., 1971-1972. (In French: *Mécanique des Milieux Continus*. Moscou, Mir. 1975)
2. Sedov L.I., 'On conditions at second-order discontinuities in the theory of gravitation', *J. Appl. Math. and Mech.* (PMM), 1972, **36**, 1, p. 1-11.
3. Sedov L.I., 'On the ponderomotive forces of interaction of an electromagnetic field and an accelerating material continuum taking into account finite deformations', *J. Appl. Math. and Mech.* (PMM), 1965, **29**, 1, p. 2-17.
4. Sedov L.I., 'Mathematical methods for constructing new models of continua', *Uspekhi Matematicheskikh Nauk*, 1965, **20**, 5, p. 121-180 (in Russian).
5. Zhelnorovich V.A., 'Models of media with internal electromagnetic and mechanical momenta', in: *Problems of Hydrodynamics and Continuum Mechanics*, Philadelphia, SIAM, 1969, p. 803-815.
6. Shtein A.A., 'Models of polarizable media and average relations which correspond to these in the case of high-frequency electromagnetic field', *J. Appl. Math. and Mech.* (PMM), 1977, **41**, 2, p. 262-272.
7. Chyorny L.T., *Construction of models of magnetoelastic continua with account of magnetic hysteresis and plastic deformations*, Scientific Papers of the Institute of Mechanics, Moscow University, 1974, No. 31, p. 100-119 (in Russian).

Special Cases of the Nonlinear Shell Equations*

*J.G. Simmonds***

Summary

The paper begins with a simple derivation, from three dimensions, of the exact integral equations of motion of a shell. This yields, in a straightforward way, definitions of two-dimensional stresses and translational and rotational variables. The addition of strain measures and stress-strain relations, consistant with a work principle, leads to various forms of the equations of nonlinear first-approximation theory, depending on the choice of the unknowns. Known, new, and conjectured simplifications of the general equations are discussed for special midsurface geometries (planes and helicoids) and special loads. In particular, the von Karman equations are extended to plates undergoing arbitrarily large rotations.

1. Introduction

Today, the equations of nonlinear, first-approximation shell theory and their attendant errors are well-established, thanks, in no small measure, to the efforts and *inspiration* of the man this symposium honors: Warner Tjardus Koiter. One might thus conclude, naively, that shell theory can now be left to the numerical analyst and pure mathematician. But Warner, I think, would be the first to protest that the work of the applied mechaniker has just begun.

The equations of motion of an elastically isotropic shell undergoing small strain but large deflection are of immense complexity. To mount a frontal assault on THE SHELL PROBLEM – to solve the equation of motion for arbitrary geometry and arbitrary initial and boundary conditions – seems patently absurd. The present paper suggests what ground has and may possibly be gained by attacking special cases of the general equations.

We begin with a derivation of the general equations of motion that leads to a natural definition of two-dimensional translation and rotation variables. After discussing, briefly, supplemental compatibility and stress-strain relations, we consider special geometries and special loads. Much of the material herein was stimulated by a number of discussions with Prof. Avinoam Libai while he was a member of the Center for Advanced Studies at the University of Virginia.

* This work was supported by the National Science Foundation under Grant MCS-76-10454.
** Dept. of Applied Mathematics and Computer Science, University of Virginia Charlottesville, VA 22901, USA

2. The equations of first-approximation shell theory

2.1. Derivation of equations of motion

The equations of motion of a shell must be exact consequences of the equations of motion of a three-dimensional continuum. To derive *differential* equations one may follow Naghdi's lead in the linear, static theory [1] and integrate the three-dimensional equations through the thickness of the shell. Alternatively, one may simple specialize, to shell-like bodies, the following *integral* equations of motion of a three-dimensional continuum:

$$\iint_{\partial V} \mathbf{S}^i N_i \,\mathrm{d}S + \iiint_V \mathbf{f}\,\mathrm{d}V = \Big(\iiint_V \rho \dot{\mathbf{R}}\,\mathrm{d}V\Big)^{\cdot} \tag{2.1}$$

$$\iint_{\partial V} \mathbf{R}\times \mathbf{S}^i N_i \,\mathrm{d}S + \iiint_V \mathbf{R}\times\mathbf{f}\,\mathrm{d}V = \Big(\iiint_V \rho \mathbf{R}\times \dot{\mathbf{R}}\,\mathrm{d}V\Big)^{\cdot}. \tag{2.2}$$

Here V represents an arbitrary volume of the reference configuration of the continuum and ∂V is its piecewise smooth boundary. Points in V are represented by a position vector of the form $\mathbf{X}(\xi^1, \xi^2, \xi^3) \equiv \mathbf{X}(\xi^i)$. $\mathbf{S}^i N_i \mathrm{d}S$ is the net force acting across an oriented differential element of surface area that initially coincided with $\mathbf{N}\mathrm{d}S$, where $\mathbf{N} = N_i \mathbf{g}^i = N^i \mathbf{g}_i$ is the outward unit normal vector on $\mathrm{d}S$ and $\mathbf{g}_i = \partial\mathbf{X}/\partial\xi^i$. $\mathbf{f}$ is a body force/initial volume and ρ is the initial mass density. Finally, $\mathbf{R}$ is the position vector of a particle that initially lay at $\mathbf{X}$ and a dot denotes differentiation with respect to time. A shell is a body whose initial configuration can be represented in the form

$$\mathbf{X}(\xi^i) = \mathbf{X}(\xi^\alpha, z) = \mathbf{x}(\xi^\alpha) + z\mathbf{n}(\xi^\alpha),\quad H_-(\xi^\alpha) \leqq z \leqq H_+(\xi^\alpha),\quad \alpha = 1,2, \tag{2.3}$$

where $\mathbf{x}(\xi^\alpha)$ defines the shell reference surface and $\mathbf{n}$ is its normal vector. Let A be an arbitrary connected area of the reference surface with a piece-wise smooth boundary ∂A. Further let νds be an oriented differential element of arc of ∂A, where $\nu = \nu^\alpha \mathbf{a}_\alpha$ is an outward normal unit vector and $\mathbf{a}_\alpha = \partial\mathbf{x}/\partial\xi^\alpha$. Then, with the aid of some well-known results from differential geometry, (2.1) may be expressed as

$$\int_{\partial A} \Big(\int_-^+ \mathbf{S}^\alpha \mu \mathrm{d}z\Big)\nu_\alpha \mathrm{d}s + \iint_A \Big\{\big[\mathbf{S}^\alpha T^\beta_\alpha H,_\beta^{\,2} + \mathbf{S}^3 \mu\big]_-^+ + \int_-^+ \mathbf{f}\mu\mathrm{d}z\Big\}\mathrm{d}A$$
$$= \Big[\iint_A \Big(\int_-^+ \rho\mathbf{R}\mu\mathrm{d}z\Big)^{\cdot}\mathrm{d}A\Big]^{\cdot}. \tag{2.4}$$

Here $\pm$ is short for $H\pm(\xi^\alpha)$ and $\mu^2 = 1 - 2zM + z^2K$, where M and K are, respectively, the mean and Gaussian curvatures of the reference surface. The tensor T^β_α depends on the geometry of the reference surface only. Its specific form may be computed readily but is not needed in what follows. Let $m = \int_-^+ \rho\mu \mathrm{d}z$ denote the mass/initial area and let $\mathbf{r} = \int_-^+ \rho\mathbf{R}\mu\mathrm{d}z/m$ denote a weighted, two-dimensional position vector.* Then with an obvious identification of variables, (2.4) takes the form of the integral equation of conservation of linear momentum for a shell:

$$\int_{\partial A} \mathbf{N}^\alpha \nu_\alpha \mathrm{d}s + \iint_A \mathbf{p}\mathrm{d}A = \Big(\iint_A m\dot{\mathbf{r}}\mathrm{d}A\Big)^{\cdot}. \tag{2.5}$$

* It is conventional but not necessary to choose H_+ and H_- so that $\mathbf{r} = \mathbf{x}$ at the initial time $t = 0$. A related definition in the linear theory of beams has been given by Cowper [2].

Now set

$$\mathbf{R}(\xi^\alpha, z, t) = \mathbf{r}(\xi^\alpha, t) + \mathbf{v}(\xi^\alpha, z, t). \tag{2.6}$$

With the aid of (2.3), (2.4), and the definitions of m and $\mathbf{r}$, it follows that (2.2) takes the form

$$\begin{aligned}
&\int_{\partial A} (\mathbf{r} \times \mathbf{N}^\alpha + \textstyle\int_-^+ \mathbf{v} \times \mathbf{S}^\alpha \mu dz)\nu_\alpha ds \\
&\qquad + \iint_A \{\mathbf{r} \times \mathbf{p} + [\nu \times (\mathbf{S}^\alpha T_\alpha^\beta H,_\beta^2 + \mathbf{S}^3\mu)]_-^+ + \textstyle\int_-^+ \mathbf{v} \times \mathbf{f}\mu dz\}\, dA \\
&\qquad = [\iint (m\mathbf{r} \times \dot{\mathbf{r}} + \textstyle\int_-^+ \rho \mathbf{v} \times \dot{\mathbf{v}}\mu dz)\, dA]^{\cdot}.
\end{aligned} \tag{2.7}$$

Let $I = \int_-^+ \rho z^2 \mu dz$ denote the moment of inertia/initial area and let $\omega = \int_-^+ \rho\nu \times \dot{\nu}\mu dz / I$ denote a weighted, two-dimensional angular velocity. Then, with an obvious identification of variables, (2.7) takes the form of the integral equation of conservation of angular momentum for a shell:

$$\begin{aligned}
&\int_{\partial A} (\mathbf{r} \times \mathbf{N}^\alpha + \mathbf{M}^\alpha)\nu_\alpha ds + \iint_A (\mathbf{r} \times \mathbf{p} + \mathbf{q})\, dA \\
&\qquad = [\iint (m\mathbf{r} \times \dot{\mathbf{r}} + I\omega)\, dA]^{\cdot}.
\end{aligned} \tag{2.8}$$

We now have exact equations of motion that involve well-defined, two-dimensional stresses and velocities, $\mathbf{N}^\alpha$, $\mathbf{M}^\alpha$, $\dot{\mathbf{r}}$ and ω. Using the divergence theorem, we may convert (2.5) and (2.8) into differential equations. These agree in form with (19) and (20) of [3] provided that, in the latter equation, we set $\mathbf{I} \cdot \omega = I\omega$. When (19) and (20) of [3] were derived, it was believed that the dynamic terms 'introduced unavoidable approximations from the start.' The above derivation shows that these terms are, in fact, exact.

2.2. **Virtual work and strain measures**

By insisting that the shell equations of motion follow from a work principle, we are lead automatically to a definition of two-dimensional extensional and bending strains. These may take several alternate forms. Budiansky [4], for example, expresses the strains in terms of the displacement vector $\mathbf{u} = \mathbf{r} - \mathbf{x}$ and its derivatives. The author and Danielson [3] take, initially, the extensional strain as a basic kinematic variable but express the bending strain (and ω) in terms of a finite rotation vector and its derivatives. The compatibility equation $\epsilon^{\alpha\beta}\mathbf{r}_{,\alpha\beta} = 0$, where the $\epsilon^{\alpha\beta}$ are the contravariant components of the permutation tensor, is then appended. Finally, Koiter [5] takes, initially, both the extensional and bending strains as kinematic variables and adjoins the Gauss-Codazzi compatibility equations of surface theory.

2.3. **Stress-strain relations**

The only errors in shell theory occur in the stress-strain relations. Thanks to the work of Hildebrand, Reissner, and Thomas, Goldenveiser, Koiter, John, N. Burger, and others (see [6] for references to these and related papers), these errors and their implications are fairly well understood.

2.4. The complete field equations of static, first-approximation shell theory

These take various forms depending on the choice of stress and/or kinematic variables. Equations (87) and (89) of [4] apply if displacements are chosen; (83)-(88) of [3] if stress resultants (or stress-functions) and rotations are chosen; and (4.14)-(4.17) of [6] if stress resultants and bending strains are chosen. For an infinite circular cylindrical shell of initial curvature k, rigid in the axial direction and subject to a constant pressure $\bar{p}$, the equations take the following one-dimensional form.

displacement components u *and* w *as unknowns* [4]

$$n' + kq + \bar{p}\beta = 0, \quad q' - kn + \bar{p}(1+e) = 0, \qquad (2.9,\ 2.10)$$

where $(\ )' = d(\ )/ds$ and

$$n = (1+e)N + 2M\beta' + M'\beta, \quad q = (1+e)M' - 2e'M + \beta N \qquad (2.11, 2.12)$$

$$e = u' + kw, \quad \beta = -w' + ku \qquad (2.13, 2.14)$$

$$N = C[e + \tfrac{1}{2}(e^2 + \beta^2)], \quad M = D[(1+e)\beta' - \beta e'] \qquad (2.15, 2.16)$$

$$C = Eh/(1-\nu^2), \quad D = Eh^3/12(1-\nu^2). \qquad (2.17, 2.18)$$

Here, E is Young's modulus, ν is Poisson's ratio, and h is the thickness of the shell, henceforth taken constant.

meridional stress resultant N *and rotation* ψ *as unknowns* [3]

$$N = N(0)\cos(ks+\psi) + D\psi''(0)\sin(ks+\psi)$$
$$+\bar{p}\int_0^s [1 + C^{-1}N(t)]\sin[k(s-t) + \psi(s) - \psi(t)]\,dt, \ \psi(0) = 0 \qquad (2.19)$$

$$D\psi'' + N(0)\sin(ks+\psi) - D\psi''(0)\cos(ks+\psi)$$
$$= \bar{p}\int_0^s \cos[k(s-t) + \psi(s) - \psi(t)]\,dt. \qquad (2.20)*$$

meridional stress resultant N *and bending strain* K *as unknowns* [6]

$$[N + D(kK + \tfrac{1}{2}K^2)]' + \underline{3C^{-1}NN'} = 0 \qquad (2.21)**$$

$$DK'' - (k+K)N + \bar{p}(1 + C^{-1}N) = 0 \qquad (2.22)$$

3. Geometrical simplifications

3.1. Plates under edge loads only

For this special case, equations (4.14) to (4.17) of [6] reduce to

* The factor $1 + C^{-1}N(t)$ in the integrand has been replaced by 1, consistant with the errors inherent in this equation.
** The underlined term may be neglected, being, on account of the N' factor, of the same order of magnitude as the inherent errors in (2.21). However, in general, terms of this type must be kept.

$$AN^{\alpha}_{\alpha}\Big|^{\beta}_{\beta} + \tfrac{1}{2}\bar{K}^{\alpha}_{\beta}K^{\beta}_{\alpha} = 0(\eta\theta^2/\lambda^2), \quad DK^{\alpha}_{\alpha}\Big|^{\beta}_{\beta} - K^{\alpha}_{\beta}N^{\beta}_{\alpha} = O(Eh^2\eta\theta^2/\lambda^2) \qquad (3.1, 3.2)$$

$$-\bar{K}^{\beta}_{\alpha}\Big|_{\beta} + \underline{A[(1+\nu)(K^{\beta}_{\alpha}N^{\lambda}_{\lambda}\Big|_{\beta} + K^{\beta}_{\lambda}N^{\lambda}_{\beta}\Big|_{\alpha}) - \nu K^{\lambda}_{\lambda}N^{\beta}_{\beta}\Big|_{\alpha}]} = O(\eta\theta^4/h\lambda) \qquad (3.3)$$

$$< N^{\beta}_{\alpha} + \underline{D(K^{\beta}_{\alpha} - \tfrac{1}{2}\delta^{\beta}_{\alpha}\bar{K}^{\kappa}_{\kappa})K^{\lambda}_{\lambda} + A\,\{2N^{\lambda}_{\alpha}N^{\beta}_{\lambda} - \tfrac{1}{2}\delta^{\beta}_{\alpha}[(1-\nu)N^{\kappa}_{\lambda}N^{\lambda}_{\kappa}}$$

$$\underline{+\nu(N^{\lambda}_{\lambda})^2]\}} > \Big|_{\beta} = O(Eh\eta\theta^4/\lambda). \qquad (3.4)$$

Here N^{α}_{β} and K^{α}_{β} are the mixed components of the symmetric stress resultant and bending strain tensors, $\bar{K}^{\alpha}_{\beta} = \delta^{\alpha}_{\beta}K^{\lambda}_{\lambda} - K^{\alpha}_{\beta}$, where δ^{α}_{β} is the Kroneker delta, $A = 1/Eh$, and a vertical bar denotes covariant differentiation based on the metric of the undeformed midplane. In the error terms on the right sides of (3.1) to (3.4), η is the maximum strain in the shell, $\theta = \mathrm{Max}(h/d, \sqrt{\eta})$, and $\lambda = h/\theta$, where d is the minimum distance to the edge. These error bounds are based on John's work [7], and are elaborated in [6] and [8]. A later example involving the buckling of a circular cylindrical shell implies that it must be possible to refine the error terms in (3.1) and (3.2) by distinguishing between the maximum bending strain η_b and the maximum extensional strain η_e. (Koiter has already put forward this suggestion in [9]). If the underlined terms in (3.1)-(3.4) are dropped, we arrive at the starting point of the celebrated von Karman equations. A well-known deficiency of these equations is that they do not contain as a special case the equations of the *elastica*. In other examples, such as the simultaneous twisting and stretching of a long strip [10], it can be shown that the von Karman equations predict the stresses quite accurately. The displacements, however, can be way off (as when the strip is twisted through a full turn).

3.2. Von Karman-like equations for arbitrarily large rotations of plates

Drawing on arguments and suggestions set forth by Koiter [5, 8, 11] and Danielson [12], we now derive a coupled set of equations for a stress function F and a curvature function W that are free from the above-mentioned shortcomings of the von Karman equations. Unfortunately, we cannot guarantee *a priori* that the new equations are always sufficiently accurate.

Consider (3.3). In both the von Karman equations and for inextensional bending, $\bar{K}^{\beta}_{\alpha}\big|_{\beta} = 0$. Moreover, the underlined terms in (3.3) are $O(\eta^2/h\lambda)$. Thus, we assume that

$$K^{\beta}_{\alpha} = W\Big|^{\beta}_{\alpha} + \underline{O(\eta^2/h)}. \qquad (3.5)$$

In general (3.5) can be validated only *a posteriori*.

Consider next (3.4). It is of the form $T^{\beta}_{\alpha}\big|_{\beta} = 0$ and hence may be satisfied identically by setting $\bar{T}^{\beta}_{\alpha} = F\big|^{\beta}_{\alpha}$. The underlined terms in (3.4), coming from stress-relations, contain errors of $O(Eh\eta^3)$. Further, as $N^{\beta}_{\alpha} = T^{\beta}_{\alpha} + O(Eh\eta^2)$, it is consistant to replace K^{β}_{α} by $W\big|^{\beta}_{\alpha}$ and $\bar{N}^{\beta}_{\alpha}$ by $F\big|^{\beta}_{\alpha}$ in these terms. This allows us to write

$$\bar{N}^{\alpha}_{\beta} = F\Big|^{\alpha}_{\beta} + \underline{D(W\Big|^{\alpha}_{\beta} - \tfrac{1}{2}\delta^{\alpha}_{\beta}\Delta W)\Delta W}$$

$$+A\{\underline{\underline{\tfrac{1}{2}\delta^{\alpha}_{\beta}[(4+\nu)(\Delta F)^2-(3+\nu)F|^{\kappa}_{\lambda}F|^{\lambda}_{\kappa}]}}$$

$$+\underline{\underline{2F|^{\alpha}_{\lambda}F|^{\lambda}_{\beta}-4\Delta FF|^{\alpha}_{\beta}\}}}+O(Eh\eta\theta^4), \tag{3.6}$$

where Δ is the Laplacian. Returning to (3.3), we may, consistant with the errors in this equation, replace the underlined terms by

$$A\{(1+\nu)[W|^{\beta}_{\alpha}\Delta F|_{\beta}+\Delta W\Delta F|_{\alpha}-W|^{\beta}_{\lambda}F|^{\lambda}_{\beta\alpha}]-\nu\Delta W\Delta F|_{\alpha}\}$$

$$=A[-(1+\nu)(\bar{W}|^{\beta}_{\lambda}\bar{F}|^{\lambda}_{\alpha})|_{\beta}+\Delta W\Delta F|_{\alpha}]. \tag{3.7}$$

(Note that $\bar{F}|^{\beta}_{\alpha\beta}=0$, but that $\bar{F}|^{\beta}_{\beta\alpha}=\Delta F|_{\alpha}$).

Let $P^{\alpha\beta}=P^{\beta\alpha}$ be any solution of

$$P^{\beta}_{\alpha}|_{\beta}=\Delta W\Delta F|_{\alpha}+\tfrac{1}{2}(1+\nu)(W|^{\lambda}_{\alpha}F|^{\beta}_{\lambda}-W|^{\beta}_{\lambda}F|^{\lambda}_{\alpha})|_{\beta}. \tag{3.8}$$

Then (3.5) takes the more explicit form

$$K^{\alpha}_{\beta}=W|^{\alpha}_{\beta}+A[(1+\nu)\underline{(\tfrac{1}{2}\bar{W}|^{\alpha}_{\lambda}\bar{F}|^{\lambda}_{\beta}+\tfrac{1}{2}\bar{W}|^{\lambda}_{\beta}\bar{F}|^{\alpha}_{\lambda}-\delta^{\alpha}_{\beta}W|^{\kappa}_{\lambda}F|^{\lambda}_{\kappa})}+\bar{P}^{\alpha}_{\beta}]+O(\eta\theta^4/h). \tag{3.9}$$

Our quest for a universally valid reduction of the general plate equations is thwarted by the term $A\bar{P}^{\alpha}_{\beta}$ that may not be $O(\eta^2/h)$, as was assumed originally. However, two simple examples will illustrate later that this term need not cause difficulties.

Finally, we insert (3.6) and (3.9) into (3.1) and (3.2) and make the following

CONJECTURE: a refined analysis of the errors in (3.1) and (3.2) that distinguishes between the maximum bending and extensional strains will show that, of all terms arising from the underlined terms in (3.6) and (3.9), it is consistant to retain in (3.1) only those coming from (3.9) and in (3.2) only those coming from the singly underlined terms in (3.6).

Thus we arrive at *the extended von Karman equations* (less error terms):

$$A[\Delta\Delta F-\underline{(1+\nu)(W|^{\alpha}_{\beta}\bar{W}|^{\beta}_{\lambda}\bar{F}|^{\lambda}_{\alpha})+W|^{\alpha}_{\beta}P^{\beta}_{\alpha}}]+\tfrac{1}{2}\bar{W}|^{\alpha}_{\beta}W|^{\beta}_{\alpha}=0 \tag{3.10}$$

$$D\{\Delta\Delta W+\underline{[W|^{\alpha}_{\beta}W|^{\beta}_{\alpha}-\tfrac{1}{2}(\Delta W)^2]}\Delta W\}-W|^{\alpha}_{\beta}\bar{F}|^{\beta}_{\alpha}=0. \tag{3.11}$$

3.3. Extension and twisting of an infinite strip

Let $Oxyz$ denote a right-hand Cartesian reference frame with associated orthonormal base vectors $(\mathbf{i},\mathbf{j},\mathbf{k})$ and consider a plate that initially occupies the region $|y|\leq b$, $|z|\leq\tfrac{1}{2}h$. Let the plate be acted upon at $x=\pm\infty$ by a net axial force P and torque T and let its sides be stress free. The strains then depend on y only and the midplane will deform into a right helicoid of the form

$$\mathbf{r} = (1+\gamma)x\mathbf{i} + [y+f(y)](\mathbf{j}\cos x\theta + \mathbf{k}\sin x\theta), \quad f(0)=0, \tag{3.12}$$

where γ and θ are constants, and f, for the moment, is arbitrary. The components of the extensional and bending strains are

$$\gamma_{12} = K_{11} = K_{22} = 0 \tag{3.13}$$

$$\gamma_{11} \equiv \gamma_x = \gamma + ½[\gamma^2 + \theta^2(y+f)^2], \quad \gamma_{22} \equiv \gamma_y = f' + ½f'^2 \tag{3.14}$$

$$\begin{aligned} K_{12} \equiv \tau &= \frac{\theta(1+\gamma)(1+f')^2}{[(1+2\gamma_x)(1+2\gamma_y)]^{½}} = \frac{\theta(1+\gamma)(1+f')}{(1+2\gamma_x)^{½}} \\ &= \theta[1+f' - ½\theta^2 y^2 + O(\eta^2)]. \end{aligned} \tag{3.15}$$

The only non-vanishing stress components are $N^1_1 \equiv N_x$, $N^2_2 \equiv N_y$, and $M^1_2 = D(1-\nu)K^1_2$, and these must be even functions of y. The Kirchhoff conditions along $y = \pm b$ reduce to $N_y(\pm b) = 0$, while over any cross section

$$\int_b^b N_x(y)\,dy = P, \quad \int_{-b}^b \{\theta[y+f(y)]^2 N_x(y) + 2D(1-\nu)\tau(y)\}\,dy = T. \tag{3.16, 3.17}$$

The last condition results from incorporating the concentrated (Kirchhoff) force perpendicular to the deformed midplane at $y = \pm b$ and from the fact that the transverse shear stress resultant along $x =$ Const. vanishes.

It may now be verified that the extended von Karman equations admit solutions of the form

$$W = \Theta xy[1 + \Gamma + O(\eta^2)] \tag{3.18}$$

$$\begin{aligned} F = ½Eh\,[\Gamma + \Theta^2 y^2/12)y^2 &+ ½(\Gamma + \Theta^2 b^2/2)^2(y^2 - x^2) \\ &+ (\Gamma^2 + \Gamma\Theta^2 y^2/b)y^2 + O(\eta^3)], \end{aligned} \tag{3.19}$$

where Γ and Θ are constants to be related to P and T. The auxiliary equations (3.6) and (3.9) yield

$$\begin{aligned} N_x = Eh\,\{(\Gamma + \Theta^2 y^2/2) \\ + (\Theta^2/2)[\Gamma(b^2 - y^2) + (\Theta^2/4)(b^4 - 3y^4)] + O(\eta^3)\} \end{aligned} \tag{3.20}$$

$$N_y = Eh(\Theta^2/2)\,[\Gamma + (\Theta^2/4)(y^2 + b^2) + O(\eta^2)](y^2 - b^2) \tag{3.21}$$

$$\tau = \Theta[1 - \nu\Gamma - (1+\nu)\Theta^2 y^2/2 + O(\eta^2)]. \tag{3.22}$$

The force-extension and torque-twist relations, (3.16) and (3.17), take the forms

$$P = 2Ehb\,[\Gamma + \Theta^2 b^2/6 + O(\eta^2)] \tag{3.23}$$

$$T = 4D(1-\nu)b\Theta[1 + (3/5)(1+\nu)\Theta^2(b^4/h^2) + 2(1+\nu)\eta(b^2/h^4) + O(\eta)], \tag{3.24}$$

which agree with the predictions of the von Karman equations [10] to within negligible errors. From (3.14), (3.20), (3.21), and the stress-strain relations $\gamma_x = A(N_x - \nu N_y)$, $\gamma_y = A(N_y - \nu N_x)$ follows

$$\gamma = \Gamma, \quad f = -\nu y\,[\Gamma + \Theta^2 y^2/6 + O(\eta^2)]. \tag{3.25}$$

Finally, a comparison of (3.15) and (3.22) gives $\theta = \Theta$. The only substantial difference between the von Karman equations and the extended von Karman equations is that the former predict the displacement vector

$$\mathbf{u} = \mathbf{i}\gamma x + \mathbf{k}\theta xy \tag{3.26}$$

while the latter give, by (3.12) and (3.25),

$$\mathbf{u} = \mathbf{i}\gamma x + y\mathbf{j}\{[1 - \nu(\gamma + \theta^2 y^2/6) + O(\eta^2)]\cos\theta x - 1\}$$
$$+ y\mathbf{k}[1 - \nu(\gamma + \theta^2 y^2/6) + O(\eta^2)]\sin\theta x. \tag{3.27}$$

In this example

$$P^\alpha_\beta = Eh\,\{O(\eta^2), \tfrac{1}{2}(1+\nu)(\Gamma + \Theta^2 y^2/2)\,[1 + O(\eta)], O(\eta^2)\}.$$

3.4. Buckling of an infinite circular cylindrical shell

Consider an infinite plate of width $2\pi R$ that is bent by pure edge couples into a circular cylindrical shell of midsurface radius R. Let the coincident edges be rigidly fixed to one another and let the shell be acted upon at its infinitely removed ends by a net axial compressive force $2\pi RP$. Finally, let the (trivial) Euler buckling mode be supressed.

The non-zero components of the stress resultant and bending strain tensors are $N^0_x \equiv N^1_1 = -P$ and $K^0_y \equiv K^2_2 = -R^{-1} \equiv -k$. The solution of the extended von Karman equations that yields these values is readily found to be

$$F^0 = -P[1 - 3AP/2 + O(\eta^2)]y^2/2$$
$$- AP^2[1 + O(\eta)]x^2/4 + Dk^2[1 + O(\eta)](x^2 - y^2)/4 \tag{3.28}$$

$$W^0 = -k[1 - A(1+\nu)P + O(\eta^2)]y^2/2. \tag{3.29}$$

Note that $P^\alpha_\beta = 0$.

Buckling equations are obtained by setting $F = F^0 + f$, $W = W^0 + w$ and linearizing. In particular the linearized form of the term $W|^\alpha_\beta P^\beta_\alpha$ in (3.10) may be taken as $k^2\Delta f$. Thus (3.10) and (3.11) yield, to within errors inherent in these equations,

$$A\Delta(\Delta+k^2)f - kw_{,xx} = 0 \tag{3.30}$$

$$D(\Delta\Delta w + k^2 w_{,yy}) + kf_{,xx} + Pw_{,xx} = 0. \tag{3.31}$$

The arguments used in [13] and [14] imply that these equations are as accurate as possible within the framework of first-approximation shell theory.

3.5. General helicoidal shells admit one-dimensional strains

For shells of revolution undergoing torsionless, axisymmetric deformation, Koiter has shown [8] that equations (4.14) to (4.17) of [6] reduce to two coupled, ordinary, second order differential equations – the Reissner-Meissner-Reissner equations. Are analogous reductions possible for other midsurface geometries? This leads to the following problem: find all surfaces with metric and curvature tensors that depend on one and the same variable. Aside from the obvious case of a general cylinder, it turns out that general helicoids are the only surfaces having this property [15]. The loading on a general helicoidal shell must, of course, be of a special type to produce one-dimensional strains. Fig. 1 illustrates such a situation. Of particular interest is the construction of functions of the form $P = \hat{P}(\delta, \psi, \overline{P})$ and $T = \hat{T}(\delta, \psi, \overline{p})$, where δ and ψ are, respectively, an overall axial extension and angle of twist per unit axial length. Clearly, the initial system of 8th order ordinary differential equations will have first integrals.

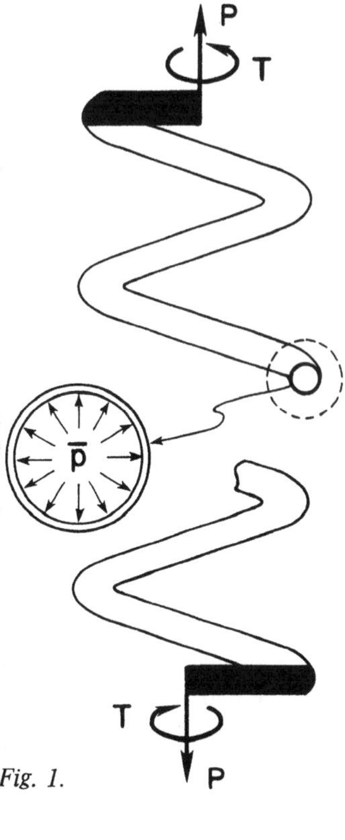

Fig. 1.

4. Special loads

If the two elastic parameters A and D are regarded as independent, it follows *formally* that if we set, respectively, $A=0$ or $D=0$, then the equations of inextensional bending or membrane stretching follow from equations (4.14) to (4.17) of [6].

Clearly, we need to consider the limit of some dimensionless parameter, say $\beta = AD/R^2$, where R is some typical dimension of the midsurface. Both theories emerge in the limit as $\beta \to 0$. The difference arises in how the stress resultants and bending strains are nondimensionalized.

In what follows assume that the surface coordinates have units of length and set

$$\xi^\alpha = R\tilde{\xi}^\alpha, \quad b^\alpha_\beta = R^{-1}\tilde{b}^\alpha_\beta, \tag{4.1, 4.2}$$

where b^α_β are the mixed components of the undeformed midsurface curvature tensor.

4.1. Inextensional deformation

Set

$$K^\alpha_\beta = R^{-1}\tilde{K}^\alpha_\beta, \quad N^\alpha_\beta = DR^{-2}\tilde{N}^\alpha_\beta, \tag{4.3, 4.4}$$

insert (4.1) to (4.4) into (4.14) to (4.17) of [6], drop all tildes, and in the resulting equations let $\beta \to 0$. The following equations (less error terms) result

$$(b^\alpha_\beta + \tfrac{1}{2}K^\alpha_\beta)\bar{K}^\beta_\alpha = 0, \quad \bar{K}^\beta_\alpha|_\beta = 0, \quad \Delta K^\alpha_\alpha - (b^\alpha_\beta + K^\alpha_\beta)N^\beta_\alpha = 0 \tag{4.5, 4.6, 4.7}$$

$$[N^\beta_\alpha + \tfrac{1}{2}(1-\nu)(b^\lambda_\alpha K^\beta_\lambda - b^\beta_\lambda K^\lambda_\alpha) + (K^\beta_\alpha - \tfrac{1}{2}\delta^\beta_\alpha K^\kappa_\kappa)K^\lambda_\lambda]|_\beta + b^\beta_\alpha K^\lambda_\lambda|_\beta = 0. \tag{4.8}$$

4.2. Inextensional deformation of plates

This brings us back to the extended von Karman equations, (3.10) and (3.11), with $A=0$ and $D=1$. Rather than attack these directly, we follow Mansfield [16] and use the fact that the deformed midplane must be a developable surface. Mansfield's analysis embodies the assumptions of the von Karman equations, but Prof. Libai and I have just finished a study of the exact equations for a cantilevered plate bounded by two straight edges and two curved ones [17, 18]. The plate is built-in along one straight edge while the other, that moves rigidly, is subject to a force and a couple. The curved edges are stress free. This is the two-dimensional analogue of the *elastica*. If the plate twists as it deforms (so that the deformed midplane has a non-degenerate edge of regression), then an independent set of 7 first order *ordinary* differential equations can be split off. These equations reduce even further for special geometries and special loads. For example, for a quadrilateral plate subject to a pure couple at its free end, there are just 4 equations involving a dimensionless load parameter ϵ. Three of the four boundary conditions at the free end are known. Mansfield's single nonlinear equation, (25) of [16], is the *uniform* limit of our equations as $\epsilon \to 0$.

4.3. Inextensional deformation of a spherical shell

The surface of simplest geometry after the plane and cylinder is the sphere. Unfortunately, there is no analogue of a developable surface for a portion of a sphere that bends without stretching. Still, (4.5) to (4.8) can be reduced and simplified considerably. With R in (4.1) and (4.2) the radius of the undeformed midsurface, the dimensionless metric and curvature tensors are identical and, as is well-known, (4.6) is satisfied identically by setting

$$K^{\alpha}_{\beta} = W|^{\alpha}_{\beta} + \delta^{\alpha}_{\beta} W. \tag{4.9}$$

Likewise, (4.8) reduces to a divergence form that may be satisfied identically by setting

$$\bar{N}^{\alpha}_{\beta} = F|^{\alpha}_{\beta} + \delta^{\alpha}_{\beta} F + [W|^{\alpha}_{\beta} - \delta^{\alpha}_{\beta}(1 + \tfrac{1}{2}\Delta W)](\Delta + 2) W. \tag{4.10}$$

Thus (4.5) and (4.7) reduce to

$$(\Delta + 2) W + \tfrac{1}{4}[(\Delta + 2) W]^2 + \tfrac{1}{2}[\tfrac{1}{2}(\Delta W)^2 - W|^{\alpha}_{\beta} W|^{\beta}_{\alpha}] = 0 \tag{4.11}$$

$$(\Delta + 2)[(\Delta + 2) W - F] + [(\Delta + 2) W + W|^{\alpha}_{\beta} W|^{\beta}_{\alpha} - \tfrac{1}{2}(\Delta W)^2](\Delta + 2) W$$

$$+ W|^{\alpha}_{\beta} F|^{\beta}_{\alpha} - \Delta W \Delta F - F\Delta W - W\Delta F - 2WF = 0. \tag{4.12}$$

These equations should be amenable to a perturbation solution in powers of the maximum amplitude of W, inasmuch as (in contrast to the plate) the lowest order equations are linear with solutions expressible in terms of Legendre functions.

4.4. An apparent paradox in the inextensional theory for arbitrary shells

For shells of constant Gaussian curvature K, (4.6) is satisfied identically by setting $K^{\alpha}_{\beta} = W|^{\alpha}_{\beta} + K\delta^{\alpha}_{\beta} W$. This representation, substituted into (4.5) yields a second order partial differential equation (PDE) for W.

For shells of nonconstant K, (4.6) may be satisfied identically by expressing K^{α}_{β} in terms of W and its derivatives up to, possibly, the 5th order, depending on the complexity of the surface. (See [19] for the analogous problem in linear membrane theory.) This representation, substituted into (4.5), yields a PDE for W of, possibly, 3rd, 4th, or 5th order. The paradox is that, in the linear theory at least, the equations of inextensional deformation, by the static-geometric analogy, may always be reduced to a 2nd order PDE by introducing either the Pucher or Vekua stress functions. Here we can offer only a clue to the resolution of this paradox.

For surfaces of revolution of nonconstant K, the differential equation for W in the linear theory is [20]

$$(1 - r'^2)[r^2 (W'/f')']' - rr''[rr'(W'/f')' - f(W'/f') + W + W^{\bullet\bullet}/f'] = 0. \tag{4.13}$$

Here r is the distance from the axis of revolution, $f = rr'' - r'^2$, $\partial W/\partial s = W'$, $\partial W/\partial\theta = W^{\bullet}$, s is distance along a meridian, and θ is the azimuthal angle. The representations for the physical components of K^α_β, from [20] and the static-geometry analogy, are

$$rK_s = [r^2 (W'/f')']', \quad T = (W'/f')^{\cdot} \tag{4.14, 4.15}$$

$$rK_\theta = rr' (W'/f')' - f(W'/f') + W + W'^{\cdot\cdot}/f'. \tag{4.16}$$

Note that $W = Cf$, C a constant, is a non-trivial solution of (4.13) that yields zero bending strains. By setting $W(s, \theta) = \Sigma w_n(s) \exp(in\,\theta)$ and $w_n(s) = f(s)v_n(s)$, we obtain a *2nd order* differential equation for v_n'. This example suggests that if the representation for K^α_β involves 4th or 5th derivatives of W, we ought to look for null strain solutions. These solutions, in some sense, should then be factored out leaving a 2nd order equation for a residual solution.

4.5. **Membrane stretching**

Set

$$N^\alpha_\beta = A^{-1} \tilde{N}^\alpha_\beta, \quad K^\alpha_\beta = R^{-1} \tilde{K}^\alpha_\beta, \tag{4.17, 4.18}$$

insert (4.1), (4.2), (4.17) and (4.18) into (4.14) to (4.17) of [6], drop all tildes, and let $\beta \to 0$. The following equations, less error terms, result:

$$\Delta N^\alpha_\alpha + (b^\alpha_\beta + \tfrac{1}{2} K^\alpha_\beta) \bar{K}^\beta_\alpha = 0, \quad (b^\alpha_\beta + K^\alpha_\beta) N^\beta_\alpha = 0 \tag{4.19, 4.20}$$

$$[-\bar{K}^\beta_\alpha - \tfrac{1}{2}(1+\nu)(b^\lambda_\alpha N^\beta_\lambda - b^\beta_\lambda N^\lambda_\alpha]\,|_\beta + b^\beta_\alpha N^\lambda_\lambda\,|_\beta$$

$$+ (1+\nu)(K^\beta_\alpha N^\lambda_\lambda\,|_\beta + K^\beta_\lambda N^\lambda_\beta\,|_\alpha) - \nu K^\lambda_\lambda N^\beta_\beta\,|_\alpha = 0 \tag{4.21}$$

$$\{N^\beta_\alpha + 2N^\lambda_\alpha N^\beta_\lambda - \tfrac{1}{2}\delta^\beta_\alpha [(1-\nu) N^\kappa_\lambda N^\lambda_\kappa + \nu (N^\lambda_\lambda)^2]\}\,|_\beta = 0. \tag{4.22}$$

We have only one, dismal observation: these equations are considerably more complicated than those of inextensional theory.

References

1. Naghdi, P.M., 'Foundations of elastic shell theory', *Progress in Solid Mechanics, Vol. IV*, Ed. Sneddon, I.N., and Hill, R., North-Holland, 1963.
2. Cowper, G.R., 'On the accuracy of Timoshenko's beam theory', *J. Engr. Mechs. Div.*, Proc. A.S.C.E., **94**, 1447-1453, 1968.
3. Simmonds, J.G., and Danielson, D.A., 'Nonlinear shell theory with finite rotation and stress-function vectors', *J. Appl. Mechs.*, **39** 1085-1090, 1972.
4. Budiansky, B., 'Notes on nonlinear shell theory', *J. Appl. Mechs.*, **35**, 393-401, 1968.
5. Koiter, W.T., 'On the nonlinear theory of thin elastic shells', *Proc. Kon. Ned. Ak. Wet.*, **69B**, 1-54, 1966.
6. Koiter, W.T., Simmonds, J.G., 'Foundations of shell theory', *Appl. Mech.* (Proc. 13th Int. Cong. Theor. & Appl. Mechs., Moscow, Aug. 1972), Ed. Becker, E., and Mikhailov, G.K., Springer-Verlag, 1973.

7. John, F., 'Estimates for the derivatives of the stresses in a thin shell and interior shell equations', *Comm. Pure & Appl. Math.*, **18**, 235-267, 1965.
8. Koiter, W.T., 'The intrinsic equations of shell theory with some applications', *Mech. Today*, **Vol. 6**, the Reissner anniversary volume, to be published.
9. Koiter, W.T., Review of [7]. *Math. Rev.*, **30**, 1030 (Review # 5569), 1965.
10. Reissner, E., 'Finite bending and twisting of thin plates', *J. Appl. Mech.*, **24**, 391-396, 1957.
11. Koiter, W.T., 'A consistant first approximation in the general theory of thin elastic shells', *Proc. IUTAM Symp. on the theory of thin elastic shells*, Delft, Aug. 1959, North-Holland, 1960, pp. 12-33.
12. Danielson, D.A., 'Simplified intrinsic equations for arbitrary elastic shells', *Int. J. Engr. Sci.*, **8**, 251-259, 1970.
13. Simmonds, J.G., 'A set of simple, accurate equations for circular cylindrical elastic shells', *Int. J. Solids & Struct.*, **2**, 525-541, 1965.
14. Danielson, D.A. and Simmonds, J.G., 'A proof of the accuracy of a set of simplified buckling equations for circular cylindrical elastic shells', *Developments in Theor. & Appl. Mechs.*, **Vol. 5**, (Proc. 5th Southeastern Conf. Theor. & Appl. Mechs., 1969), Univ. N.C. Press, 1971, pp. 1015-1028.
15. Simmonds, J.G., 'Surfaces with metric and curvature tensors that depend on one coordinate only are general helicoids', *Q. Appl. Math.*, **37**, 82-85, 1979.
16. Mansfield, E.H., 'The inextensional theory for thin flat plates', *Q. J. Mechs. & Appl. Math.*, **8**, 338-352, 1955.
17. Simmonds, J.G. and Libai,A., 'Exact equations for the inextensional deformation of cantilevered plates', to appear in *J. Appl. Mechs.*
18. Simmonds, J.G. and Libai, A., 'Alternate exact equations for the inextensional deformation of arbitrary, quadrilateral, and triangular plates', to appear in J. Appl. Mech.
19. Truesdell, C.A., 'Invariant and complete stress functions for general continua', *Arch. Rat. Mechs. & Anal.*, **4**, 1-29, 1959.
20. Simmonds, J.G., 'Six different membrane stress functions for shells of revolution', *Mech. Today*, **Vol. 6**, the Reissner anniversary volume, to be published.

Some Recent Advances in the Application of Nonlinear Elastostatics to Singular Problems *

*E. Sternberg***

Summary

This survey outlines two interrelated sequences of recent studies devoted to singular problems in the finite equilibrium theory of elastic solids: one of these deals with locally unbounded – the other with discontinuous – deformation gradients.

Introduction

Although the finite theory of elasticity was complete in its essential features before the turn of the last century, its application to specific problems awaited a revival of interest in this subject around the middle of the current century. The work ensuing at that time – prompted in part by a concern with rubber elasticity and partly impelled by issues in the field of elastic instability – yielded a number of exact solutions to special problems involving large deformations of incompressible isotropic elastic solids and produced various results pertaining to small – superposed on large homogeneous – deformations.[1]

Since then the literature on applications of nonlinear elasticity theory has grown steadily in size and scope. A comprehensive review of recent progress in connection with such applications is beyond the limitations of this brief survey and of its author. Instead, my purpose is at once less ambitious and more narrow-minded. I shall merely attempt to convey the gist of a string of studies within finite elastostatics in which I have been engaged during the past few years in collaboration with my colleague J.K. Knowles, as well as touch upon some closely related work carried out independently by Knowles and by others. These studies have a common theme in that all of them deal with singular problems which by their very nature are either far removed from the range of validity of the linearized theory of elasticity or have no counterpart at all in the linear theory. Thus our efforts have been directed at certain inherently strong nonlinear effects, rather than at second-order amendments of predictions based upon the linear theory.

Singular problems involving locally unbounded deformation gradients

One class of problems in the category to be considered here concerns geometrically induced singular stress concentrations, such as those arising in crack problems, which have

* Some of the work described in this paper was done under Contract N00014-75-C-0196 with the Office of Naval Research in Washington, D.C.
** California Institue of Technology, Pasadena, California 91125, U.S.A.
[1] The notes preceed the references

long played a pivotal part in fracture mechanics, or in punch and identation problems, which are also of engineering importance. In problems of this kind, the linearized theory of elasticity – in conflict with its underlying basic approximative assumption – gives rise to locally unbounded infinitesimal strains, regardless of the size of the loads. Consequently the solutions alluded to, though presumably realistic at finite distances from the singular points at sufficiently *small* loads, cannot possibly be uniformly valid in the immediate vicinity of such points. This local behavior, in turn, is bound to depend upon the material response to severe deformations. The foregoing observations motivated our interest in exploring the corresponding implications of the finite theory, which allows for arbitrarily large deformations and takes account of constitutive nonlinearities as well.

Recent work in finite elastostatics bearing on locally unbounded deformation gradients is reviewed in survey articles by Knowles [3], [4], [5]. We return to this work here in a highly cursory manner, partly in order to supply a motivation for a second sequence of investigations, to be described in the succeeding section of this paper.[2] On the other hand, we rely on the earlier surveys for references to the related literature in plasticity theory, most of which is tied to the assumption of infinitesimal deformations.

Apparently the first analysis of a crack problem within the framework of nonlinear elastostatics is due to Wong and Shield [6], who obtained an approximate global solution to the problem of a finite crack in an all-around infinite thin *incompressible* sheet of a *Neo-Hookean* material, which is subjected to bi-axial tension at infinity. The approximative scheme employed in [6] presupposes the deformations to be large throughout the sheet.

The analogous crack problem for *finite plane strain*, for a crack in a *uni-axial tension field* at right angles to the plane of the crack and a class of *compressible* elastic materials, is the object of a study reported in [7], [8], which aims directly at the elastostatic field induced near the crack-tips. The underlying constitutive hypothesis, which characterizes the governing elastic potential in the presence of severe extensional deformations, permits the tensile stress in large homogeneous uni-axial tension to be proportional to a continuously adjustable power of the axial principal stretch. Depending on the value of the exponent determining this power (hardening parameter), both hardening and softening materials are accommodated.

The method used in this investigation is an adaptation to the finite theory of an asymptotic scheme first introduced into linear elastostatics by Knein [9] and later on systematically exploited by Williams [10]. Its point of departure is the assumption that each of the two cartesian components of displacement in the vicinity of either crack-tip is asymptotic to the product of an initially unknown power of the distance from this point and an originally undetermined function of the angular coordinate around the tip. This Ansatz is found to be asymptotically consistent with the governing field equations and boundary conditions, provided both displacements, but not all displacement gradients, remain bounded as the crack-tip is approached and if the range of the hardening parameter is suitably restricted.

In this manner one is led to a nonlinear eigenvalue problem for an ordinary second-order differential equation to be satisfied by one of the two unknown functions initially introduced and containing the associated power-exponent as the eigenvalue parameter. The solution to this subsidiary problem is deducible in closed elementary form and – but for an arbitrary amplitude parameter – renders fully determinate the dominant behavior at the crack-tips of the displacement and normal stress parallel to the applied loading. Further, for small load intensities a precise estimate of this amplitude parameter is obtainable on the basis of the available global solution to the analogous linearized crack problem with the aid of a conservation law discovered by Eshelby [11] and first applied by Rice [12] in a similar context.[3]

The determination to dominant order of the displacement *parallel* to the crack-faces and of the remaining two in-plane components of true stress turns out to be a task of surprising complexity, which necessitates higher-order asymptotic considerations.

The results obtained in this study reveal that the nature and specific structure of the stress singularities arising at the crack-tips depend crucially on the value of the hardening parameter. In particular, for the range of this material constant admitted in [7], [8], the actual normal stress in the load-direction is always unbounded at the crack-tips; also, when this stress is referred to local spatial (Eulerian) coordinates, it is found to be more strongly singular than its counterpart predicted by the linear theory if the material is hardening in large uni-axial tension, and less so for a softening material.

The results of an unpublished analysis of the analogous *plane-strain* crack problem for an *incompressible* elastic material of the *Mooney-Rivlin* type, which furnish asymptotic estimates up to the second order for the local displacements and stresses near the crack-tips, are cited in [3].

The nonlinear plane-strain crack problem for a *transverse shear* loading at infinity (loading Mode II, in the terminology of fracture mechanics) remains to be explored. On the other hand, Knowles [18] considered the problem of a crack in a field of *longitudinal shear* (Mode III) for a class of *incompressible* isotropic materials whose elastic potential is completely determined by their response to a homogeneous simple shear deformation. Incompressible materials of this kind are capable of sustaining non-homogeneous anti-plane shear deformations.[4] The crack problem treated in [18] is thus reducible to a nonlinear boundary-value problem governed by a single quasi-linear partial differential equation of the second order for the scalar displacement parallel to the edges of the crack; the remaining two components of displacement vanish identically. The constitutive law adopted in [18] is then particularized for a sub-class of materials whose response in simple shear obeys a power law involving a hardening parameter that accommodates both hardening and softening behavior.[5] If the shearing stress in simple shear increases steadily beyond bounds with the amount of shear applied, the present problem is amenable to an asymptotic approach strictly parallel to that followed in [7], [8]. In fact, for resasons yet to be understood but hardly accidental, one is led to precisely the same eigenvalue problem encountered there. The two non-vanishing shear stresses are found to become unbounded at the crack-tips, the order of these singularities being dependent on the value of the hardening parameter. Here again hardening aggravates, while softening of the material mitigates the analogous singularities prevailing in the linearized problem. In contrast, the axial normal stress, which vanishes according to the linear theory, becomes infinite like the reciprocal of the distance from the singular points for all admitted values of the hardening parameter, the same being true of the strain-energy density.

The limiting case in which the shearing stress in simple shear rises monotonically to a *finite* value, as the amount of shear is increased beyond bounds, no longer yields to the assumption that the displacement is a locally separable function of the polar coordinates at the crack-tips. This transition case is treated in [18] by an entirely different approach – limited to small amounts of shear at infinity. In this asymptotic scheme the crack of finite width is replaced by a semi-infinite one, while the *far* field is required to match the elastostatic field *near* the crack-tip predicted by the solution to the original problem according to the linearized theory. The emerging global 'small-scale nonlinear crack problem', which is an analogue of what is known as the small-scale yielding problem in the related plasticity literature, is solved in [18] by means of the hodograph method. The local results established in this manner furnish *finite* shear stresses at the crack-tip, but the induced axial normal stress, as well as the strain-energy density, remain unbounded there.

Although the solution to the foregoing small-scale nonlinear crack problem is entirely smooth, it reveals rapid changes in the angular distribution of the elastostatic field around the crack-tip that presage a breakdown in smoothness once the response curve in simple shear undergoes a reversal in slope. An example of this kind is dealt with in a more recent analysis [21], which will be discussed later on.

We turn now to an asymptotic study [22] of a different sort, which aims at the singular plane-strain field behavior in the vicinity of a point that separates a load-free from adjoining fixed portion of the boundary of an elastic body. A familiar illustration of such mixed boundary conditions is supplied by the particular problem of a rigid flat indentor that is bonded to the straight edge of a semi-infinite elastic slab and subjected to an axial load. The well-known solution to this problem in linear elastostatics predicts displacements and stresses that are oscillatory near the punch-corners, the stresses becoming unbounded as such a corner is approached.

In [22] the corresponding implications of the finite equilibrium theory are explored with limitation to *compressible* elastic materials of *'harmonic type'*, which were introduced by John [23]. For this special class of materials a *non*-oscillatory behavior of the appropriate local elastostatic field is found to be consistent with the nonlinear theory, although the stresses remain unbounded at the singular boundary point. This conclusion at once suggests the elusive question as to the transition from the finite to the linearized theory in the present instance and as to the precise approximative status within the nonlinear theory of the solution to the linearized punch problem mentioned above.

It should be remarked that in [22] the desired local elastostatic field is assumed *a priori* to involve large extensional deformations because of certain physical anomalies in the behavior of harmonic materials at sufficiently large contractions.[6] The applicability of the results of [22] to the problems of the adhering punch is thus contingent upon a 'pull-out' – rather than an indenting – load. For this and other reasons it would be useful to re-examine the issue dealt with in [22] for other elastic materials.

Finally, we refer to a paper by Varley and Cumberbatch [24], who employ the harmonic material in considering the plane problem of a slab with an elliptic hole (for a homogeneous deformation at infinity) and give some attention to the limiting case of a crack.

Singular problems involving discontinuous deformation gradients. Loss of ellipticity in finite elastostatics.

The constitutive assumption introduced in connection with the plane-strain study [7], [8] of a crack in a tension field is satisfied in particular for an elastic potential proposed by Blatz and Ko [25] on the basis of experiments with a highly deformable rubberlike material. Our attempts to adapt the asymptotic scheme described earlier to this elastic potential were frustrated by difficulties which suggested that the global nonlinear crack problem may not admit a solution of unlimited smoothness near the crack-tips in this instance and led us to exclude this special case from our analysis. The complications thus encountered pointed to a possible breakdown in the ellipticity of the corresponding elastostatic field equations in the presence of sufficiently severe local deformations.

The foregoing conjecture was borne out in [26], where we showed that the displacement equations of equilibrium appropriate to this special isotropic elastic material are elliptic at a solution and at a material point if and only if the local principal stretches are confined to the interior of a convex hexagonal pyramid in the principal-stretch space. The conclusions reached in [26] in particular imply a loss of ellipticity at severe enough local states of three-dimensional uni-axial stress, plane-strain uni-axial stress, and simple shear. In contrast, ellipticity can never fail at a state of three-dimensional or plane isotropic extension.

Necessary and sufficient conditions for ordinary and strong ellipticity of the displacement equations of equilibrium governing *plane* deformations of *arbitrary compressible isotropic elastic* materials are established in [27]. These conditions are in the form of inequalities that restrict the local principal stretches and involve the first and second gradients of the strain-energy density (elastic potential) with respect to the principal stretches. The results of [27] include the local characteristic directions associated with a failure of ellipticity. Unfortunately, the inequalities just referred to do not appear to admit a convenient physical interpretation. Comparably explicit ellipticity conditions for *not necessarily plane* deformations of general compressible isotropic elastic solids are apt to be difficult to obtain.

Ellipticity conditions for *plane* elastostatic deformations of *incompressible* isotropic materials can be found in a doctoral dissertation by Abeyaratne [28]. The relevant system of equations in this instance consists of the appropriate displacement equations of equilibrium together with the incompressibility requirement, and the resulting inequalities have a fairly simple physical meaning. An even more transparent situation arises in *anti-plane shear* deformations of *incompressible* isotropic elastic solids of the kind admitted in [18]. Here the governing displacement equation of equilibrium is found to be locally elliptic at a solution if and only if the response curve in simple shear associated with the material has a positive slope at an amount of shear equal to the magnitude of the corresponding local displacement gradient. The small-scale nonlinear crack problem analyzed in [18] and discussed in the preceding section of the present paper thus refers to the limiting elliptic case within the class of power-law materials introduced there.

A breakdown of ellipticity in the elastostatic field equations suggests the possible emergence of solution fields with finite jump discontinuities in the first displacement gradients. Such discontinuities are encountered in connection with localized shear failures and are discussed by Rudnicki and Rice [29] in a different constitutive setting.[7]

Elastostatic fields possessing discontinuities of the type described above are referred to as 'elastostatic shocks' in [31] because of their more than passing resemblence to gas-dynamical shocks in steady transonic flows. The study summarized in [31] is confined to *plane* deformations of *compressible* (not necessarily isotropic) elastic materials and aims at certain local aspects of elastostatic shocks in this context. To that end the shock-line is assumed to be straight and the deformation on either side of it is taken to be homogeneous. If one of the foregoing two homogeneous deformations is regarded as given, the existence of a corresponding 'piecewise homogeneous elastostatic shock' is contingent upon the existence of another such deformation and an orientation of the shock-line compatible with the required continuity of the displacements and tractions across this line. As shown in [31], a necessary condition for the existence of *any* such elastostatic shock is that the displacement equations of equilibrium associated with the material at hand suffer a loss of *strong* ellipticity at *some* homogeneous deformation. The intuitively plausible conjecture that the existence of elastostatic shocks necessitates a failure of *ordinary* ellipticity remains as yet unproven. One confirms easily, however, that a loss of ordinary ellipticity is indeed both necessary and sufficient for the existence of shocks in *linear* elastostatics of homogeneous and isotropic materials. This precludes the emergence of shocks in the linearized equilibrium theory at physically acceptable values of the elastic constants and identifies the issue under discussion as inherently nonlinear.

The analysis in [31] leads one to anticipate that once there exists a piecewise homogeneous shock for a given deformation on one side of the shock-line, there is a one-parameter family of such shocks. Such a family of shocks may be traced in a variety of ways. The particular tracing parameter adopted in [31] is the relative change in the mass density across the shock-line. This parameter, which vanishes in the absence of a shock, is a natural measure of the shock-strength for *compressible* elastic materials and coincides with the shock-strength commonly employed in gas dynamics.

Taking for granted the availability of a smooth family of shocks in a neighborhood of zero shock-strength, we deduce in [31] lowest-order asymptotic estimates – pertaining to 'weak shocks' – for the jumps across the shock-line in certain kinematical field quantities, as well as for those in the stresses and the strain-energy density. The existence of the underlying family of shocks is found to imply a loss of ordinary ellipticity at the pre-assigned homogeneous deformation, which can be shown to be necessarily dynamically unstable. Further, as the shock-strength tends to zero, the limiting shock-line tends to a characteristic line associated with the governing displacement equations of equilibrium at the given deformation. In addition to weak shocks, an example of a shock of *finite* strength – based on the elastic potential of Blatz and Ko [25] – is explored in detail. The results reveal that all finite shocks admitted by this material are accompanied by a loss of ordinary ellipticity at least on one side of the shock-line.

In gas dynamics the sign of the shock-strength is determined by the entropy inequality, which reflects the dissipative character of shocks arising there (in the absence of viscosity) and necessitates an increase in the mass density as the shock-surface is traversed from the upstream to the down-stream side. An entropy-like inequality for elastostatic shocks is deduced in [31] from the proposed requirement that time-dependent quasi-static shocks of the kind under consideration be 'dissipative' in an analogous sense. The resulting dissipation inequality, which involves Eshelby's [14] energy-momentum tensor, in general restricts the range of the elastostatic shock-strength but does not lead to a universal decision as to its sign.

Knowles [32] returns to the energetics of elastostatic shocks in a considerably broader setting. Here general *three*-dimensional elastostatic fields with finite jump discontinuities across *curved* surfaces are admitted and *compressible* as well as *incompressible* materials are encompassed. It is shown in [32] that if the requirement of dissipativity is imposed in these general circumstances one is led to precisely the same dissipation inequality previously arrived at in [31]. Also, for compressible elastic solids, a continuum-thermodynamical argument is produced in support of the dissipation inequality. Finally, the implications of this inequality are studied for *weak* shocks in plane strain of *compressible* – not necessarily isotropic – materials and for shocks of an *arbitrary* suitably defined strength in *anti-plane* shear of the class of isotropic *incompressible* materials introduced in [18].

The results of [31] concerning piecewise homogeneous plane elastostatic shocks are extended by Abeyaratne [28] to *incompressible* materials.[8] For this purpose a definition of shock-strength applicable to locally volume-preserving plane deformations is adopted. The treatment of finite shocks in [28] is confined to isotropic solids but is otherwise general, rather than merely illustrative.

In addition, [28] contains a detailed analysis of a displacement boundary-value problem exemplifying the occurrence of elastostatic shocks in a plane deformation of an incompressible isotropic elastic material. The specific problem treated is that of a hollow circular cylinder whose inner boundary is subjected to a rigid rotation (twist) about its axis, while the outer boundary is held fixed. In order to render fully determinate the in-plane response of the material to *any* plane deformation it suffices here to prescribe its response in simple shear. The latter is assumed in [28] to be such that the corresponding response curve rises steadily to a maximum as the amount of shear increases from zero, then declines to a minimum, and thereafter risen monotonically beyond bounds as the amount of applied shear continues to grow indefinitely. This postulated constitutive behavior allows for the eventuality of plane deformations that are accompanied by a local loss of ellipticity.

Although the foregoing problem enables one to restrict the unknown displacement field to one of pure axisymmetric torsion and is thus reducible to a one-dimensional boundary-value problem for an ordinary differential equation, the possibilities as to the character and multiplicity of its solutions are remarkably rich. In particular, it is demonstrated in [28] that the present problem admits a unique smooth solution, which is wholly elliptic, for all sufficiently small and all large enough values of the given twist. Further, there are always twists in the intermediate range that preclude the existence of a smooth solution, although there may exist wholly non-elliptic smooth solutions at *some* intermediate twists. Finally, for *each* twist in the intermediate range there exists an infinity of mixed solutions involving a cylindrical shock-surface, all of which conform to the dissipation inequality, but only one of which is stable with respect to axisymmetric torsional perturbations of arbitrary size.

We turn now to an investigation [21], cited earlier, which illustrates the emergence of elastostatic shocks in a genuinely two-dimensional boundary-value problem. The example treated in [21] is a particular small-scale nonlinear crack problem for a crack in an anti-plane shear field. As mentioned before, such a problem is analyzed in [18] by means of the hodograph method for the limiting elliptic case of an isotropic incompressible material whose response in simple shear obeys a power-law.

In [21], which is a pilot study suggested by – and complementary to – the investigation reported in [18], a possible loss of ellipticity is provided for. Here the shearing stress induced by a simple shear deformation is assumed to rise linearly to a finite peak at a critical amount of shear and is required to diminish steadily to zero in a definite manner as the amount of shear is further increased. Accordingly, the appropriate scalar displacement equation of equilibrium suffers a loss of ellipticity at any solution whose local gradient in magnitude exceeds the critical amount of shear.

When the ellipticity of the required solution field is not assured in advance, the mapping from the physical to the hodograph plane need no longer be uniquely invertible. Even in these circumstances, however, the hodograph method may be used formally to generate solutions of the nonlinear second-order partial differential equation at hand. For the special constitutive law adopted in [21] one is thus led to exact elliptic and hyperbolic solutions of closed elementary form in certain subdomains of the relevant physical domain, which is the exterior of the semi-infinite crack. These solutions, in turn, may be pieced together so as to yield a global solution of restricted smoothness to the boundary-value problem under consideration.

The explicit solution constructed in this way exhibits two curved shock-lines, which are symmetrically situated with respect to the crack-axis, issue from the crack-tip, and terminate in the interior of the body – ahead of the tip. Both the displacement gradient and the stresses experience finite jump discontinuities across these lines, but the displacements and tractions remain continuous there. Further, the emerging elastostatic shocks are found to be consistent with the dissipation inequality, if the amount of shear applied at infinity is an increasing function of 'time'. The solution obtained is hyperbolic in a bulb-shaped domain whose boundary consists of the two shock-lines and a circular arc connecting their terminal points; it is elliptic outside this region. All stresses remain finite at the crack-tip with the exception of the induced axial normal stress, which becomes unbounded if the tip is approached from the interior of the hyperbolic region. The question as to the non-existence of a smooth solution to this problem remains open, as does the issue concerning the stability of the solution described above, in which shocks are admitted.

In [21] we also discuss modifications in the solution to the problem that come about if the linearly rising portion of the response diagram for simple shear is retained but the response past critical shear is generalized to allow for different rates of decay. Of particular interest in this connection is the limiting *non*-elliptic case in which the shear stress remains constant beyond the critical amount of shear. In this instance the hodograph method leads to a solution free of shocks, whose asymptotic behavior near the crack-tip is identical with that arrived at in [18] for the limiting *elliptic* case of a power-law material. The results pertaining to the limiting non-elliptic case alluded to above are also closely related to a solution obtained by Hult and McClintock [33] to the anti-plane shear problem for an elastic-perfectly plastic body with a notch, which degenerates into a crack as its opening angle tends to zero. This solution too is based on the hodograph method.

Further generalizations of the response curve for simple shear underlying [21] are currently in progress. These include a rather broad class of *smooth* response curves that rise to a maximum shear stress and then decline to a *positive* ultimate stress.

A problem of greater physical interest is the plane-strain small-scale nonlinear crack problem for a crack in a tension field, under constitutive assumptions that entail a potential loss of ellipticity. Unfortunately this problem, which is governed by a *fourth*-order system of partial differential equations, is no longer accessible by means of the hodograph transformation.

Acknowledgment

I am indebted to J.K. Knowles and R.C. Abeyaratne, who read the manuscript and offered helpful suggestions.

Notes

1. An account of such investigations, up to 1968, may be found in the treatise by Green and Zerna [1]. See also Green and Adkins [2].
2. Some of the work delineated there is included in [5].
3. A derivation of this and related conservation laws from Noether's theorem on invariant variational principles is given in [13]. See also Eshelby [14], Günther [15], Chadwick [16], as well as Chen and Shield [17].
4. Necessary and sufficient conditions for such to be the case are established elsewhere by Knowles [19].
5. A closely related constitutive law is used by Lo [20] in dealing with the same nonlinear crack problem.

6 As observed by John [23], in these circumstances harmonic materials violate the Baker-Ericksen inequalities. See also the related discussion in [22].

7 See also Rice [30].

8 As noted earlier, the dissipation inequality is shown by Knowles [32] to remain unaltered in this case.

References

1. A.E. Green and W. Zerna, *Theoretical elasticity*, Second Ed., Oxford, 1968.
2. A.E. Green and J.E. Adkins, *Large elastic deformations*, Second Ed., Oxford, 1970.
3. J.K. Knowles, 'Finite elastostatic fields with unbounded deformation gradients', in *Finite Elasticity*, Appl. Mech. Div., Amer. Soc. Mech. Eng., **Vol. 27** (1977), 23.
4. J.K. Knowles, 'Crack problems in finite elastostatics', *Proc., Symp. on Math. Problems in Fracture*, Amer. Math. Soc. and Soc. Ind. Appl. Math., New York, March 1978.
5. J.K. Knowles, 'On some inherently nonlinear singular problems', *Proc., Eighth U.S. Nat. Cong. Appl. Mech.*, Amer. Soc. Mech. Eng., June 1978.
6. F.S. Wong and R.T. Shield, 'Large plane deformations of thin elastic sheets of Neo-Hookean material', *Zeit. Angew. Math. Physik*, **20**, (1969), 176.
7. J.K. Knowles and Eli Sternberg, 'An asymptotic finite-deformation analysis of the elastostatic field near the tip of a crack', *J. of Elasticity*, **3** (1973), 2, 67.
8. J.K. Knowles and Eli Sternberg, 'Finite-deformation analysis of the elastostatic field near the tip of a crack: Reconsideration and higher-order results', *J. of Elasticity*, **4**, (1974), 3, 201.
9. M. Knein, 'Zur Theorie des Druckversuchs', *Zeit. Angew. Math. Mech.*, **6** (1926), 414.
10. M.L. Williams, 'Stress singularities resulting from various boundary conditions in angular corners of plates in extension', *J. Appl. Mech.*, **19** (1952), 526.
11. J.D. Eshelby, 'The continuum theory of lattice defects', *Progress in Solid State Physics*, Vol. 3, Academic Press, New York, 1956, 79.
12. J.R. Rice, 'A path-independent integral and the approximate analysis of strain concentrations by notches and cracks', *J. Appl. Mech.*, **35** (1968), 2, 379.
13. J.K. Knowles and Eli Sternberg, 'On a class of conservation laws in linearized and finite elastostatics', *Arch. Rat. Mech. Anal.*, **44** (1972), 3, 187.
14. J.D. Eshelby,'The elastic energy-momentum tensor', *J. of Elasticity*, **5** (1975), 3-4, 321.
15. W. Günther, 'Über einige Randintegrale der Elastomachanik', *Abhandlungen, Braunschweiger Wissenschaftliche Gesellschaft*, **14** (1962), 53.
16. P. Chadwick, 'Applications of an energy-mementum tensor in nonlinear elastostatics', *J. of Elasticity*, **5** (1975), 3-4, 249.
17. F.H.K. Chen and R.T. Shield, 'Conservation laws in elasticity of the J-integral type', *Zeit. Angew. Math. Physik*, **28** (1977), 1.
18. J.K. Knowles, 'The finite anti-plane shear field near the tip of a crack for a class of incompressible materials', *Int. J. Fracture*, **13** (1977), 5, 611.
19. J.K. Knowles, 'On finite anti-plane shear for incompressible elastic materials, *J. Australian Math. Soc.*, **XIX** (1976), Ser. B, 4, 400.
20. K.K. Lo, 'Finite-deformation crack in an infinite body under anti-plane simple shear', *Int. J. Solids and Structures*, **13** (1977), 1219.
21. J.K. Knowles and Eli Sternberg, 'Discontinuous deformation gradients near the tip of a

crack in finite anti-plane shear: an example', Technical Report No. 40, Contract N00014-75-C-0196, Calif. Inst. of Tech., January, 1979 (to appear in *J. of Elasticity*).
22. J.K. Knowles and Eli Sternberg, 'On the singularity induced by certain mixed boundary conditions in linearized and nonlinear elastostacs', *Int. J. Solids Structures*, **11** (1975), 1173.
23. F. John, 'Plane strain problems for a perfectly elastic material of harmonic type', *Communications Pure Appl. Math.*, **XIII** (1960), 2, 239.
24. E. Varley and E. Cumberbatch, 'The finite deformation of an elastic material surrounding an elliptic hole', in *Finite Elasticity*, Appl. Mech. Div., Amer. Soc. Mech. Eng., Vol. 27 (1977), 41.
25. P.J. Blatz and W.L. Ko, 'Application of finite elasticity to the deformation of rubbery materials', *Trans., Soc. Rheology*, **6** (1962), 223.
26. J.K. Knowles and Eli Sternberg, 'On the ellipticity of the equations of nonlinear elastostatics for a special material', *J. of Elasticity*, **5** (1975), 3-4, 341.
27. J.K. Knowles and Eli Sternberg, 'On the failure of ellipticity of the equations for finite elastostatic plane strain', *Arch. Rat. Mech. Anal.*, **63** (1977), 4, 321.
28. R.C. Abeyaratne, *Discontinuous deformation gradients in plane finite elastostatics of incompressible materials*, Doctoral dissertation, Calif. Inst. of Tech., June 1979.
29. J.W. Rudnicki and J.R. Rice, 'Conditions for the localization of deformation in pressure-sensitive dilatant materials', *J. Mech. Phys. of Solids*, **23** (1975), 371.
30. J.R. Rice, 'The localization of plastic deformation', *Proc., 14th IUTAM Congress*, Delft, The Netherlands, 1976, North Holland, Amsterdam, 1976.
31. J.K. Knowles and Eli Sternberg, 'On the failure of ellipticity and the emergence of discontinuous deformation gradients in plane finite elastostatics', *J. of Elasticity*, **8** (1978), 4, 329.
32. J.K. Knowles, 'On the dissipation associated with equilibrium shocks in finite elasticity', Technical Report No. 38, Contract N00014-75-C-0196, Calif. Inst. of Tech., December 1977 (to appear in *J. of Elasticity*).
33. J.A.H. Hult and F.A. McClintock, 'Elastic-plane stress and strain distributions around sharp notches under repeated shear', *Proc., 9th IUTAM Congress, Brussels, 1956*.

W.T. Koiter Public Lecture

Forty Years in Retrospect, the Bitter and the Sweet
Omzien in verwondering maar niet in wrok

W.T. Koiter

The highly valued presence today of friends from abroad justifies my speaking mostly in English, but you will forgive me, if I start with a few words of a more personal and local character in my native language.

U zult niet voor de eerste maal hebben gehoord, dat het afscheid van een dierbare taak met gemengde gevoelens gepaard gaat. Ook al is er geen bindend voorschrift, althans nog niet, voor zulk een afscheid bij het bereiken van de leeftijd van 65 jaar, meer dan één overweging heeft mij tot dit besluit geleid. Een opsomming van deze overwegingen zal ik U besparen maar de belangrijkste grond wil ik U niet onthouden, omdat hij naar menselijke berekening niet alleen voor mij zal gelden. Ik doel op de moeilijkheid met het klimmen der jaren de stormachtige ontwikkeling van de wetenschap te volgen, ook op mijn toch tamelijk beperkte vakgebied. De verworven ervaring maakt het soms nog wel mogelijk een bijdrage tot verdieping en synthese van inzichten te leveren, maar de opgave om aan de wetenschappelijke ontwikkeling voor de toekomst op de juiste wijze mede richting te geven wordt bijkans onuitvoerbaar.

Onder de gemengde gevoelens die ik al noemde speelt een grote dankbaarheid voor hetgeen het leven bracht duidelijk de hoofdrol, bijgestaan door een onmiskenbare weemoed om het verleden, en helaas ook een zekere bitterheid over teleurstellende ervaringen; deze bitterheid wordt echter duidelijk overstemd door dankbaarheid en gedempt door gelatenheid.

Het was een onnoemelijk voorrecht te mogen opgroeien in een omgeving waarin waarden van geest en hart hoger stonden aangeschreven dan stoffelijke welvaart. De levensstaat van het ouderlijk leraarsgezin met vier opgroeiende kinderen was uiteraard heel bescheiden, zeker in de crisis van de jaren dertig. Onze ouders waren echter meesters in de letterlijke zin in het stimuleren van onze belangstelling voor literatuur in de eerste plaats, maar ook van onze verwondering over de dode en levende natuur en van de daaruit voortvloeiende dorst naar kennis. Voor hen en voor mij was het een grote voldoening hen bij mijn promotie en later bij mijn ambtsaanvaarding in Delft te mogen begroeten.

Een tweede zeldzaam voorrecht werd mij verschaft in de personen van een aantal voortreffelijke leermeesters in de wiskunde en de mechanica. Al in de eerste jaren op de HBS liep ik in mijn meccano-gids aan tegen het rechtgeleidingsmechanisme van Peaucellier, dat ik niet vermocht te doorgronden. Mijn wiskundeleraar, de heer Hageman, nam zonder aarzeling de moeite een goniometrische behandeling te vertalen in de moeizamer analyse met behulp van eenvoudige planimetrische eigenschappen, waarbij ook de essentiële voorwaarde te voorschijn komt, die aan de dimensionering van het mechanisme moet worden gesteld. In het laatste jaar op school gaf de directeur, de heer Lem, op de vrije woensdagmiddag een facultatieve cursus 'hogere wiskunde' die het me mogelijk maakte het eerste jaar in Delft goeddeels te luieren, een vermoedelijk niet bedoeld neveneffect. Mijn wat sluimerende belangstelling werd pas echt wakker in het derde studiejaar bij de colleges over theoretische en toegepaste mechanica van de uiterst precieze en droge (schoon niet altijd nuchtere) Schuh en van mijn onvolprezen leermeester Biezeno met zijn magistrale en ook vaak theatrale betoogtrant. Dankbare herinneringen bewaar ik ook aan de wat moeilijker maar inspirerende colleges van Burgers over aero- en hydrodynamica en aan het ontwerp van een wringingsvermoeiingsmachine onder leiding van Koch, van wie ik in een dagelijks praatje van tien minuten gedurende drie maanden meer over construeren heb geleerd dan van alle specifiek constructieve docenten in mijn hele studietijd. Het vervult me met misschien ongepaste trots, dat deze machine na meer dan veertig jaren nog altijd in bedrijf is, evenals een grotere latere opvolger volgens hetzelfde beginsel. Weemoed bekruipt me bij het terugdenken aan de studietijd maar dankbaarheid voert weer de boventoon, mede door de aanwezigheid vandaag van mijn studievriend Piet Kroon die hiervoor een verre reis heeft moeten maken.

Mijn goede gesternte in de ontmoeting van inspirerende leermeesters heeft me ook na de studie niet verlaten. Bij de toenmalige Rijks-Studiedienst voor de Luchtvaart, het huidige Nationaal Lucht- en Ruimtevaart Laboratorium, mocht ik onder leiding van Van der Neut mij verder verdiepen in de toegepaste mechanica en vooral in haar concrete toepassing op constructies. Hij heeft me geleerd een rapport te schrijven, een aspect dat destijds bij de studie vrijwel geen aandacht kreeg, en ook nu misschien nog wel minder dan wenselijk is. Goede herinneringen bewaar ik verder aan de vele leerzame gesprekken met mijn helaas te vroeg verscheiden vrienden Koning en Plantema.

Dankbaar in nog bredere zin gedenk ik ook mijn betreurde goede vriend De Winter wiens moed, onkreukbaarheid, plichtsbetrachting en industrieel inzicht mij een lichtend voorbeeld zijn geweest in mijn ambtelijke loopbaan bij de Rijksluchtvaartdienst. De onstuimige wederopbouw van luchtvaart en vliegtuigindustrie in de eerste jaren na de oorlog liet vrijwel geen ruimte voor wetenschappelijke activiteit of studie, een gemis dat door mij wel als pijnlijk maar ook als onvermijdelijk werd ervaren. Dank zij de brede blik van mijn chefs mocht ik echter toch de internationale mechanicacongressen van 1946 in Parijs en van 1948 in Londen bijwonen, waardoor de verbinding met de wetenschap niet geheel te loor ging. Ik heb het ook altijd als een voorrecht beschouwd te mogen medewerken aan de totstandkoming van het Nederlands Instituut voor Vliegtuigontwikkeling, dat thans ook de ruimtevaart in zijn vaandel heeft opgenomen, een onmisbare schakel tussen overheid en industrie voor de handhaving van een Nederlandse vliegtuigbouw volgens eigen ontwerp. Onder de bezielende leiding van Van der Maas heb ik de eerste jaren als lid van het dagelijks bestuur deel mogen nemen aan de werkzaamheden van dit Instituut. Sinds mijn benoeming in Delft heb ik deze werkzaamheden op wat grotere afstand mogen volgen als gewoon bestuurslid. Daarnaast heeft mijn periode als voorzitter van de Wetenschappelijke Commissie NLR-NIVR in de jaren 1955-1968 mij veel voldoening geschonken.

In 1949 teruggekeerd in Delft op een leerstoel voor toegepaste mechanica ging zo niet mijn roeping dan toch wel mijn ideaal in vervulling van een gecombineerde taak als onderwijzer en onderzoeker. Velen, zeer velen, curatoren, collega's en medewerkers in alle geledingen ben ik grote dank verschuldigd voor de steun in mijn werk die ik allerwegen heb mogen ondervinden. Zonder zulk een steun van collega-docenten, wetenschappelijke en technische medewerkers, secretaressen en medewerkers van algemene diensten van de Afdeling der Werktuigbouwkunde zou het onmogelijk zijn een zware onderwijs- en onderzoektaak naar behoren te vervullen. Hun aantal is veel te groot om namen te noemen.

Het is altijd verstandig de oorzaak van teleurstellingen in het leven in de eerste plaats te zoeken in eigen tekortkomingen. Het kleine aantal promovendi die ik tot hun doctoraat heb mogen begeleiden behoort tot deze teleurstellingen die ik wel in hoofdzaak aan mijzelf te wijten heb. Ik kom hierop straks wat uitvoeriger terug. Toch mag ik ook niet verhelen dat mijn geheel andere en nog ernstiger teleurstelling over de z.g. universitaire democratisering in wezen te wijten is aan de karakterloze houding van mijn eigen generatie, en in Delft in het bijzonder van de vroegere senaat, die in dit opzicht zijn jammerlijk falen in de oorlogsjaren nog eens heeft bevestigd. Met grote dankbaarheid maak ik hierbij echter ook gewag van het onbaatzuchtige initiatief van vele collega's, op grond waarvan de regering tot een bevredigende oplossing is gekomen voor mijn persoonlijke problemen in dit verband. Dank zij de mogelijkheid mijn werk voort te zetten op een persoonlijke grondslag, zonder het dwangbuis van de door mij volstrekt verworpen nieuwe structuur van de hogeschool, is de bittere nasmaak van de gebeurtenissen in de vroege jaren zeventig goeddeels verdwenen. Met recht kan ik de terugblik bij mijn afscheid dan ook parafraseren met de dichterlijke woorden 'Omzien in verwondering maar niet in wrok'.

Staat U mij thans toe, dames en heren, dat ik vervolg in een taal die ook voor onze buitenlandse gasten toegankelijk is.

I apologize that I have spoken so far in a language which is inaccesible to most of our foreign guests. I shall continue in English with a review under the title 'Forty years in retrospect, the bitter and the sweet', a free translation of the motto ending my words in Dutch. I have selected this title because it is the appropriate counterpart of my optimistic, and occasionally perhaps even naive, inaugural lecture thirty years ago 'Prospects for the development of applied mechanics'. Moreover, it would be presumptuous on my part to attempt a sketch of present trends and possible future developments, in view of the authoritative papers by so many outstanding experts at this week's symposium 'Trends in solid mechanics 1979'. The bitter aspects of my retrospect are largely due to my failure to achieve many things I had in mind at the start, but they are dominated by the sweet, mainly thanks to so many friendly contacts both at home and abroad.

The first decade of my work in Delft was in several respects the most difficult one. True enough, the dual task of teaching and research strongly appealed to me but the arrears in my knowledge loomed like an insurmountable barrier. The five years of Nazi occupation had isolated us from the immense research activity in the free world, and my intense preoccupation with urgent practical business in the Department of Civil Aviation in the immediate post-war years had effectively prevented me from a serious study of the enormous stream of literature after the war. Some remarks in my inaugural lecture testify to my inadequate understanding of the theory of plasticity, developed so extensively during the war years and in the post-war era in Britain, the United States and the Soviet Union. I was also hardly aware of the modern theory of plane strain and generalized plane stress, developed first in the U.S.S.R. and during the war independently in Britain. The English translation of Muskheliskvili's famous book in 1953 and the publication of the treatise by Green and Zerna a year later were eye openers to me.

As the youngest member of a foursome, consisting of Biezeno, Koch, Legger and myself, I carried a fairly heavy teaching load. In the associated burden of grading of papers I gratefully acknowledge the unselfish help by Legger, and later also by Floor and younger collaborators. The actual lecturing to undergraduates has always been a pleasant and even stimulating duty to me, but the marking of examination papers was an unpleasant and unthankful chore.

The end of my first decade in Delft was marked by the retirement of Biezeno after no less than 44 years of service as a full professor, for many years alone, later supported by Hencky and after a few years by Koch. His imprint on the Dutch school of solid mechanics will never fade, and we all remember with gratitude his monumental work 'Technische

Dynamik', written in close collaboration with his good friend Grammel. A.D. de Pater was appointed in Biezeno's place, and a year later Besseling also returned from the United States to join our professorial staff. These changes had a significant effect on my own activities. Whereas I had been dabbling in elasticity, plasticity and vibrations during the first decade, a concentration on a more restricted field now became desirable. De Pater was and is an expert in vibration theory and problems of dynamics in general, and Besseling had won his spurs already in plasticity. It was no more than natural for me to return to my old loves: elastic stability and the theory of thin elastic shells.

Koch retired in 1963, and Lekkerkerker was appointed as his successor, thus like Koch crowning his academic career in our university. The end of the second decade was marred by the introduction of the unfortunate university administration reform bill, in a terrified response to the so-called student revolt in Europe in the late sixties. Neither the university senate nor the government seemed to have learnt their lesson from wartime experience that fear is the worst possible adviser. Gradually it became clear to me that I would be unable to maintain my academic integrity under the new organization, and early in 1973 I decided to relinquish my chair of engineering mechanics. On this occasion it was an unexpected and agreeable surprise that so many colleagues and institutions intervened on my behalf. I am also grateful to the minister and the secretary of state for higher education who granted me a year's leave of absence and reappointed me on a personal chair, free from the interference by incompetent committees. The at the same time stimulating and relaxing atmosphere at Caltech enabled me to regain my balance and composure, and to return to academic work in Delft, both in research and graduate teaching. I am afraid that my stand on this issue is not (yet) generally approved, and I interpret the newspaper paragraph that I am now retiring from the chair for the theory of *liberty*, strength and stability of structures as a Freudian mistake by the reporter.

It is one of the serious shortcomings in my professorial career that I have accompanied only a few students to their doctor's degree. True enough, this degree has always been more an exception than the rule in engineering education in the Netherlands, but this is no adequate reason for my failure in this respect. A partial explanation may be that prospective candidates may have been frightened away be exaggerated rumours of the high standard of my requirements. A further reason is undoubtedly that I had not much need of assistance by younger associates in my own research. Hence I had no urge to encourage many of them to a prolonged period of research under my guidance unless they took the initiative themselves. Moreover, all my doctoral students were exceptionally qualified, and their stimulation of my own thoughts and research often continued long after they had obtained their degree. For example, the revival of my interest in the general theory of thin elastic shells was activated by my first doctoral student Cohen. His thesis on helicoidal shells and wide propeller blades revealed a number of open questions in shell theory which kept me busy for quite a few years. Alblas performed a thorough study of the three-dimensional stress distribution in a flat plate with a circular cylindrical hole. His asymptotic result for thin plates proved again most valuable nearly twenty years later as a check on Gol'denweizer's and Van der Heijden's modified boundary conditions at a free edge of a plate. Incidentally, the even more difficult asymptotic case of very thick plates in the Alblas problem was finally solved by Youngdahl and Sternberg. The two-way traffic between teacher and student was again emphasized by Van der Heijden's master thesis, a forerunner to his doctoral work. He opened my eyes to the significant advances achieved by Gol'denweizer in the nineteensixties by his asymptotic integration of the three-dimensional equations of elasticity theory, leading to a justification and subsequent refinement of classical plate and shell theory.

Lekkerkerker's thesis on the stress concentration around a circular hole in a cylindrical shell was perhaps, with one exception, the most independent work by my doctoral students. His theory represents an essential refinement of earlier work by Lure'e. It may be of some interest to note that the same problem was dealt with at the same time by Van Dijke by means of a different approach in the Division of Engineering and Applied Science at Harvard University. The results were, as they should be, in full mutual agreement, and they were also confirmed by Lekkerkerker's careful experiments. The comprehensive analytical study of stress distributions in perforated plates by Meijers was largely based on my earlier work on the two-dimensional theory of elasticity in domains with a doubly-periodic set of holes of arbitrary shape. His careful and detailed numerical evaluation soon found its way to design codes. A worthwhile complement was provided by Slot in the form of three-dimensional corrections obtained by finite element techniques. Slot's thesis was a good example of doctoral work performed at a long distance from the university. This possibility of obtaining a doctor's degree on the basis of independent research outside the university, more or less by correspondence, is a fortunate tradition in our country. We may only hope that it will not be eroded like so many other ancient good traditions.

It is not only to colleagues and associates in my own departments, Mechanical Engineering and up to my resignation in 1973 also Aeronautical Engineering, that I owe a significant debt in connection with my research. Apart from experimental techniques which I have never been able to master, due to the possession of two left hands, mathematics is the most important tool in mechanics. It would have been far more difficult, if not impossible, to achieve anything without the stimulation and help received from a number of colleagues in our Department of Mathematics. Two of them have passed away, Timman sadly enough quite prematurely and Van Veen long after his retirement. To both of them I owe more than I can summarize in a few words. Van Veen's deep and extensive knowledge of analysis, both classical and contemporary, has been invaluable in tracing more or less hidden sources of information, whereas Timman's perceptive mathematical intuition has often stimulated my own way of thinking. Until his retirement some years ago Bottema contributed the teaching and research of rational mechanics to several engineering departments. Even if I have long felt the need of the incorporation of this branch of mechanics in engineering mechanics, Bottema's brilliant teaching convinced us that this move should be postponed as long as our students might have the benefit of his lectures. The courses of rational mechanics are now the responsibility of the group of engineering mechanics. We are fortunate, however, that Bottema is still quite active in research, as testified by his recent original monograph on kinematics, written jointly with the younger colleague Roth of Stanford University.

Even more than to my colleagues, associates and students in Delft I am indebted to the world scientific community in mechanics for their encouragement and stimulation throughout my career. It starts with the debt I owe to an outstanding scientist and teacher whom I never met because he died rather shortly after my graduation in Delft. I refer to Trefftz whose fundamental thinking about elastic stability was never properly recognized outside his own country but whose influence has permeated my own work more than anything else. Shortly after the war it was my privilege to make some life-long friends at the Paris congress of mechanics, and I am happy that one of them, Jim Stoker, has confirmed our friendship once again by his presence here this week. The list of outstanding colleagues and friends whose generous hospitality and equally generous stimulation I have enjoyed is too long to read here. As I once had occasion to remark, I have more friends abroad than in my own country, possibly because The Netherlands is so small, but more likely because my weaknesses are better known at home. At the risk of doing an injustice to others I shall mention only two names. First of all Eli Sternberg, because our association has been extended over no less than three academic years, one in Delft, and later two in the United States where I enjoyed first the hospitality of Brown University and twelve years later of Caltech. It is difficult to say to what aspect I owe more, to Eli's clarity and precision in formulation or to his wonderful sense of humour. It will be my eternal regret that we have authored together only one paper, but I remember this joint effort with gratitude and affection. The institutes abroad with which I feel most closely associated professionally are the Courant Institute in New York and the Division of Engineering and Applied Science at Harvard University, although for quite different reasons. Fritz John of the Courant Institute represents to me the ideal of an applied mathematician. Our frequent discussions of shell theory and its mathematical justification during the nineteensixties have influenced my thinking more than I can say. On this topic the interaction between sound mathematical reasoning and engineering intuition has been most fruitful, hopefully mutually so. The ties with the Harvard group lie more on the common ground of engineering science, and I fondly hope that the interaction with this powerful group has also been mutually beneficial. I am happy that this group has been represented here so strongly this week.

Up to the early nineteensixties my personal contacts abroad had their centre of gravity in the United States. One obvious reason is the important role played by that large country in the development of solid mechanics, and in particular of its applications to engineering which were already somewhat familiar to me from my aeronautical experience. Political and linguistic barriers hampered similar contacts with colleagues in the Soviet Union, although this enormous country was perhaps even more powerful in the more mathematical approach to our field. The language barrier has always remained a serious obstacle, primarily by my own fault, but I am happy that differences in political philosophy have not prevented a gradual growth of ties of friendship with many colleagues in the U.S.S.R., and I have warmly welcomed the participation of some of them in this week's symposium. The slower growth of my scientific relations in Western Europe may be partly due to a certain provincialism in most European countries where scientists tend to some national complacency in scientific matters, in a marked contrast to the more open minds of our American colleagues. A further even more cogent reason for a closer relationship with American science is to be found in the unparalleled generous hospitality offered by universities and other scientific institutions in the United States. Travel across the Atlantic was thus made easier for the impoverished Europeans than visits within Europe. Increased prosperity and in particular the activities of Euromech have led to a significant improvement of regional contacts including both East and West. More than one hundred European mechanics colloquia have now been held since their inception at the initiative of George Batchelor in 1964. The International Centre of Mechanical Sciences in Udine with its many courses in solid mechanics under the inspired leadership of Waçlaw Olszak has also made a significant contribution to improved contacts.

It has been my good fortune that quite soon after my appointment in Delft I was drawn into the circle of international scientific organizations. On a proposal by Burgers I was involved in the organization of the first IUTAM symposium on nonlinear vibrations on the island Porquerolles, as secretary of the scientific committee chaired by Pérès. I am still saddened by the reason for my appointment as secretary of the International Committee for the Congresses of Applied Mechanics in 1952, my good friend Carel Koning's fatal illness, but I recall with gratitude the numerous contacts in this office with very many colleagues, and in particular first with my own teacher Biezeno and his close friend Southwell, and later with George Batchelor, Ernst Becker and Wladek Fiszdon. Some years later I also became involved with the Bureau of the International Union of Theoretical and Applied Mechanics

(IUTAM) which I was priviliged to serve subsequently as treasurer, as president and vice-president. I still consider it a great honour to have been allowed to serve in the latter offices as a successor to my great predecessors with such diverse personalities: Joseph Pérès, Hugh Dryden, Folke Odqvist, George Temple and Maurice Roy. It may be true that the far from trivial administrative burden of these activities detracted time and energy from the pursuit of research, but the stimulation of ideas by contacts with so many outstanding colleagues more than compensated for this loss.

The dominant interest in my scientific life has lain in the somewhat restricted field of elastic stability, buckling and post-buckling behaviour, in particular with reference to the so diverse and often precocious buckling behaviour of plate and shell structures. The general theory of thin elastic shells was included in this main interest as an indispensable background. This interest arose from my earliest activity in aeronautical engineering after graduation from Delft. The advent of thin-walled metal structures in the nineteenthirties had drawn this problem area into the foreground. After a few years of more or less routine work in checking of stress analyses by aircraft companies, combined with some structural research, I came to the conclusion that it would be worthwhile to write a doctoral thesis in the evenings. My first approach to Biezeno with this idea in 1938 met with a reply that was not quite encouraging. I raised the question about a suitable topic, and he replied that I might come back when I had found one. The next year I imagined I had indeed found a suitable basis, namely the theorem that the stable domain in load space of an elastic structure with a linear pre-buckling state is always convex. It was quite a disappointment when I discovered some time later that Papkovich had published this theorem in 1934 at the Cambridge Congress of Applied Mechanics. In the spring of 1940 I had the misfortune to be laid up with pneumonia and a pleurisy, and I was hardly able to get out of bed when the Nazi invasion came on 10-th May. In bed, however, I read an exciting article by Leslie Cox in the Journal of the Royal Aeronautical Society of March 1940, and his simple account of the behaviour of a column with a nonlinear elastic support suggested to me the basic idea that instability *at* the critical load might be the explanation of the disastrous effect of imperfections.

This conjecture led to the development of the general nonlinear theory of elastic stability in my thesis. My ideas came into full focuss in the summer of 1940, and a fairly detailed outline, contained in a letter of December 1940, was approved in principle by Biezeno. The basic work, including the application to the baffling problem of buckling of cylindrical shells under axial compression, was completed in February 1942, and the first draft of the manuscript was sent to Biezeno before the end of the year; his comments led to some minor changes in June 1943. I have a vivid recollection of New Year's eve in December 1942 when my wife and I sat in front of a very modest fire and discussed, in addition to the most acute war-time problems, what would happen to my work which I considered to be significant but questioned whether it would be recognized as such. Since German was the only foreign language permitted by the Nazis for publications my manuscript could obviously only be written in my native language, not easily accessible to most colleagues abroad. In January 1943 I entered into an agreement with the publishing firm H.J. Paris in Amsterdam with the intention of completing the printing by September and having the oral examination in the fall of 1943. We had to stop this, however, in August 1943 because the University was forced by the Nazis to require a declaration of loyalty to the Nazi occupation authorities from all students who wished to pass an examination. I remember my indignation at the University Senate for yielding to this pressure, the cowardice I referred to earlier in connection with this body. I replied I was no longer interested in their degree, I am afraid not with

Biezeno's approval. At that time the publisher took a very generous initiative which I remember with gratitude and respect. He proposed to complete the typesetting of my manuscript, to reserve paper for printing, and to wind up the matter (including my payment to him) only after the war. Both of us implicitly assumed the end to be victorious. This enabled me to get my degree very soon after the University opened its doors again in September 1945.

I have related this history of my first and most basic work in elastic stability in some detail, going back in time over more than forty years, not only to show that commercial firms are occasionally more humane and courageous than universities, but also because the delay in my return to this topic closest to my heart requires some explanation. I have already referred to my intense preoccupation with urgent more mundane problems in the immediate post-war years and to the serious arrears in my general knowledge in the field of engineering mechanics in the first years of my professorship. A further reason was that I considered I had already published my basic thinking on elastic stability, and that it would be improper to reiterate on the same topic, a curious mixture of modesty and immodesty, the latter because I took it for granted that my published work was accessible in principle to anyone actively interested in the field. This aspect is most obvious by the paper I read at the Paris congress of mechanics in 1946, the first paper I presented anywhere. I did not even think of returning to a topic which I considered to have been dealt with adequately elsewhere, and I gave a paper on the behaviour of flat plates in the advanced post-buckling regime, published only in report from in the last years of the war. I cannot resist the temptation to record several amusing aspects in connection with this paper. The work was performed at the Aeronautical Research Institute in Amsterdam, where I had obtained a refuge from the German occupation authorities, under a contract from the Department of Civil Aviation, drawn up by myself before I went into refuge. In this way I was able to evade work for the Germans. The congress session in which I presented the paper, with some unexpected numerical results, was chaired by Von Mises who opened, or rather cut short the discussion by saying that he did not believe a word of it, a most healthy, if perhaps not quite welcome cold shower to a young man. Finally, the proceedings of the Paris congress were never published, although I had the privilege of correcting the proof sheets of my paper. Up to this time the annual report of IUTAM refers to this situation quite properly and concisely by stating in its list of proceedings for the Paris congress 'was given in the hands of Gauthier-Villars, Paris', and there it rests.

My unhurried return to the field of nonlinear elastic stability during the nineteenfifties was accelerated by a course I gave at Brown University in the academic year 1961-62. The audience included my good friend Eli Sternberg, and I shall always remember his friendly, constructive and penetrating comments in our discussions of the topic in question. Interest in this field also mushroomed at that time, in particular at Harvard University where I fondly believe my work had some influence, and at University College in London where a similar approach for discrete elastic systems was developed more or less independently, as described so eloquently in the preface by Thompson and Hunt to their monograph 'A general theory of elastic stability'. Particularly stimulating to me have been Hutchinson's investigation of spherical shells under external pressure, Budiansky's reformulation of buckling theory in the modern language of functional analysis, the paper by Budiansky and Hutchinson in the Van der Neut anniversary volume, and now again the exciting joint paper by Budiansky and Hutchinson at this week's symposium. I also realize the obligation I have incurred by Van der Heijden's excellent lecture notes of the course I gave this year in Delft!

In connection with the application of stability theory to shell buckling in my thesis I was in need of a nonlinear theory of shells, which did not exist at the time. Chapter 5 of the thesis is thus devoted to the topic of nonlinear shell theory up to the point of an adequate expression for the elastic energy, in terms of the middle surface strains and the changes of curvature, which quantities are in their turn expressed in terms of the displacement components of the middle surface and their derivatives. This theory also included a procedure to allow for the presence of geometric imperfections in the shell shape.

As mentioned before, Cohen's 1955 thesis had exhibited a number of open questions in the evaluation of linear shell theory in the literature, and this stimulated my interest in the general theory of thin elastic shells, both in the linear and the nonlinear domain. It was my privilege, after the opening lecture by Biezeno at the first IUTAM symposium on the theory of thin elastic shells in Delft in 1959, to present the first paper in which I outlined the most basic, and I believe irrefutable part of the results I had obtained at the time. The resulting equations of linear theory turned out to be the tensor equivalent of the Sanders equations published earlier that year and derived along different lines. The follow-up with more details in the nonlinear range was discussed in a number of lectures on both sides of the Atlantic in the following years, but their publication was delayed for several reasons. First of all, I was not yet quite satisfied with the results obtained, and secondly Sanders published in 1963 an outstanding paper on the nonlinear theory which obviated a further publication by me at that time. Fritz John's fundamental paper in 1965 put me on the right track for a further study. His justification of the classical assumptions of shell theory, with concrete error estimates, represents a mile stone in the history of this theory. However, it will not be the last step. Error estimates for the *solutions* of the shell equations are even more important than error estimates for the *equations*. Here we have still a long way to go. Apart from a very special case we have only root mean square error estimates for the solution of linear shell problems. Here again I have been privileged to take part in this development in the company of Danielson and Simmonds in the United States, and more recently of Ladevèze in France.

Rereading my inaugural lecture I am dismayed by my oversight of several significant developments which were at that time in their incubator. I shall mention only two of these, fracture mechanics and the impact of electronic computers. In the first of these fields our group has been able only to make a contribution in the domain of crack problems in the theory of elasticity. I remember with pleasure my joint effort with Benthem to compile available information on asymptotic approximations to crack problems. I also take pride in referring to Benthem's fundamental and penetrating three-dimensional analysis of cracks ending at a free surface. This is perhaps the most significant contribution of elasticity theory to fracture mechanics since the formula for cracks in plane strain, derived by Inglis in 1913. I am also grateful to Besseling for taking up the finite element method, with modern computers perhaps the most powerful tool at present in the hands of mechanical scientists. His significant contributions in this field, notably in the nonlinear range, more than compensate for my lack of enthusiasm to enter this new field myself.

In order to end with a more optimistic note I am glad to observe that I have remained true to my basic philosophy that mechanics requires in its further development the employment of more and more abstract and sophisticated mathematics. Functional analysis, for example, is now an effective tool in solid mechanics, and I am pleased to note that I have employed this tool, at least in a heuristic manner, long before it had come into common use. I hope I have also kept in mind the adage that research in engineering science cannot be considered to be completed until its full numerical evaluation has been obtained, and the results have been presented in a form accessible to other engineers.

I had not intended to dwell on educational problems in this lecture. The basic reason for this intentional omission was again my reluctance to reiterate on a topic on which I have expressed my views in unmistakable words elsewhere. Quite recent developments, however, have forced my hand, and I feel compelled to say at least a few words on this curcial issue.

Since the original proposal for a shortening of the basic academic course in engineering more than ten years ago I have warmly and publicly supported this reform under the essential proviso that an adequate post-graduate curriculum for research and development is introduced at the same time. Under this proviso I do not share the prevalent view in university circles that such a shortening of the basic course would be harmful. Nor do I see any serious objection against maintaining the same civil effects as up to now for graduates from a four-year course. We should not fool ourselves, however, that their degree would imply higher qualifications than a bachelor's degree in the Anglo-Saxon countries. Post-graduate education up to a level comparable to a master's degree from leading universities abroad will require a *minimum* of two years. It appears now that the Government plans to restrict such post-graduate curricula to at most one year, primarily for budgetary reasons, but obviously also influenced by the current tide of anti-science and anti-intellectualism. If these governmental plans are carried through, the result will be disastrous. In that case my swan song may well be followed by a 'marche funèbre' for science and engineering in the Netherlands. We can only hope that sanity will prevail.

You will have noticed, before I did so, that I have spent quite some time on reminiscences, a telling symptom of advancing years, and one more reason not to postpone retirement. I should add that I had planned to spare you this ordeal and to leave the university without a farewell lecture. In a way it is the wages of sin, be it a lovely sin of my colleagues in the group of engineering mechanics who have organized this week's symposium in connection with my birthday tomorrow. My heart was deeply touched when Besseling and Van der Heijden announced to me their preparations for this meeting some time ago, and I realized immediately that there would be no escape from a lecture today. The list of organizers, speakers and participants is too long to read here. Let it suffice therefore to express my heartfelt thanks to all of you for the great honour and deep pleasure you have bestowed upon me in this memorable way. It is an additional joyful aspect to have seen among you so many old and good friends, both from nearby and far away, including those friends who have so generously sponsored my degree at their foreign universities.

Having said all this about my feelings of gratitude to so many colleagues and friends, both at home and abroad, I have to single out one very special person to whom I owe more than to any one else in the world, a debt I can never hope to pay in full. My inaugural lecture, Lous, was deeply saddened by your forced long absence. The more joyful is your presence today which allows me an opportunity to recognize in public, without fear of contradiction, that it is *your* care and love which have enabled me to work, and live a happy life. It is you, and during some years your beloved mother, who has reared our children and inspired them by your example with the right ideals and human values. I am proud to see all of them true to these basic ideas. One of our friends paid you, in your absence, the greatest possible and true compliment in words which I shall never forget: 'Lous has not even a grain of egotism in her character'. We have not been spared sadness in our common life, but there has been far more joy. In fact, looking back over forty years with you they seem to me to have passed by like a single sunny summer day. Bless you!

MIX
Papier aus verantwortungsvollen Quellen
Paper from responsible sources
FSC® C105338

If you have any concerns about our products,
you can contact us on
ProductSafety@springernature.com

In case Publisher is established outside the EU,
the EU authorized representative is:
Springer Nature Customer Service Center GmbH
Europaplatz 3, 69115 Heidelberg, Germany

Printed by Libri Plureos GmbH
in Hamburg, Germany